Reconfiguring Knowledge Production

Reconfiguring Knowledge Production

Changing Authority Relationships in the Sciences and their Consequences for Intellectual Innovation

Edited by
RICHARD WHITLEY,
JOCHEN GLÄSER,
and
LARS ENGWALL

OXFORD
UNIVERSITY PRESS

OXFORD
UNIVERSITY PRESS

Great Clarendon Street, Oxford OX2 6DP

Oxford University Press is a department of the University of Oxford.
It furthers the University's objective of excellence in research, scholarship,
and education by publishing worldwide in

Oxford New York

Auckland Cape Town Dar es Salaam Hong Kong Karachi
Kuala Lumpur Madrid Melbourne Mexico City Nairobi
New Delhi Shanghai Taipei Toronto

With offices in

Argentina Austria Brazil Chile Czech Republic France Greece
Guatemala Hungary Italy Japan Poland Portugal Singapore
South Korea Switzerland Thailand Turkey Ukraine Vietnam

Oxford is a registered trade mark of Oxford University Press
in the UK and in certain other countries

Published in the United States
by Oxford University Press Inc., New York

© Oxford University Press, 2010

The moral rights of the author have been asserted
Database right Oxford University Press (maker)

First published 2010

British Library Cataloguing in Publication Data

Data available

Library of Congress Cataloging in Publication Data

Data available

Typeset by SPI Publisher Services, Pondicherry, India
Printed in Great Britain
on acid-free paper by
MPG Group, Bodwin and King's Lynn

ISBN 978-0-19-959019-3

3 5 7 9 10 8 6 4 2

Preface

This book developed from discussions in 2006 between Sigrid Quack, then Chair of the European Group for Organization Studies (EGOS), Richard Whitley, and Lars Engwall about how the changing governance of the public sciences in many countries was affecting the direction of organization studies. These led to the forming of a subtheme group on 'The Changing Organisation of the Sciences and the Changing Sciences of Organisation' at the 2008 EGOS Colloquium in Amsterdam. It generated enough interrelated papers to form the basis of a coherent volume about the effects of the restructuring of public science systems on the authority relationships governing research activities and, consequentially, on intellectual innovations in different scientific fields. Revised versions of these contributions, together with a few new invited papers were discussed at a workshop held at the Royal Swedish Academy of Letters, History and Antiquities in Stockholm in February 2009. We are most grateful to the Academy for its support and assistance in hosting this workshop. We are also indebted to the Bank of Sweden Tercentenary Foundation for financial support for this event. Finally, we want to thank all the participants at the two meetings for their contributions to our discussions, and to the authors of the chapters in this book for responding so effectively to our editorial suggestions.

Richard Whitley
Jochen Gläser
Lars Engwall

Contents

Contents

List of Contributors

Martin Benninghoff is senior researcher at the Observatory Science, Policy and Society (OSPS) of the University of Lausanne. His work focuses on issues of academic identity, power, policy and legitimacy, and the social organization of scientific activities. Currently, he is involved in a European research project of the EUROCORES Programme on scientific innovation. Martin Benninghoff has published books on science policy (with J. P. Leresche, *La Recherche: Affaire d'État* (PPUR); with Martina Merz *et al.*, *La Fabrique des sciences* (PPUR)) and has also contributed to research mandates from Swiss higher education and research bodies, like the Swiss University Conference, the Rectors' Conference of the Swiss Universities, and the Swiss Science and Technology Council.

Harry de Boer is a senior researcher and research coordinator at the Center for Higher Education Policy Studies (CHEPS) at the University of Twente, the Netherlands. His research interests in the field of higher education studies concern governance and steering models, public policy analysis, institutional governance and management, and strategic planning and decision-making. He frequently publishes his articles in the journals *Higher Education*, *European Journal of Education*, *Higher Education Research and Development*, *Public Administration*, *Leadership*, and *Tertiary Education and Management*, and over the years he has published many book chapters on these topics.

Dietmar Braun is Professor of Political Science at the Institut d'Études Politiques et Internationales of the University of Lausanne. He has also been research councillor at the Swiss National Science Foundation since 2004, and was editor of the *Swiss Political Science Review* between 2001 and 2004. He has worked on topics linked to the science and technology field since the beginning of the 1990s, especially on research funding policies and funding agencies. His habilitation thesis is on the political guidance of science. His other research interests are studies on comparative federalism and political theory.

Jürgen Enders is Professor at the School of Management and Governance and Director of the Center for Higher Education Policy Studies (CHEPS) at the University of Twente, the Netherlands. He also serves as one of the leaders of the research programme on 'Governance of Innovation, Technology, Higher Education, and Research' of the Institute for Governance Studies (IGS) at the University of Twente. His research interests are in the areas of the political sociology of education, science, and innovation; the governance and management of higher education, research, and knowledge transfer; organizational change in higher education; higher education and the world of work; and the academic profession. He is member of the editorial board of the book series 'Higher Education Dynamics' and the journal *Higher Education*. Jürgen Enders has spent periods as a visiting scholar at the University of California, Berkeley, USA, the University of Beijing, China, the Foundation Nationale de Sciences Politiques in Paris, France, and the Social Science Research Center (WZB), Berlin, Germany.

Lars Engwall has been Professor of Business Administration at Uppsala University since 1981, and has held visiting positions in Belgium, France, and the United States. His research has been directed towards structural analyses of industries and organizations as well as the creation and diffusion of management knowledge. Among his books are *Mercury Meets Minerva* (Pergamon Press, 1992; EFI, 2009), *Management Education in Historical Perspective* (Manchester University Press, 1998, ed. with Vera Zamagni), *Management Consulting: The Emergence and Dynamics of a Knowledge Industry* (Oxford University Press, 2002, ed. with Matthias Kipping), and *The Expansion of Management Knowledge: Carriers, Flows and Sources* (Stanford Business Books, 2002, ed. with Kerstin Sahlin-Andersson). He is a member of several learned societies, among them the Royal Swedish Academy of Sciences and Academia Europaea.

Jochen Gläser is a senior researcher at the Center for Technology and Society of the Technical University, Berlin. His major research interest is the interaction of epistemic and institutional factors in the shaping of conduct and content of research at the micro-level of individuals and groups and at the meso-level of scientific communities. He has also published on qualitative methods, research methods in science studies including bibliometrics and interviewing, and methods of research evaluation. Current empirical projects concern national systems of research evaluation and funding in an internationally comparative perspective, the responses of German universities to evaluations, and the impact of changing authority relations on conditions for scientific innovation. Key publications include: *Wissenschaftliche Produktionsgemeinschaften: Die Soziale Ordnung der Forschung,* (Frankfurt a.M.: Campus, 2006) and, ed. with Richard Whitley, *The Changing Governance of the Sciences: The Advent of Research Evaluation Systems* (Dordrecht: Springer, 2007).

Matthias Kipping is Professor of Strategic Management and Chair in Business History at the Schulich School of Business of York University in Toronto, Canada. His main research interest concerns the international transfer of management knowledge—in particular the evolution and role of management consultants and management education. He has published widely on these topics in business history and management journals. He is the co-editor (with Timothy Clark) of *The Oxford Handbook of Management Consulting* (OUP, 2010) and is working on a monograph about *The Management Consultancy Business in Historical and Comparative Perspective.* Together with Lars Engwall he is currently conducting research on the internationalization of business schools.

Robert Kneller is a professor in the University of Tokyo's Research Center for Advanced Science and Technology (RCAST). His research focuses on university–industry cooperation, the role of venture companies in innovation, the discovery and commercialization of biomedical technologies, and conflicts of interests associated with academic entrepreneurship. Amongst many other publications, he is the author of *Bridging Islands* (Oxford, 2007), which compares the environments for new high technology companies in Japan and the USA, and the role these companies play in innovation.

Stefan Lange is a senior researcher at the Endowed Chair for Science Organization, Higher Education, and Science Management, German University of Administrative Sciences, Speyer. He is the editor-in-chief of the German journal *Hochschulmanagement* (Higher Education Management) and principal investigator in the research project *External Evaluations as Stimuli for Internal Governance Innovations in Universities.* Recent publications in English include: (with G. Kruecken) 'German Universities in the New Knowledge Ecology: Current Changes in Research Conditions and University–Industry-Relations', in Craig Calhoun and Diana Rhoten (eds.), *Knowledge Matters: The Public Mission of the Research University* (New York: Columbia University Press, 2010); (with U. Schimank) 'The German University System: A Late-Comer in New Public Management', in Catherine Paradeise *et al.* (eds.), *University Governance. Western European Comparative Perspectives* (Dordrecht: Springer, 2009); 'The Basic State of Research in Germany: Conditions of Knowledge Production Pre-Evaluation', in Richard Whitley and Jochen

Gläser (eds.), *The Changing Governance of the Sciences: The Advent of Research Evaluation Systems* (Sociology of the Sciences Yearbook, 26; Dordrecht: Springer, 2007).

Grit Laudel is a senior researcher at the Rathenau Institute in The Hague. Her research focuses on the influence of institutions on the production of scientific knowledge. She has investigated the impact of funding schemes of the German Deutsche Forschungsgemeinschaft on interdisciplinary collaboration, the impact of project funding in Germany and Australia on the content of research, and the adaptation of Australian university research to the indicator-based block funding of Australian universities. Currently she is investigating the mechanisms by which national systems of institutions shape academic careers and, through careers, the content of research. Her key publications are: 'Collaboration, Creativity and Rewards: Why and How Scientists Collaborate', *International Journal of Technology Management*, 22(7–8) (2001): 762–81; 'The Art of Getting Funded: How Scientists Adapt to their Funding Conditions', *Science and Public Policy* 33(7) (2006): 489–504; 'Interviewing Scientists', *Science, Technology and Innovation Studies*, 3 (2007) (with Jochen Gläser).

Liudvika Leišytė is a research associate at the Centre for Higher Education Policy Studies, University of Twente. She is involved in research and teaching in the areas of higher education and research policy, higher education, research governance and management, research commercialization in a comparative perspective. Her major theoretical interests lie in the sociology of science and the sociology of organizations. In particular, she is interested in the role of institutions in shaping academic identities and practices in different scientific and organizational fields. She has published two monographs, a number of chapters in edited books, and peer-reviewed articles in *Higher Education, Higher Education Policy, Public Administration* on the issues of academic practices, university governance, and national higher education and research policies in selected European countries.

Séverine Louvel is assistant professor in sociology at the Grenoble Institute for Political Studies (IEP Grenoble) and director of the master programme 'technology, science and politics'. She is also a researcher at the Department for Innovation Studies at the PACTE research centre (CNRS and Grenoble Institute for Political Studies). Her Ph.D. showed how changes in the governance of the French public science system affected the implementation of scientific strategies, the career management of academics, the access to financial resources, and work relationships. Her current research interests include academic careers (glass ceiling, career scripts of senior academics, early academic careers), politics of higher education and research in Europe (comparison of governance structures in the European university field, rise of research evaluation), and the institutionalization of nanosciences and nanotechnologies (emergence of a job market for 'nanoscientists' and 'nanoengineers', development of new curricula).

Ben Martin, Director of SPRU (Science and Technology Policy Research) at the University of Sussex from 1997 to 2004, has carried out research for thirty years in the field of science policy. With John Irvine, he pioneered the notion of 'foresight' as a process for looking into the longer-term future of science and technology with the aim of identifying research areas and emerging technologies likely to yield the greatest benefits. Recently, he has carried out research on the benefits from government funding of basic research, the changing nature and role of the university, the impact of the Research Assessment Exercise, and the evolution of the field of science policy and innovation studies. In 2004–5 he was Deputy Chair of the EU High-Level Expert Group that put forward the rationale for establishing a European Research Council in order to pursue 'frontier research'. He is currently Chair of an ESF panel reviewing the possibilities for creating a research output database for the social sciences and humanities, and a member of the Royal Society 'Fruits of Curiosity' Group on the economic and social value of science. He has published seven books, eight monographs, and official government reports, and fifty journal articles. He is editor of *Research Policy*, and the 1997 winner of the de Solla Price Medal for Science Studies.

Frank Meier is a researcher at the Institute of Sociology at the University of Bremen. His research interests include organization studies, science studies, higher education governance, and sociological theory. Recent publications: *Die Universität als Akteur: Zum Institutionellen Wandel der Hochschulorganisation* (Wiesbaden, 2009); 'Linkages to the Civil Society as "Leisure Time Activities"? Experiences at a German University' (with Georg Krücken and Andre Müller), *Science and Public Policy*, 36: 139–44; 'Turning the University into an Organizational Actor' (with Georg Krücken), in Gili Drori, John Meyer, and Hokyu Hwang (eds.), *Globalization and Organization: World Society and Organizational Change* (Oxford, 2006).

Norma Morris is a research fellow in the Science and Technology Studies Department at University College London. Through a previous career as a senior manager at the UK Medical Research Council, and a period as

Chairman of the General Chiropractic Council (then a new statutory regulatory body for chiropractice), she has acquired hands-on experience of public policy-making and government/science relationships in the biomedical research and professional regulatory sectors, which has to varying degrees informed her subsequent research. Through a series of grants from the UK Economic and Social Research Council (ESRC) she has been able to pursue research on a number of inter-related topics. These include science policy and research governance (with special reference to developments in the life sciences), the role and contribution of volunteers in biomedical research, and issues arising in social science/natural science collaborations with particular reference to conflicting expectations around the role of 'embedded social scientists' in scientific institutions.

Uwe Schimank is Professor of Sociology at the faculty of social sciences at the University of Bremen. His research interests include sociological theory, theories of modern society, organizational sociology, science studies, governance theories, and sociology of sport. Recent publications: *Die Entscheidungsgesellschaft. Komplexität und Rationalität der Moderne* (Wiesbaden, 2005); 'Germany: A Latecomer to New Public Management' (with Stefan Lange), in Catherine Paradiese *et al.* (eds.), *University Governance: Western European Comparative Perspectives* (Dordrecht, 2009), 51–75; 'Research Evaluation as Organisational Development: The Work of the Academic Advisory Council in Lower Saxony (FRG)' (with Christof Schiene), in Richard Whitley and Jochen Gläser (eds.), *The Changing Governance of the Sciences: The Advent of Research Evaluation Systems* (Dordrecht: Springer, 2006), 171–90.

Behlül Üsdiken is currently Professor of Management and Organization at Sabanci University, Istanbul, Turkey. He has previously taught at Boğaziçi University and Koç University. His research has appeared, in addition to Turkish academic journals, in outlets such as *Organization Studies, Management Learning, Journal of Management Inquiry, Strategic Management Journal, British Journal of Management, Scandinavian Journal of Management, Business History*, and *International Studies of Management and Organization*. He served as a co-editor of *Organization Studies*, 1996–2001, and is currently the European editor of the *Journal of Management Inquiry*. His research interests are in organization theory, history of managerial thought, and history of management education.

Richard Whitley is Professor of Organisational Sociology at Manchester Business School, University of Manchester, and has held visiting professorships at universities in Hong Kong, Japan, and the Netherlands. Recent authored and edited books include: *The Oxford Handbook of Comparative Institutional Analysis* (2010), *Business Systems and Organizational Capabilities* (2007), *Changing Capitalisms?* (2005), *The Multinational Firm* (2001), *Divergent Capitalisms* (1999) (all published by Oxford University Press), and *Competing Capitalisms* (Edward Elgar, 2002). He has also edited two special issues of *Organization Studies*, one on 'The Dynamics of Innovation Systems' (2000) and one on 'Institutions, Markets and Organisations' (2005), as well as one of the *Journal of Management Studies* on 'The Changing Multinational Firm' (2003). In 1998–9 he served as the Chair of the European Group for Organizational Studies, and in 1999–2000 was the President of the Society for the Advancement of Socio-Economics. In 2007 he was elected a Foreign Member of the Royal Swedish Academy of Letters, History and Antiquities.

Introduction

1

Reconfiguring the Public Sciences

*The Impact of Governance Changes on
Authority and Innovation in
Public Science Systems*

Richard Whitley

Introduction

Over the past six or so decades since the end of the Second World War the
formal system for producing, evaluating, and coordinating published scientific knowledge—the public science system (PSS)—has undergone a
number of major changes in most industrialized societies. In addition to
the rapid and large-scale expansion of certified public knowledge published
in refereed journals, books, and other scientifically legitimate media, and
the proliferation of specialized research fields, the conditions governing the
production, coordination, and control of formal scientific knowledge have
altered considerably.

In particular, the state and other collective actors have become more
proactive in seeking to steer the direction of academic research through the
implementation of formal science policies and the incorporation of public
policy goals into funding agencies' evaluation criteria. Towards the end of
the twentieth century, such steering became more focused on the economic
payoffs from public investment in academic research, especially its impact

Earlier versions of this chapter were presented to a workshop held at the Erasmus Research
Institute of Management, University of Rotterdam, 31 Oct. 2008, and to a workshop on
Reconfiguring the Public Sciences held at the Royal Swedish Academy of Letters, History and
Antiquities, 18–20 Feb. 2009. I am grateful for the comments and suggestions made on those
occasions.

on the development of technological innovations and their associated industries (Biegelbauer and Borras 2003; Jasanoff 1997; Pavitt 2001), and led to public research institutions such as universities being encouraged to be more entrepreneurial and strategic in their behaviour (Clark 1998; Marginson and Considine 2000). These changes represent significant shifts in the overall size, organization, and governance of the public sciences and the context in which research is carried out, which have been seen by some as leading to 'post-academic' science (Ziman 2000: 67–82).

The changing role of the state in organizing the public sciences in the post-war period has been especially marked with regard to the organization of support for academic research and the governance of universities. The funding of such research has undergone, and in some countries is still undergoing, a transformation from being allocated on a predominantly recurrent, block grant, basis for institutes and universities to being dependent on success in competitive bidding for project grants, as discussed in many of the contributions to this volume.

Additionally, universities as organizations have been encouraged to become more significant strategic actors, with increasing control over 'their' resources, and more accountable to the state for their performance through various research evaluation systems (Chapters 2 and 8 below; Musselin 2007; Whitley 2008; Whitley and Gläser 2007). The combination of declining public financial support for scientific research in real terms since the 1970s in many countries, and widespread belief that academic knowledge should be a significant resource for economic competitiveness, has also intensified pressures for public research organizations (PROs) to seek revenues from the commercialization of research results and to collaborate with private companies, leading to considerable disquiet in a number of countries, especially the USA (Bok 2003; Croissant and Restivo 2001).

We can summarize these general changes in the governance of the public sciences as six important developments in the organization and control of scientific research. First, there has been a rapid expansion of the number of qualified scientists and resources for research followed by a period of much more limited growth in the public funding of scientific research in what Ziman (1994) has termed a *dynamic steady state*. This reduction in the rate of growth of state funding has often been accompanied by, second, a shift away from relatively stable recurrent support for research institutes and universities towards more *competitive project-based funding*. Third, states have developed a series of relatively proactive policies for *steering* the direction of research as part of a more general recasting of science–society relations (Drori *et al.* 2003; Guston 2000).

Additionally, many states have undertaken a substantial restructuring of higher education systems following the rapid expansion of students and staff (Braun and Merrien 1999; Clark 1983, 1995). This has involved, fourth, the formal *delegation* of some administrative and financial authority to the managers of universities and other PROs, as well as, fifth, the institutionalization of various procedures for *assessing* PRO performance and auditing their outputs. Finally, sixth, there has been a reorganization of relationships between the public sciences and private business, which has led many PROs to become more actively concerned with the *management* of research commercialization and the encouragement of academic entrepreneurship, especially in biomedical fields (Cohen *et al.* 2002; Hughes 2001; Kleinman and Vallas 2001; Owen-Smith *et al.* 2002).

These developments have changed authority relationships governing the selection of scientific goals and evaluation of results in many OECD countries. In particular, the nature of the groups and organizations that are legitimately involved in the choice and definition of research problems and intellectual approaches has altered considerably as science has become more 'collectivized' (Ziman 2000: 69–71; see also Chapter 8 below). Both the variety of authoritative agencies able to influence intellectual priorities, and their willingness to steer knowledge development proactively, have increased since the 1950s, as has the specialization and internationalization of the epistemic communities selecting research results for incorporation into the formal corpus of scientific knowledge and recognizing the intellectual merits of contributions.

Depending on the way that the public science system is organized, funded, and coordinated in each country, many of these changes have enhanced the strategic capacity and influence of funding agencies, scientific communities, and employment organizations over research strategies and performance standards. In some cases, they have also encouraged researchers to take more account of commercial possibilities in selecting problems to study and publication strategies, particularly in the USA (Colyvas and Powell 2006; Cooper 2009).

Such shifts in the authority relationships governing research priorities and evaluation criteria can be expected to have a number of consequences for intellectual coordination and innovation, mediating the impact of general changes in the governance and funding of the public sciences on patterns of intellectual development. In particular, the intensity of competition for disciplinary prestige, the extent of national and international coordination of research strategies, the time span of discretion within which researchers are expected to produce significant results, and the ease with which new scientific research programmes and fields can be established as distinct reputational organizations (Whitley 2000), are all affected by the changing balance of interests and power of these kinds of agencies. To understand how PSS have changed in the postwar period and what these changes mean for scientific development, it is important, then, to analyse variations and shifts in the importance, structure, and behaviour of these authoritative agencies.

Additionally, since these general changes in the size, governance, and organization of PSS have occurred in different ways in differently organized systems of higher education, state funding regimes, and labour markets (Clark 1983, 1995; Whitley 2003; Whitley and Gläser 2007), it is critical to take account of the major differences between national public science systems and variations in how such changes have been implemented. Despite the growing 'denationalization' of many sciences in the late nineteenth and twentieth centuries (Crawford *et al.* 1993), and the increasing influence of international scientific elites on reputational judgements and reward allocation processes, the coordination and control of research goals and results remains dominated by nationally organized higher education systems and patterns of financial support.

How states manage their relationships with universities and faculties, how authority is allocated between different groups and administrative levels within universities, and how they are connected to different groups and organizations in the wider society, continue to vary greatly between countries (Trow 1993, 1999; Wittrock 1993), as does the general financing of PROs for different purposes (Lenoir 1997; Torstendahl 1993). Such variations affect both the ease with which new scientific fields become established as legitimate areas of research, and how they do so, as Engwall *et al.* illustrate in their account of business studies in Europe and the USA.

In order, then, to analyse how recent governance changes can be expected to affect scientific research in different countries, we need to compare the key characteristics of national PSS and consider how differences in these are likely to influence both their implementation and outcomes for intellectual coordination and innovation. This comparison can usefully be done by identifying the critical features of different ideal types of PSS that highlight the major ways in which PSS have varied between states in the twentieth century. In particular, it is important to identify the different roles of state agencies, employing organizations, and scientific elites, the stratification of academic institutions and the nature of research funding arrangements, as well as changes in these, in different types of PSS.

Accordingly, in this chapter I suggest how the major changes that have taken place in the organisation and direction of PSS since the end of the Second World War have altered authority relationships governing research priorities and the assessment of results in different kinds of PSS, and how these shifts in authority have had varying effects on intellectual innovation and integration. I first summarize the key differences between six ideal types of PSS in terms of the relative authority of the state, intellectual elites, and employers in guiding intellectual goals and evaluating approaches. Second, I outline how the six shifts in governance of public science systems have affected the authority of six different groups and organizations over research activities.

Next, I suggest how these changes in the relative authority of different groups and agencies can be expected to influence patterns of intellectual coordination and innovation in the public sciences in general. Finally, I examine how these connections between changes in authority and the generation and selection of intellectual innovations in different societies are likely to be affected by the key features of the six different kinds of PSS. Overall, while some of these changes have increased the authority of research funding agencies, scientific elites, and university administrators at the expense of research team autonomy and discretion in some PSS, this is not always the case, and the degree of stratification of universities and other PROs and diversity of funding sources for research are key intervening factors in such relationships.

Authority Relations in Six Ideal Types of Public Science System

A central difference between PSS in OECD countries concerns the relative authority over research goals and standards of three sets of authoritative agents: the state, scientific elites, and the research organizations where scientific work is conducted. Variations in the relationships between these agencies, and in their internal organizational cohesion and structure, help to identify the key contrasts between different kinds of PSS and the ways in which they are changing as a result of the governance shifts outlined above. At least three distinct pairs of ideal types of PSS can be identified in terms of the relative dominance of state agencies, scientific elites, and employing organizations: *state-dominated, state-delegated, and employer-dominated.*

The first pair consists of PSS where the state retains considerable levels of control over employment and resource allocation, but differ in the degree to which states share authority with intellectual elites. In *state-centred* PSS the state not only employs researchers, allocates resources, and determines reward policies, but also incorporates scientific elites into political patronage networks that limit their autonomy and discretion. In *state-shared* PSS, in contrast, scientific elites are more independent and able to determine their own research priorities and standards, often to the extent of constituting distinct academic 'oligarchies' (Clark 1983) that control personnel decisions and the criteria for allocating resources.

In the state-delegated pair of PSS, scientists are employees of universities and other PROs, but these organizations remain largely funded and chartered by the state. The two types of state-delegated PSS differ in terms of the amount of independent discretion that researchers have over research goals and approaches relative to funding agencies, scientific elites, and PRO managers. This is lower in *state-delegated competitive* PSS than in *state-delegated discretionary* ones because in the former they have to compete intensively to gain research resources from a small number of research foundations, and so are highly dependent on the decisions of a few peer review panels and foundation priorities.

The last pair of PSS consist of situations where employers are much more able to determine employment conditions, resource allocation, and organizational structures independently of the state, but have to obtain most of their resources competitively from diverse sources. Universities and other employers of researchers in *employer-competitive* PSS are much more concerned to become scientifically prestigious by making major contributions to intellectual goals than are PROs in *employer-centred* ones, and so share considerable authority with scientific elites in making many decisions.

In considering how these types differ in terms of the relative authority of key agencies, and how this is changing as a result of the governance shifts mentioned above, it is useful to compare the extent of authoritative influence of six distinct agencies. First, there is the *state* and its associated organizations that typically fund much, if not most, public scientific research, control the framework within which universities and other PROs operate, and establish public policy priorities. Second, there is a variety of public and private research *funding agencies* that allocate funds and other resources with differing degrees of independence from the state and varied dependence on peer review for allocating resources. Third, universities and other *PROs* that provide facilities for, and sometimes employ, researchers can exercise authority over resource allocation between scientific fields as well as over the recognition and reward of scientists. Their strategic autonomy and capabilities as administratively integrated organizations vary considerably between countries and over time, as do their internal authority structures and variety of purposes (see, for example, Chapters 5 and 7–9 below).

Fourth, local *organizational scientific elites* can exercise considerable authority over research priorities, resource allocation, and personnel decisions in some PSS. This is especially the case where academic oligarchies play a major role in running research institutes and laboratories (Clark 1983, 1995), but is much less in PSS where researchers are able to gain research resources directly and manage their own project teams (Chapter 6 below). Fifth, national and international *scientific elites* exercise varying degrees of authority over the evaluation of scientific merit, competence, and significance and access to resources and material rewards. Their national cohesion and prestige also differ greatly between states and sciences, as does their ability to influence state policies and coordinate research goals across PROs, which has been considerable in the UK (as Morris shows in Chapter 8).

Finally, sixth, a considerable variety of *private commercial interests* have some influence on research priorities, including established and new firms, R&D consortia, trade associations, and commercialization agents. Important characteristics of these interests and how they influence research strategies concern: (*a*) the diversity of their goals and ways of using published scientific knowledge, (*b*) their willingness to support research for publication, and (*c*) the extent to which researchers and/or their employers control the flow of resources and purposes for which research results are used.

The relative authority of these six sets of agencies over research strategies and standards in the six ideal types of PSS are summarized in Table 1.1 and will now be further discussed, before seeing how they can be expected to alter as a result of the general changes in the governance of PSS discussed above. The extent of authoritative influence and their impact on research direction and organization in each ideal type are characterized by a three-point scale (low, medium, and high).

Beginning with the state-centred PSS, this is characterized by high levels of state authority over: (*a*) the employment and reward of researchers, (*b*) the allocation of resources between scientific fields and laboratories, and (*c*) the kinds of research that are most highly valued, usually because universities and other PROs are integrated parts of the state. Such integration means that that they have little or no formal authority as independent organizations. It also usually means that the formal heads of individual laboratories and institutes combine considerable administrative and

Table 1.1. Variations in authority in six ideal types of public science systems

Relative authority of	State-dominated		State-delegated		Employer-dominated	
	State-centred	State-shared	Discretionary	Competitive	Competitive	Employer-centred
The state	High	Medium	Medium	Medium	Low	Low
Research funding foundations	Low	Low	Low	High	High	Low
PRO administrative centres	Low	Low	Medium	Medium	Medium	High
Organizational scientific elites	Medium	High	Low	Medium	Low	High
National scientific elites	Low	Medium	Medium	High	High	Low
Private interests	Low	Low	Low	Medium	Medium	Varies

intellectual authority and may be able to develop their own long-term research programmes, although some may choose to delegate a degree of intellectual discretion to individual researchers.

In practice, of course, the actual extent of political and bureaucratic influence on research priorities and individual projects can vary considerably between states and over time, and the authority of co-opted scientific elites may be considerable, as Clark (1983) suggests has been the case in post-war Italy. However, such intellectual patrons exercise their power through the state machinery and by virtue of their being selected by state officials in these kinds of PSS, rather than as leaders of relatively independent reputational communities. They are therefore susceptible to political shifts and often change when the party in government alters, as discussed by Musselin (1999) in the French case.

In state-shared types of PSS, the state shares more authority with research institute directors and department heads—organizational scientific elites—but retains substantial formal authority over resource allocation, employment status, and facilities. Researchers remain state employees, and senior appointments in universities and other PROs still require state approval, but individual institute heads have considerable autonomy in setting research goals and managing 'their' staff to achieve them. Such autonomy is enhanced by the state providing substantial research support through the recurrent grant to PROs and departments, so that the need to obtain additional funding on a short-term and frequent project basis from research foundations is limited, as still seems to be the case in Switzerland (Liefner 2003; Chapter 3, below). The authority of national reputational elites over research goals and standards exercised through peer review of grant applications relative to that of institute heads is thus constrained in these kinds of PSS.

In more decentralized PSS, the state delegates control over employment, resource allocation, and facilities to PROs and scientific elites, while retaining the right to exercise ultimate authority over the nature and structure of PROs, the award of qualifications, and processes of resource allocation. The crucial difference between these types and the two just discussed is that these state-chartered higher education systems grant degree-awarding rights to universities as separate public corporations that are able to hire their own staff and manage their own financial affairs. They are therefore formally independent of the state, despite being largely funded by it, particularly in the twenty-first century. Additionally, in the Anglophone world the degree of vertical authority over research projects and approaches within departments and institutes of PROs is lower than in more state-dominated PSS. Thus, organizational scientific elites have less authority in these kinds of PSS than where they constitute powerful academic oligarchies.

State-delegated discretionary PSS, in particular, are characterized by high levels of researcher autonomy from both state agencies and local administrative hierarchies. Additionally, through a relatively generous block grant funding system that provides substantial and predictable recurrent funding of universities for both teaching and research activities, scientists are here able to pursue their own goals without needing to justify them very frequently to either department heads or disciplinary elites. In these kinds of PSS, individuals' scope of discretion is considerable as they compete for reputations in their specialist fields over the medium term, although it is important not to exaggerate the degree of individual autonomy in the postwar PSS most resembling this ideal type, such as the UK between roughly the 1950s and 1980, as Martin and Whitley point out in Chapter 2. As long as research facilities and materials do not require extensive external support in these kinds of PSS, we would expect them to generate considerable variety of research ideas and results with researchers not being greatly inhibited by established departmental and disciplinary boundaries, as exemplified by the work of Watson and Crick at the Cavendish laboratory in the 1950s, albeit they were working in a unit of the Medical Research Council rather than being employees of Cambridge University.

In contrast, state-delegated competitive PSS constrain the work of individual researchers much more by making them more dependent on gaining resources from a few funding agencies reliant on peer review procedures for allocating resources. In these ideal types of PSS, closely resembling perhaps the Australian case in recent decades (Gläser *et al.* 2007, and Chapter 10 below), authority over research priorities and approaches is concentrated in the hands of public research foundations and their advisers from the different sciences. While, then, PROs ostensibly have considerable autonomy from the state and their employees are not highly constrained by administrative hierarchies, in practice both are quite dependent on success in highly competitive contests for limited resources governed by current scientific elites' standards for deciding which projects can be expected to make significant contributions to intellectual goals.

The last two types of PSS to be considered here are characterized by limited levels of state authority, usually because states readily grant charters to universities and other PROs without much regulation about what they can do or how they are organized. In the case of the newly independent USA, for instance, 'the granting of charters came to be regarded as an aspect of political patronage and the spoils system.... (it) carried no promise, explicit or implicit, of financial support from the state government' (Thelin 2004: 43), and after the 1819 Supreme Court decision in the Dartmouth College case, few states challenged the powers of college presidents and 'their' trustees (ibid. 70–3). As employing organizations, PROs in employer-dominated PSS are usually able to establish their own policies and procedures for recruiting, managing, and rewarding researchers, to establish their own research priorities, and to shift resources between different fields and topics. They do, though, have to compete for these resources from a variety of public and private sources and to provide a range of services in order to gain them. In these kinds of PSS, then, employers are key authoritative agents affecting scientific research through their investment decisions, provision of support and rewards, and organization of research and teaching activities.

Two different types of such employer-influenced PSS can be distinguished in terms of: (*a*) their willingness to share authority with reputational elites in pursuing scientific prestige, (*b*) the strength of research foundations and other funding agencies, and (*c*) the ability of researchers to control their own research resources. Employers in employer-competitive PSS compete for social and intellectual reputations by attracting and supporting research leaders in different fields, as well as for students and funds. They therefore invest in the researchers and faculties that they think will contribute most to disciplinary and specialism goals, and seek resources from a wide range of different agencies and organizations. As Geiger (1986: 10) suggests in his account of the rise of American research universities, towards the end of the nineteenth century there was 'above all competition for prestige—an ineffable combination of publicity, peer esteem and pride ... For the academic development of research universities the most significant competition was for research-minded faculty.'

As long as there are a considerable number and variety of funding sources, competition here encourages organizational and intellectual flexibility and diversity of goals and approaches. Furthermore, the more those individual scientists are able to access research support funds on their own, as quasi-independent entrepreneurs, the more they will be able to pursue different objectives and become relatively independent of their immediate employer. This is especially likely when those funds include substantial contributions to employers' overhead costs, as in the post-war USA (Geiger 2004). In these kinds of PSS, then, employers' ability to allocate resources between competing investments and project teams, and thereby achieve particular kinds of organizational goals, is constrained by the specialist knowledge, skills, and resource-controlling powers of individual scientists and groups.

In more employer-centred PSS, in contrast, both the level of organizational competition for intellectual prestige and the availability of resources from external agencies are less. Here, scientists are dependent on their employer for funds and facilities required to conduct research. This seemed to have been the case in many US universities in the pre-war period (Geiger 1986), as is illustrated by Terman's encouragement of research activities at Stanford through obtaining funds from local businesses and supporting new firm formation based on research results (Hughes 2001; Leslie 2000; Lowen 1991).

Such dependence on employers' resources limits both researchers' autonomy and the authority of reputational elites, especially if universities have to compete to attract the best students through the provision of extensive student support services, not to mention the investments needed to be competitive in college athletics and other sporting contests in the USA (Bok 2003). The search for external funding may, though, enable some private foundations with an interest in science, such as the Rockefeller in the inter-war period, to exercise considerable influence on research priorities and approaches, and facilitate the growth of new fields such as molecular biology (Kohler 1979). In general, the relatively limited control over research priorities and significance standards exercised by reputational elites in these kinds of PSS restricts the degree of intellectual coordination around particular disciplinary goals, and allows researchers in different PROs to establish distinct schools of thought that are only weakly integrated.

The Impact of Governance Changes on Authority Relations in Science

The relative authority of the state, scientific elites, and employers in these six ideal types can be expected to alter as a result of the governance changes summarized above. In Table 1.2, I suggest how the roles of six key agencies exercising authority over research priorities and approaches are likely to have changed in the post-war period. First, the rapid growth of (*a*) certified researchers producing knowledge for publication, (*b*) published research reports and, (*c*) journals and books devoted to them between the 1950s and the 1980s, produced PSS that are massively larger than those typical of the first half of the twentieth century, and encompass many more areas of study. In terms of the internal organization of scientific fields, this

Table 1.2. Expected effects of governance changes in public science systems on the authority of different agencies

Changes in the influence of authoritative agencies	Steady state funding	Increased project-based funding	State steering of research priorities	Delegation of authority to PROs	Consequential monitoring of PROs	Increased PRO management of commercialization
State ministries	Increased	Increased where state guides project priorities	Increased	Some reduction	Increased	Reduced
Research foundations	Increased	Increased	Increased		Increased where performance standards depend on project funding success	Reduced
PRO managers				Increased where they control discretionary resources and employment	Increased	Increased where controlled by PRO centres
Organizational scientific elites				Reduced	Reduced	Reduced where controlled by PRO centres
National scientific elites	Increased where funding depends on peer review	Increased where funding depends on peer review	Reduced		Increased where funding depends on peer review judgements	Reduced
Commercial interests	Increased		Increased where focus is on technological innovation		Increased where performance assessment involves commercial success	Increased

expansion has accelerated the specialization of the reputational organizations coordinating published research as well as strengthening the prestige hierarchy of many publications (Weingart 2003; Whitley 2000).

This intensification of scientific specialization was facilitated by the post-war growth of public support for scientific research and higher education in many countries, which made it easier for young scientists to establish new research goals and approaches without having to supplant existing intellectual elites (Hagstrom 1965). Particularly in higher education systems where epistemic authority was not tied to positions of organizational authority and individual researchers were able to pursue their own concerns without having to follow the goals and approaches of institute and department heads, as in many Anglophone countries, this meant that they had considerable discretion and autonomy from both intellectual elites and administrative managers.

However, when this public support ceased to grow at the same rate as the production of qualified researchers, the move to a more steady-state pattern of funding increased scientists' dependence on the reputational elites that govern intellectual standards, as well as on research funding agencies where they required external financial support. Overall, then, the slowdown in growth of state funding for public scientific research can be expected to increase the authoritative influence of state ministries, public and private foundations, and commercial interests on research priorities. As public support declines, scientists seek resources from other organizations with different goals, as seems to have happened in the USA in the 1980s and 1990s, although the federal government here remains by far the greatest source of academic research support (Geiger 2004: 134–40).

In addition to the level of such funding being reduced in real terms and sometimes nominally as well, many governments have also reduced the proportion allocated on a block grant basis relative to that awarded competitively to research projects. Depending on how such assessments are made, and on the proportion of funding for published research that comes from the state, this shift in the prevalent mode of allocating research resources seems likely to intensify the consequences of moving to a steady-state funding regime.

Making scientists compete for resources to conduct discrete projects whose outcomes can be reliably predicted can be expected to increase researchers' dependence on established disciplinary elites. The more that scarce resources are allocated through peer review of the scientific merits of project proposals, the more scientists will propose projects for dealing with problems that are considered significant in terms of current standards and fit in with established perceptions of predictable outcomes (Ziman 1994: 107). As many informal anecdotes and conventional wisdom suggest, the chances of gaining project funding are greatly increased when much of the preliminary work has already been done and the requested funding is fairly sure to produce interesting, but not too radical, results (Laudel 2006).

Additionally, of course, increased dependence on project grant funding from research foundations and similar agencies results in their administrators and advisers acquiring more influence over research goals, not least through their selection of referees and management of the evaluation process. While such influence depends on foundations being able to develop their own strategies for achieving policy goals, it can become quite considerable and affect the direction of whole fields, as in the case of the Rockefeller foundation under Warren Weaver and the growth of molecular biology (Kohler 1979; Yoxen 1982). Moving away from recurrent block grant to project-based funding of research also facilitates state steering of research priorities in some PSS, as ministries are more able to influence the selection criteria governing resource allocation on a project basis than is usually the case with recurrent funding of chairs and institutes.

Dependence on established scientific elites' standards for assessing the merits of project proposals may be modified by the third major change in the governance of the public sciences since the end of the Second World War: the increasing willingness of state agencies and policy elites to steer public scientific research towards particular policy objectives. As many commentators have suggested, relationships between politicians, bureaucratic elites, and scientists have undergone several changes since 1945 in most OECD countries, but most have involved greater state attempts to ensure policy payoffs from public investment in research (see, for example, Guston 2000; Meulen 2007; Martin 2003). From the initial post-war compact between policy-makers and scientific elites, whereby increased public funding was provided for research on the basis that it would eventually lead to both public and private benefits, many governments have intensified their efforts to ensure that publicly funded scientific research contributes to specific policy goals.

In some cases, this has led to an increased emphasis on project-based funding from public research foundations being tied to particular policy-focused programmes that are initiated, developed, and coordinated by public officials, such as those of the US National Institutes of Health (Cozzens 2007; Stokes 1997). As many states have reduced the amount of research support provided through block grants to PROs, and thus increased the dependence of scientists on research foundations and their peer review panels, this has further encouraged the growth of research combining the search for fundamental mechanisms with contributing to social purposes in what Stokes (1997) has termed 'Pasteur's quadrant'.

Depending on how states and research foundations manage the implementation of such policy objectives in resource allocation procedures and decisions, and the general diversity and munificence of sources supporting research in the public sciences, this kind of state steering of research priorities can increase researchers' independence from current scientific elites and broaden their choice of research problems to investigate. However, it does of course also increase the potential influence of state ministries on intellectual priorities, as well as the importance of commercial considerations where technological payoffs are expected in return for state funding.

The importance of such public policy goals and their incorporation into research foundation procedures and practices has grown with the fourth and fifth sets of changes to many PSS in the post-war period: the restructuring of universities and other PROs, especially their governance, financing, and evaluation. As many governments increased their expectations of how these organizations could contribute to the social and economic welfare of society, states initiated a variety of reforms that altered their formal status, powers, and responsibilities. These became more systematic and widespread in their impact as the expansion of state funding of higher education began to slow down and change its basis.

In some countries, such as the UK and Australia, these new demands and complexities have led to increasing state coordination and direction of universities in exchange for public funds, while in much of Continental Europe and Japan they have encouraged states to separate them organizationally from the civil service and to delegate more administrative authority to their management (Chapters 4 and 7 below). In both cases, though, there has been a considerable increase in political and bureaucratic monitoring of PRO performance and the establishment of novel procedures for evaluating this (Chapter 9 below; Whitley and Gläser 2007). In principle, such delegation of authority to the central administrations of universities and other PROs should encourage them to develop some strategic autonomy and capabilities, but this varies greatly between states and organizations, as well as being inherently limited by the uncertainties involved in producing new knowledge and academic teaching processes and the influence of scientific elites (Musselin 2007; Whitley 2008).

Often premised on the assumptions of the 'new public management', and the view that universities producing new scientific knowledge could be managed and assessed in the same way as other publicly supported organizations (Schimank 2005), such demands for accountability, transparency, and 'excellence' (Weingart and Maasen 2007) have encouraged many university managers to imitate what are considered to be best practices in the private sector. In particular, they have invested in formal monitoring and evaluation procedures to meet state demands, developed organizational research strategies, and attempted to manage resource allocation between fields and faculties as a strategic activity.

As van der Meulen (2007) has emphasized in the case of Dutch universities, various state initiatives to link funding to group research programmes, and to encourage national evaluations of research performance, have provided the basis for university administrations to standardize the unit of analysis in comparing the achievements of departments and faculties in relation to the resources consumed. Such rationalization of scientific research activities—at least formally—has occurred quite widely amongst OECD countries and facilitates the management of PROs as project-based organizations in which managers can allocate resources between competing components of investment portfolios (Whitley 2006).

In principle, such standardization and evaluation can lead to increased employer coordination of research goals and control over resource allocation, with a consequent reduction in the powers and independence of senior academics, especially in higher education systems previously dominated by academic 'oligarchies' (Clark 1983; Chapter 5 below; Schimank 2007; Whitley 2007). This strengthening of the middle organizational layer between state ministries and researchers has often been seen by policymakers as a way of emulating certain characteristics of the post-war US high-education system and thereby gaining some of its perceived advantages for the wider society. It has also enabled politicians to delegate responsibility for managing the consequences of reduced public funding of higher education to universities and similar organizations (Trow 1999).

In addition to such delegation and performance monitoring potentially increasing the strategic autonomy and capabilities of PROs, they can also enhance the influence of established scientific groups on research priorities where evaluations depend greatly on the judgements of intellectual elites to evaluate quality, as in many of the UK's research assessment exercises (Chapters 2 and 8 below). Depending on the significance of technological and commercial criteria in such assessments and their impact on resource allocation, they may additionally increase the influence of business interests on research choices.

An important feature of this restructuring of state–university relationships in the last few decades has been the active encouragement by many governments of closer university links with private industry and greater academic involvement in commercial activities (see, for example, Geiger and Sa 2005; Woolgar 2007). While not uncommon in many countries such as France, Germany, Japan, and the USA before the Second World War (Metlay 2006; Odagiri 1999; Homburg 1992; Shinn 1979), such connections have become more institutionalized at the organizational level, with many universities and other PROs playing a more systematic and strategic role in commercializing research results since the 1970s.

One of the more visible signs of this increased institutional commitment is the widespread establishment of technology transfer offices and similar administrative units in many research universities in the Americas, Europe, and Japan towards the end of the twentieth century, despite their limited success in contributing to university funds in most cases (Kruecken 2003; Kruecken and Meier 2006, Kneller 2007*a*; Siegel *et al.* 2003). Another has been the expansion of university patenting activity, particularly in the USA since the passage of the Bayh-Dole Patent and Trademark Act Amendments of 1980 (Powell *et al.* 2007). As Mowery *et al.* (2004) and others have found, though, such patents vary greatly in their commercial payoffs, with many failing to recover the costs involved (Geuna and Nesta 2006).

This growing emphasis on the commercialization of research outcomes, especially in biomedical fields, has led some to see the increasing role of private business interests in guiding research strategies and university policies as a significant shift in their governance, and to criticisms of 'academic capitalism' and the commercialization of higher education (see, for example, Bok 2003; Croissant and Restivo 2001; Krimsky 2003; Owen-Smith 2003; Slaughter and Leslie 1997). Depending on how such growing influence of private interests and concerns takes place in different national contexts, it could reduce the power of established reputational elites to control research priorities and the coordination of research. It may also restrict the powers of universities as employers to coordinate research strategies when individual star scientists become able to establish successful spin-off companies that generate significant revenues (Powell and Owen-Smith 1998), just as the growth of federal funding in the USA increased the bargaining power of Dr 'Grant Swinger' (Greenberg 1966).

These shifts in the relative influence of different agencies on research strategies and resource allocation as a result of the governance changes discussed above can be summarized as five changes in authority relations. First, researchers have become much more dependent on peer-reviewed project-based funding of proposals and so on established scientific elites. Second, public policy goals and funding agencies are much more influential in setting research priorities. Third, PROs in some countries have developed greater strategic autonomy as organizations. Fourth, PROs have become more subject to systematic, formal, and consequential state monitoring of their performance. Fifth, there are increasing opportunities for, and encouragement of, research commercialization by scientists and PROs, potentially leading to greater influence of both PROs and private business interests on research goals.

Overall, the extent of direct state control over intellectual priorities and personnel, and resource allocation, seems to be declining in many PSS, but so too is researcher discretion over problem formulation and time horizons, particularly in state-delegated discretionary PSS. The authority of academic oligarchies has also been reduced in some PSS, while that of PRO managers, funding agencies, and international scientific elites has probably grown.

Impact of Changing Authority Relations on Intellectual Coordination and Innovation in the Public Sciences

Many of these changes in governance and authority relations can be seen as encouraging conservative research strategies as scientists compete for limited funds and have to demonstrate more frequently to epistemic elites how their research contributes to collective intellectual goals in order to gain recognition and rewards (Geuna and Martin 2003; Schiene and Schimank 2007; Tabil 2001). In particular, the combination of steady-state funding, project-based resource allocation and formal, public monitoring of PROs' performance may well restrict intellectual novelty and encourage scientists to work on mainstream topics with established techniques and concepts in preference to tackling interdisciplinary problems with a variety of novel methods and approaches that challenge established boundaries and identities.

In very general terms, the two key aspects of scientific development that are likely to be most affected by the governance changes just outlined concern, first, the generation of new ideas and results, especially the variety of different ones, and, second, the interpretation and selection of these contributions as being significant advances to established knowledge. If the modern sciences are indeed evolutionary systems of knowledge production that blindly produce variations to be selectively retained, then their key components are those that (*a*) generate variety in intellectual innovations and (*b*) select the best supported knowledge claims for inclusion in the institutionalized body of formal knowledge (Ziman 2000: 276–82).

The more that the *ex ante* design of projects to fit public policy goals and produce expected results is emphasized over the *ex post* selection of diversely produced knowledge claims, the more limited the range of blind variations may become (Braben 2004).

However, not all of these changes in authority relations can be expected to have the same impact on intellectual variety and selection, and the mechanisms through which they operate on scientific development are affected by variations in the organization of PSS between countries, particularly the stratification of higher education systems, the diversity of funding agencies and their goals, and the roles of state ministries and PROs in exercising authority over research priorities and rewards. To consider their likely implications for novelty and selection more systematically, it is useful to distinguish between the following four aspects of intellectual coordination and innovation.

1. The intensity of competition for disciplinary prestige based on contributions to the dominant intellectual goals of established scientific disciplines. This affects the extent to which scientists concentrate on problems and approaches that are central to their discipline, as distinct from pursuing a wider variety of concerns with a range of concepts and methods from different fields.

2. The strengthening of national and international intellectual coordination of research goals and approaches across universities and national public science systems, which reflects the interdependence between researchers in different locations and so their ability to develop idiosyncratic research programmes in separate institutes and departments (Whitley 2000).

3. The shortening of the period within which research is expected to lead to interesting results, and so the feasibility of undertaking long-term commitments to major intellectual problems.

4. The ease with which scientists are able to establish new fields that draw on novel concepts and approaches, and particularly their ability to obtain resources and gain intellectual legitimacy.

   The likely connections between changes in authority relations and these aspects of intellectual innovation and integration are summarized in Table 1.3 and will now be more fully discussed.

The combination of reduced growth in public support for scientific research and increasing reliance on peer-review judgements of the quality of project proposals for allocating such support can be expected to intensify

Table 1.3. Expected effects of changing authority relations on intellectual innovation and coordination

Expected effects on innovation and coordination	Increased researcher dependence on epistemic elites for project-based funding and reputations	Increased influence of public policy goals and foundations on research funding	Increased PRO strategic autonomy and capability	Increased state monitoring of PRO performance	Increasing opportunities and encouragement of research commercialization by PROs
Competition for disciplinary prestige	Intensified	Reduced	Intensified in low prestige PROs, weakened in elite ones	Intensified in low prestige PROs where evaluations have major financial consequences	Reduced
National and international coordination of research goals and results	Strengthened	Reduced	Strengthened in low prestige PROs	Intensified where dominated by peer-review judgements of quality	Reduced
Researchers' time horizons for producing significant results	Shortened	Shortened	Reduced in low prestige PROs, probably increased in elite ones	Shortened, except in elite PROs with access to discretionary resources	Shortened
Ease of establishing new scientific fields	Reduced	Increased where state provides new resources	Increased where elite PROs have access to discretionary resources	Reduced	Increased for fields close to technological opportunities

competition for disciplinary prestige within established fields as scientists seek to persuade their senior colleagues of the significance of their projects' contributions to disciplinary and specialism goals. This is especially likely when funding panels and their advisers are organized around established disciplines and the variety of research support agencies is limited, as in Australia (Gläser and Laudel 2007; Chapter 10 below).

Such peer-reviewed and project-based support should also strengthen the *national and international coordination of research goals and results* across departments and institutes, as applicants have to demonstrate how their projects contribute to the collective intellectual purposes of national and international epistemic communities. By making access to public support dependent on departments' performance as judged by international standards, states reduce the variety of idiosyncratic research goals and approaches adopted by groups in different universities, as happened in Dutch philosophy departments in the 1980s (Meulen and Leydesdorff 1991).

Correlatively, increased competition for project-based funding shortens *the time horizons for producing significant results* as researchers design projects that can be reliably expected to generate publishable outcomes within two or three years (Ziman 1994: 107). Where project funding contributes to university overheads such pressures are likely to narrow the scope of many research problems, especially if they cannot readily be decomposed into separate projects and/or require the coordination of different kinds of expertise and knowledge. Scientists are less likely to work on large-scale issues that appear intractable in the short to medium term in these circumstances, particularly where they involve the development of novel research skills and technologies at the expense of existing ones.

They will also be discouraged from investing considerable amounts of energy in *developing new research fields* around novel kinds of goals and approaches. Where the reputational benefits of such innovations appear unlikely to be achieved within the usual project funding cycle, researchers will be reluctant to attempt to establish new areas that transgress established intellectual boundaries and draw upon methods and skills from different disciplines. This is especially so when resources are limited and are concentrated in a small number of research foundations reliant on a narrow group of advisers.

Some of these tendencies may be counterbalanced by the second major change in authority relations: the growing influence of public policy goals and foundation officials on the allocation of resources for research. To the extent that such influence steers research funding away from established disciplinary priorities and significance standards, it should reduce the intensity of intra-disciplinary competition and also weaken the need for applicants to integrate their projects around disciplinary objectives. By broadening the range of purposes for which research support is available, the incorporation of policy objectives into research foundation objectives could increase the variety of intellectual activities. Given the reliance on project-based funding and the need to demonstrate useful results from such support, though, it seems unlikely that this will greatly reduce pressures to fund research that provides fruitful results in the short to medium term.

Depending on how states and research foundations manage the incorporation of public policy goals into funding procedures and decisions, as well as on the general diversity and munificence of organizations supporting the public sciences, this kind of state steering of research priorities can encourage the establishment of new scientific fields, as arguably has happened in the case of US post-war investment in computer science and health sciences. While this encouragement of policy-related research goals can increase scientists' independence from current scientific elites and broaden their choice of research problems to investigate, it seems less likely to happen when resources are more limited and researchers are highly dependent on the decisions of a small number of foundations. In these latter PSS, the variety of intellectual innovations and willingness to pursue what might be regarded as deviant goals are likely to be quite constrained—particularly those that focus on long-term fundamental processes.

Considering next the impact of increasing the autonomy and authority of PRO central administrations on intellectual innovation and selection, this is likely to differ between types of public science systems—particularly their stratification, the availability of discretionary resources to particular PROs, and the nature of the competitive environment. In principle, delegating administrative and financial authority to PRO managers could enable them to pursue a variety of strategies and invest in different kinds of science in order to distinguish themselves from competitors. This might facilitate the establishment of new fields and encourage scientists to undertake more radically innovative research. Depending on the availability of resources and how these are allocated, some PROs might be willing to encourage researchers to study wide ranging problems that may only pay off reputationally in the long term.

However, less prestigious institutions that have few of their own resources available for such risky strategies are unlikely to encourage scientists to challenge disciplinary orthodoxies unless they are strongly supported by state agencies pursuing public policy goals. Conversely, elite universities in highly stratified academic systems, which are able to commit substantial resources that are not so dependent on peer review, may feel able to support staff investigating unfashionable issues and/or complex interdisciplinary problems. Such organizations are also less likely to respond directly to state attempts to steer research towards public policy goals and resist the establishment of new fields that do not seem to contribute to traditional 'high science' ideals, such as engineering and business studies at the universities of Cambridge and Oxford. Rejection of professional education and research is, though, less likely in countries where universities are more dependent on market demand than on central state support, as in the USA (Chapter 11 below; Geiger 1986).

Similarly, the effects of the fourth major change in authority relations, the intensified performance monitoring and sanctioning of PRO performance, can be expected to vary between organizations and funding regimes (Whitley 2007). Where such monitoring has significant consequences for university finances and is closely tied to success in gaining external, peer-reviewed project funding, as has been the case in Australia (Gläser and Laudel 2007), there seems little scope for pursuing deviant strategies or challenging established orthodoxies. In these conditions, the formal organizational autonomy from the state and considerable central managerial authority of PROs are more than counterbalanced by the tightly constrained availability of research funding and its concentration in a few foundations.

In contrast, where universities are not so dependent on a limited range of state funding councils for discretionary resources and/or state assessment of their performance is not so consequential for their finances, they may not feel constrained to follow the results of such evaluations and can continue to cross-subsidize weaker departments and support long-term investments. This is obviously easiest for the most prestigious organizations that can rely on endowment income and similar discretionary resources, as may have happened in the early days of the British RAE (Chapter 2 below).

Finally, turning to consider the likely effects on intellectual coordination and innovation of encouraging more commercialization of research results, this could reduce the intensity of disciplinary competition for intellectual prestige and coordination of research strategies as the power of established reputational elites to control research priorities becomes limited by the broader availability of funds. It might, then, increase researcher autonomy from mainstream priorities to pursue unfashionable avenues and so add to intellectual variety. However, commercialization pressures are perhaps more likely to narrow research strategies and focus scientists' attention on technologically relevant problems that are resolvable in the short to medium term. In any event, the stronger such pressures become, the more they can support the development of new research fields that appear to offer useful knowledge for private interests and be seen to orient problem choice towards such purposes, as in the USA (Cooper 2009).

Consequences of Changing Authority Relations in Different Public Science Systems

These general connections between changing authority relationships and patterns of intellectual coordination and innovation in the public sciences can be expected to be affected by differences between the types of PSS discussed above. Many of the changes in state–university relations, for instance, will have limited effects on scientists' behaviour in employer-dominated PSS where the state already has delegated considerable powers to PROs and there is a strong tradition of organizational independence from national authorities.

In Table 1.4 I suggest how the five kinds of changes in authority relationships discussed above are likely to have contrasting effects on scientific integration and innovation in differently organized types of PSS, focusing especially on the different effects expected between these types. It is worth emphasizing here that the changes in authority are themselves not independent of the nature of different PSS, and the way that they occur does to some extent vary between, say, state-centred and state-shared ones.

Considering first the impact of increasing researcher dependence on scientific elites in different kinds of PSS, the general intensification of disciplinary competition and reduction in diversity of intellectual goals seems likely to occur in both state-dominated and state-delegated PSS. As scientists have to compete more for publicly provided resources that are primarily allocated on the basis of intellectual contributions to collective goals, they become more dependent on the judgements of current reputational elites and funding agencies. They are therefore less likely to invest in 'deviant' research strategies and attempt to establish new kinds of scientific fields that use unorthodox techniques.

These effects are less likely to be so strong in state-centred PSS where the state delegates only limited authority to scientific elites, and/or where there is a considerable variety of different kinds of funding agencies pursuing diverse objectives. Particularly with regard to the intellectual integration of research goals and approaches around current disciplinary significance

Table 1.4. Effects of changes in authority relations on intellectual coordination and innovation in different types of PSS

Effects on intellectual coordination and innovation	Type of PSS	Increased researcher dependence on epistemic elites for project-based funding and reputations	Increased influence of public policy goals and foundations on research funding	Increased PRO strategic autonomy and capability	Increased state monitoring of PRO performance	Increasing opportunities and encouragement of research commercialization by PROs
Intensification of competition for disciplinary prestige	State-dominated	Increased	Reduced in S-centred PSS, limited impact in S-shared ones	Increased in S-shared PSS with strong disciplinary elites	Increased when tied to resource allocation in S-shared PSS	Reduced
	State-delegated	Increased	Reduced in S-delegated competitive PSS with weak academic oligarchy	Increased in S-delegated competitive PSS	Increased when tied to resource allocation	Reduced in S-delegated competitive PSS
	Employer-dominated	Limited impact	Reduced in E-Comp PSS, limited impact in E-Centred PSS	Limited impact	Limited impact	Limited impact
National and international reputational coordination of goals and results	State-dominated	Limited impact in S-centred PSS. Increased in S-shared PSS with stratified HE system	Reduced	Increased where PROs' prestige and resources depend on intellectual reputations	Increased in both types of state PSS where dominated by peer-review judgements of quality that affect funding	Reduced
	State-delegated	Increased, especially with stratified HE system and limited variety of funding sources	Reduced	Increased in S-delegated competitive PSS Limited impact in S-delegated discretionary and employer dominated PSS		Reduced in S-delegated competitive PSS
	Employer-dominated	Limited impact	Reduced, with limited impact where there is a variety of funding sources		Limited impact	Limited impact

Researchers' time horizons for producing significant results	State-dominated	Reduced	Reduced	Limited impact in S-centred PSS, reduced in S-shared and S-delegated PSS	Reduced	Reduced in all types, unless leading to endowment-type funding that supports long-term projects, probably in elite universities
	State-delegated Employer-dominated	Reduced	Reduced		Reduced	
	Employer-dominated	Limited impact	Limited Impact	Limited impact	Limited impact	
Ease of establishing new fields and goals	State-dominated	Reduced in all state PSS, except for direct state investment	Increased, except where scientific elites are powerful	Increased where PROs have access to discretionary funding, otherwise limited impact	Reduced for all types, especially in stratified systems where reputational elites are strong	Increased for all types where research commercialisation supports public research
	State-delegated Employer-dominated	Limited impact	Increased			

29

standards, the authority of scientific elites may be less enhanced by such changes in state funding in state-centred and employer-dominated PSS.

It is perhaps worth mentioning a further 'intervening variable' in such relationships, which affects the differentiation of PROs and the reproduction of elite status: the stratification of universities and other PROs. Where this is high, we could expect the effects of reduced funding and growing reliance on competition for project proposals to be strongest for those researchers based in relatively low-prestige institutions. Since the intellectual elites determining the criteria according to which research projects are assessed as significant and competent are likely to work in the more prestigious universities where they are also likely to have better facilities, they are also more likely to be successful in gaining resources under the new, more competitive, regime.

Again, this outcome will be mitigated in PSS where there are diverse funding agencies with varied goals and groups of advisers, as in the USA where some universities have developed strategies to 'circumvent the tradition of scholarly review boards … [by persuading] a supportive member of Congress to attach a "rider bill" to establish a research project as an obscure part of some larger federal works legislation' (Thelin 2004: 356). However, in state-dominated PSS, the same elite is likely to be closely connected to political and bureaucratic elites which may well favour the more prestigious institutions, as seems to have happened in Japan where the former imperial universities have disproportionately become centres of excellence (Kneller 2007b; Chapter 4 below).

Second, while the incorporation of public policy goals into the allocation of research funding is generally likely to mitigate many of these effects of researcher dependence on scientific elites, it is particularly likely to do so when those elites are not so dominant that the reputational risks of pursuing research contributing more to such goals than to purely intellectual ones are very high. Thus, competition for disciplinary prestige may well be reduced by such public policy goals in state-centred PSS and state-delegated ones where disciplinary elites are not dominant. However, in PSS where there is a strongly entrenched academic oligarchy, this effect is less likely to be noticeable.

Establishing new fields that contribute both to policy goals and to intellectual understanding should be easiest in state- and employer-dominated PSS, and rather less so in state-delegated ones, where scientific elites are cohesive and prestigious. As in the previous cases, the effects of such changes in state funding regimes are likely to be limited when resources are available from a variety of non-state agencies and foundations.

Turning next to consider, third, the probable effects of increasing PRO autonomy in different kinds of PSS, this seems likely to have greatest impact on the more state-dominated higher education systems. It is doubtful if states that have developed highly centralized PSS over many centuries will in fact delegate substantial decision-making control to university heads, at least in a few decades, or that many administrators will believe in the permanence of such formal delegation when political and/or financial pressures become significant, as in the case of France (Merrien and Musselin 1999; Musselin 1999). The likelihood of universities developing much strategic actorhood and influence over the research priorities of 'their' staff as a result of such shifts is, then, limited in state-centred PSS. Where the state shares authority with academic elites to a greater extent, this restructuring of state–university relationships is often intended to grant more responsibility for managing resources and activities to universities and to encourage their central administrations to exercise more influence over academics, especially institute heads (see, for example, Chapters 4 and 7 below). To the extent that it does in fact generate and enhance such organizational capability to manage resources and change activities in state-shared PSS, it could lead to more variety between universities as they compete for resources and prestige, including perhaps their establishment of new departments and research areas.

This increased heterogeneity would, though, depend on the general availability of resources for such initiatives and the extent to which current reputational elites dominate intellectual standards and resource allocation criteria. Where PROs are highly dependent on their staff obtaining resources for research from a small number of foundations dominated by disciplinary elites, they are unlikely to invest in supporting radically novel projects and skills. This is even more probable when the PSS is highly stratified so that the formal separation of universities from state ministries and of academics from state employment does not really change the dominance of a few elite institutions over the higher education system, as in Japan (Chapter 4 below).

Furthermore, the willingness of the newly empowered deans and presidents of universities to exercise their authority over departments and institutes remains quite limited in many of the European PSS where some formal delegation has taken place (see, for example, Muller-Camen and Salzgeber 2005; Schiene and Schimank 2007). Given the novelty of these changes, and their susceptibility to revision as politicians and bureaucrats change their minds, this is perhaps not too surprising. Such centralization of organizational authority could increase when new appointments to these roles are not so dependent upon collegial support, and universities begin to function as independent employers, as well being more able to obtain resources from a wider range of funding agencies and organizations.

In the case of more state-delegated PSS where PROs (*a*) already have some autonomy from the state, (*b*) are able to shift resources between departments and activities, and (*c*) to make their own employment decisions, the effects of any increase in state delegation seem likely to enhance their strategic actorhood at the expense of individual researchers and departments. This is especially probable where the political rhetoric associated with such delegation encourages university administrators to act more like private company heads and exercise control over their employees' activities, as in the UK (Chapter 2 below).

The development of organizational research strategies and formal procedures for making trade-offs between investment alternatives at central university, faculty, and departmental levels that has become such a feature of many PROs in the Anglophone world reflect, such pressures, and can be expected to limit the discretion of researchers and groups. While some of these strategies remain largely formal, and are often vague lists of aspirations rather than systematic attempts to allocate resources preferentially to realize specific objectives, they do represent a reduction in the overall independence of scientists from their employers and in the possible diversity of research goals and approaches, especially in state-delegated competitive PSS.

The extent to which such increasing state delegation and encouragement of strategic autonomy additionally enables PROs to make strategic investment decisions that conflict with current disciplinary priorities depends on the concentration of resource control in one or a very few public research foundations and their reliance on peer-review advice from scientific elites. Where scientists and universities as a whole are highly dependent on research funds from a small number of foundations pursuing intellectual goals, the realistic level of autonomy from such elites will be limited. If, on the other hand, universities and other PROs are able to attract funds from a variety of foundations, private interests, licensing, and so on that enhance their discretion over resource allocation, they may well increase their strategic capabilities, at least as far as making differential investments in favoured fields and researchers is concerned (Owen-Smith 2001).

Considering, fourth, the impact of increased state surveillance and assessment of PRO performance on intellectual coordination and innovation in different types of PSS, this is likely to be most marked in societies where the state continues to provide the bulk of resources for research, and legitimates the public status of universities and other PROs, while sharing considerable authority with scientific elites over the determination and implementation of the standards used to assess performance. It is, then, in state-shared and state-delegated PSS, that the impact of what can be termed strong research evaluation systems should be greatest (Whitley 2007). These combine standardized procedures for assessing the quality of research outputs with the publication of results for each 'unit of assessment', as the British Research Assessment Exercise (RAE) construed knowledge production groupings, and with the allocation of resources on the basis of such evaluations.

The more such performance auditing is conducted publicly according to formal and standard procedures that enable policy-makers and the public to rank PROs in terms of their research 'excellence' (Weingart and Maasen 2007) and leads to differential funding for research according to such *ex post* evaluations, the more researchers come to depend on the verdicts of the scientific elites that make such judgements. Such dependence can be expected to make it more difficult to establish new fields that combine new techniques from different areas and to pursue radically novel intellectual goals.

This will be especially so in PSS that are so highly stratified that the standards governing evaluations tend to be established and policed by scientific elites based at the most prestigious and well-endowed universities. The stronger the research evaluation systems are in these kinds of society, the more likely that they will reinforce such elite standards and the Matthew effect become more marked. To some extent, the insistence on making qualitative judgements of research outputs according to national and international norms of scientific significance in the evolution of the British RAE can be seen as reinforcing the standards and goals of the scientific elite and the institutions where they are mostly located (see Chapters 2 and 9). As a result, it may well have reduced the likelihood of establishing new kinds of research fields and developing novel techniques drawing on ideas and methods from different sciences.

It is important to note here, though, that scientific elites are not necessarily passive recipients of these kinds of governance changes, and in some case may be active participants in their implementation. As Morris emphasizes in her chapter, and Wilson (2008) has graphically described in the case of the restructuring of the biological sciences at Manchester University, the British scientific elite had formed a powerful coalition with some Research Councils and science advisers to encourage greater selectivity in the award of state funds for scientific research well before the RAE was developed, and some organizational scientific elites used the opportunities provided by external pressures to comprehensively reorganize university departments, teaching programmes, and research priorities.

Considering finally the likely impact of the growing encouragement of research commercialization on scientific integration and innovation in different kinds of PSS, this depends greatly on the relative influence of individual scientists, PROs, and state agencies on (*a*) the flow and use of funds from such activities and (*b*) the ability of private interests to control the production and dissemination of knowledge, materials, and technologies. Where resources from companies are primarily provided to individuals and their research teams for work that contributes to both certified knowledge and private purposes on an informal and largely personal basis, as has been the case for many of the chemical and technological sciences since the last decades of the nineteenth century, they reduce the authority of disciplinary elites and enable scientists to undertake a wider range of projects.

On the other hand, if such contacts and resource flows become managed by state agencies and/or by employers, they may reduce the discretion of individual researchers. This would enable the state to pursue particular technology policies and employers to invest in particular scientific fields and groups. This last is obviously dependent on universities and PROs having some discretion over resource allocation and being able to act as employers. In state-delegated and employer-dominated PSS, then, the active exploitation of commercialization opportunities by PROs can develop their strategic capabilities and undertake investments in new fields that would be too risky if they were highly dependent on public funds from foundations using peer review to allocate resources.

In PSS that are not very strongly stratified into a stable hierarchy of prestige and resources, gaining resources through working with companies and commercializing research results and technologies additionally enables PROs to invest in the recruitment of new and established research stars and provide high levels of support for their work as a means of increasing their social and intellectual standing (Owen-Smith 2001). This is especially so in employer-dominated PSS, such as in the USA, where they can compete in a relatively fluid scientific labour market and develop distinctive strategic capabilities as separate organizations pursuing varied kinds of intellectual projects. However, the more that they try to control intellectual property rights and use them to raise revenues through licensing and restricting the use of materials, ideas, and instruments to organizations that are prepared to pay for them, the more they become similar to commercial organizations and threaten both the collaborative ethos of the public sciences and their own legitimacy as non-profit charitable organizations (Nelson 2004).

In summary, this analysis of how the major post-war changes in the governance and funding of public scientific research are likely to affect intellectual competition, innovation, and coordination in different kinds of PSS highlights five points.

1. These changes have led to a significant shift in the authority relations governing both the definition and choice of research problems, strategies and approaches and the evaluation and integration of research outcomes into established public knowledge. These involve the development of new kinds of authoritative agencies, such as public research foundations, ministries of science and technology policy, and agencies commercializing intellectual property rights as well as the restructuring of existing ones, such as reputational communities, education ministries, universities, and other PROs.

2. It is important to recognize that the scale and significance of the changes discussed in this chapter have varied considerably across the OECD economies, as shown by the contributions to this book. How they were introduced and continue to develop differ between states such that many features of their national PSS remain quite distinct. In particular, the recent shifts in state–university relations in many countries are more likely to be far-reaching and significant where PROs were fully integrated parts of the state than in societies where the state has only a limited role in their governance, funding, and legitimation. Despite states continuing to 'learn from abroad' in their educational and science reforms, as they have done to some extent since the seventeenth and eighteenth centuries, the organization of scientific research and its governance still vary greatly between countries in ways that affect intellectual competition, innovation, and coordination.

3. Such shifts in authority patterns can be expected to have varied consequences for intellectual innovation and evaluation depending on the degree and rigidity of stratification of universities and other PROs, on the one hand, and the diversity of funding sources for published scientific research, on the other hand. When PSS are highly stratified and funding is dominated by a small number of research foundations, attempts to institutionalize a competitive market for resources based on the excellence of individuals' and departments' contributions to collective intellectual goals are likely to reinforce both existing prestige hierarchies

and researchers' dependence on the standards and goals of current scientific elites.

Conversely, where there is a variety of different providers of research resources for a range of intellectual and social purposes that rely on the judgements of different sets of peer reviewers and other experts, the effects of many of these changes on intellectual innovation and integration may be less restricting. Furthermore, if such stratification is relatively fluid and not institutionalized through state hierarchies and policies, considerable diversity of research funding and goals may encourage non-elite universities to support the development of new kinds of research and approaches, including those transgressing current intellectual boundaries and norms.

To an extent, the ability to gain revenues from a variety of sources, including research commercialization, has enabled some US universities to invest in the pursuit of novel intellectual goals as well as more conventional attempts to poach research stars (Brint 2005; Owen-Smith 2003). This may facilitate the establishment of new intellectual enterprises and research specialisms, particularly when the state commits substantial resources to them, as more generally does the incorporation of different kinds of scientific goals and audiences into the activities of US research universities and the relative fluidity of academic and business labour markets in the USA (Casper 2007).

4. This discussion has emphasized that these changes in the organization and governance of PSS often have contradictory effects on intellectual competition, innovation, and coordination. While some can be expected to narrow the scope of intellectual novelty and researcher discretion, others could mitigate such consequences depending on the context in which they are introduced. Similarly, while formal state delegation of authority to PROs could enhance their strategic capability in some PSS, this might be constrained by research stars being able to attract significant funds for their own purposes and establishing new firms that provide substantial revenue streams for universities. It may also, of course, be inhibited by intensified state monitoring of performance and reduced public funding to the extent that PROs have effectively very limited freedom of action to determine their futures as independent organizations.

Finally, it is worth emphasizing that the ability of research foundations, state agencies, and PROs to exert authority over the direction of research projects and the assessment of their outcomes is limited by the inherent uncertainty of most scientific research and the diversity of peer-group judgements. As many chapters highlight, research teams retain considerable autonomy over how research goals are pursued and projects conducted, as well as over the interpretation of results. Although authority over project selection, investment priorities and research strategies may be more formally assumed by PROs, foundations, and state agencies than before, it remains shared with peer reviewers and researchers who collectively reconstruct and coordinate results, albeit to varying degrees between scientific fields and different PSS.

The Structure of the Book

These points are explored in more detail in the next three parts of this book. The first part deals with the reorganization of academic systems as a whole, and focuses on how three very different PSS are being restructured by national governments in varying ways with contrasting consequences. In Chapter 2, Ben Martin and Richard Whitley reconstruct the history of the British Research Assessment Exercise in its various incarnations, and show how it has largely enhanced the authority of established scientific elites rather than orienting academic research towards economic goals and what the state considered to be user needs. Once research quality as judged by disciplinary elites was established as the key criterion for evaluating this research, it became very difficult to insist on its short-term economic impact as a central factor in allocating central state resources to universities. Furthermore, the greater variety of funding sources for much academic research in the UK as compared to Australia has meant that state steering of university science towards user needs has been less influential than that reported by Gläser and his colleagues in Chapter 10.

A very different picture is presented by Martin Benninghoff and Dietmar Braun in their account of the growing importance of the Swiss National Science Foundation (SNSF). In what closely resembled the state-shared type of PSS outlined above, the post-war establishment and expansion of the SNSF heralded a more competitive funding system in which researchers increasingly have had to compete for external support if they wish to carry out projects in many sciences. More recently, university managers have begun to acquire new competences and develop distinctive research strategies, thus constraining individuals' problem choices somewhat. However, the nature of Swiss federalism and continuation of substantial block grant support for universities have limited the effects of these changes on the Swiss PSS, which remains more similar to the state-shared type than the state-delegated highly competitive variant found in the UK and Australia.

In the case of Japan, Robert Kneller shows how a highly stratified and largely state-shared academic system in which patronage has historically played a major role in appointments and promotions is beginning to change in response to governance shifts and reduced budgets. In particular, he sees some weakening of the patronage powers of professors at the dominant universities, although the state's focus on creating a relatively small number of internationally 'excellent' universities has reinforced the concentration of resources and prestige in a few elite institutions. The intensification of competition for resources between newly empowered universities seems likely to further exaggerate the stratification of the system rather than encourage greater intellectual and organizational variety.

The second part of the book focuses on the changing authority relationships within research organizations in different PSS—especially the expansion of central administrative controls over resource allocation, research strategies, and profiles, mostly at the expense of senior professors and institute heads. In Chapter 5, Jochen Gläser, Stefan Lange, Grit Laudel, and Uwe Schimank consider how the development and use of different kinds of research evaluation systems have both legitimized and technically assisted managerial control in five academic systems. Analysing these RES in terms of their timeliness, richness, comparability, and validity of the information they provide, and their legitimation of central strategic control, they show how the cheapness and timeliness of indicator-based RES are outweighed in terms of enabling strategic change and limiting distortions by the greater validity, richness, and legitimacy of peer-review-based RES. Overall, their use has increased the authority of both scientific elites and university managers at the expense of individual researchers.

Governance changes in a traditionally state-dominated PSS have encouraged significant authority shifts within French research organizations, and perhaps in the PSS as a whole where funding agencies and PRO administrative centres have gained authority largely at the expense of laboratory directors. To some extent, the growing importance of competitive project-based funding has also resulted in research groups gaining some autonomy and separate identities as authority relations move from a patronage to a partnership model, as described by Severine Louvel. As in Japan, there has been some encouragement of career mobility and external recruitment with a weakening of organizational scientific leaders' power, but hierarchical authority within research units remains considerable.

The importance of local organizational contexts and intellectual traditions in affecting the impact of national governance changes on intra-organizational authority relationships, highlighted in Louvel's study, is also evidenced by Frank Meier and Uwe Schimank's study of authority changes in the Lower Saxony academic system. Here, the dual domination of the state ministry and the professoriate is gradually being supplanted—or perhaps, more accurately, complemented—by a new power structure dominated by coalitions between university managers and national scientific elites. Through both delegated powers and the use of research evaluation systems, some universities in Lower Saxony are developing strategic capabilities in research profiling, which enable them to reallocate positions and concentrate resources. In so doing, they could lead to greater organizational specialization in particular sciences, as arguably existed in the heyday of the nineteenth-century German university system, although, as Meier and Schimank show, this is a long way from realization at the moment and is beset by contradictory state policies and actions.

In Part III, contributors analyse the impact of differences between scientific fields on the consequences of governance changes in different PSS. In many, if not indeed most, cases, these changes have been implemented across the sciences as if they are essentially the same, or, if not, then they should become more similar to the current icon of scientificity. However, as these chapters make plain, the major differences between the sciences have greatly affected how scientists have responded to funding and authority changes, just as variations in the organization of largely national PSS have continued to structure both the nature of governance changes and their effects. In particular, scientific elites in some fields were by no means passive bystanders in the development and implementation of these changes, as Norma Morris indicates in her discussion of the limited impact of the British RAE on biologists' research practices. In the post-war biosciences, she suggests:

Scientists' extreme resource-dependency, long-term exposure to research council competitive funding systems, and espousal of group-based work systems generated its own need for research management and monitoring. Thus the prevailing patterns of authority relations developed in science departments (which had little in common with the unchallenged individual autonomy once enjoyed by many academics in other parts of the university), provided structures and strategies ready for deployment in the face of perturbations produced by the RAE.

Parts of the British scientific elite had already developed strategies for ensuring continued state support for the best research in a dynamic steady state, which could accommodate many features of the RAE once peer-reviewed quality became the central assessment criterion. As a result, its direct impact on research practices was less marked in these fields than it seems to have been in sciences that had experienced less change in work organization in the post-war period.

The importance of external resource dependence and group-based research organization in mediating the impact of funding and governance changes on research problem selection is highlighted by Liudvika Leišytė, Jurgen Enders, and Harry de Boer's study of biotechnology and medieval history researchers in England and the Netherlands. While some adaptation to the changing environment took place in both countries, this was more marked in the English cases and in biotechnology, where external resource dependence is high. However, most respondents reported symbolic compliance rather than the more severe shifts found by Gläser and Laudel (2007) in Australia.

In Chapter 10, Jochen Gläser, Stefan Lange, Grit Laudel, and Uwe Schimank extend their earlier studies of the effects of research evaluation systems on scientists' research strategies to examine more closely the ways in which differences between scientific fields affect the impact of changing authority relations on research goals and approaches. Comparing proximate and remote epistemic properties on research in six disciplines in the sciences and humanities, they show how variations in resource dependence, research portfolio diversity, and other factors have impinged upon scientists' responses to resource scarcity and state restructuring in academic governance in Australia, as well as how these factors in turn reflect more deep-seated features of research styles, such as the role of personal interpretation in problem formulation, decomposability of research problems, and mode of access to empirical evidence. Such analysis highlights the considerable variety of epistemic properties across the sciences and the consequential dangers of imposing a standard model of scientific practises of all fields through RES and related measures.

Differences in the authority of scientific elites between differently organized PSS are shown to have been critical to the contrasting patterns of institutionalization of business studies as a research field in Europe and North America by Lars Engwall, Matthias Kipping and Behlul Usdiken. The relative weakness of the professoriate in most US universities before the post-war expansion of federal research funding enabled presidents and trustees to incorporate new fields focused on professional education, such as business studies, into leading universities at an earlier stage than in most European countries where scientific elites commonly disdained such vocationally oriented fields and new commercial schools were often established outside the public academic system. This case exemplifies the path-dependent nature of both governance changes and innovation in different PSS, and their unlikely convergence to a single model as seems to be desired by some policy-makers in an idealized view of the contemporary US research university system.

Indeed, despite many formal similarities in the governance changes developed by state agencies and other actors in Japan, Europe, Australasia, and parts of North America, the chapters in this book demonstrate quite convincingly that their impact on the relative authority of scientific elites, academic oligarchies, university managers, and other groups varies across PSS, and is often contradictory in the consequences for intellectual innovation. Like any set of socio-political innovations, how public knowledge production is being reconfigured and its effects on scientific development are greatly influenced by established institutional systems governing the public sciences. Additionally, the changing distribution and direction of authority over research goals and judgements of results is central to our understanding of how changing forms of academic governance are influencing the public sciences. In the final chapter, Jochen Gläser considers how the analyses of changing authority relationships in the public sciences presented in this book help to make research on shifts in governance over PSS more clearly and precisely connected to patterns of intellectual change.

References

Ben-David, Joseph (1972) *The Scientist's Role in Society* (Englewood Cliffs, New Jersey: Prentice-Hall).

Biegelbauer, P., and Borras, S. (eds.) (2003) *Innovation Policies in Europe and the US: The New Agenda* (Aldershot: Ashgate).

Bok, Derek (2003) *Universities in the Marketplace: The Commercialisation of Higher Education* (Princeton: Princeton University Press).

Braben, Donald W. (2004) *Pioneering Research: A Risk Worth Taking* (Chichester: Wiley).

Braun, Dietmar, and Merrien, F.-X. (eds.) (1999) *Towards a New Model of Governance for Universities? A Comparative View* (London: Jessica Kingsley Publishers).

Brint, Steven (2005) 'Creating the Future: "New Directions" in American Research Universities', *Minerva*, 43: 23–50.

Casper, Steven (2007) *Creating Silicon Valley in Europe: Public Policy towards New Technology Industries* (Oxford: Oxford University Press).

Clark, Burton (1983) *The Higher Education System: Academic Organization in Cross-National Perspective* (Berkeley, Calif.: University of California Press).

—— (1998) *Creating Entrepreneurial Universities: Organizational Pathways of Transformation* (Oxford: Pergamon Press).

—— (1995) *Places of Inquiry: Research and Advanced Education in Modern Universities* (Berkeley, Calif.: University of California Press).

Cohen, Wesley, Nelson, Richard, and Walsh, John (2002) 'Links and Impacts: The Influence of Public Research on Industrial R&D', *Management Science*, 48: 1–23.

Coleman, Samuel (1999) *Japanese Science: View from the Inside* (London: Routledge).

Colyvas, Jeannette, and Powell, Walter (2006) 'Roads to Institutionalization: The Remaking of Boundaries between Public and Private Science', *Research in Organizational Behavior*, 27: 305–53.

Cooper, Mark H. (2009) 'Commercialization of the University and Problem Choice by Academic Biological Scientists', *Science, Technology and Human Values*, 34: 629–53.

Cozzens, Susan (2007) 'Death by Peer Review? The Impact of Results-Oriented Management in U.S. Research', in Whitley and Gläser (eds.), *The Changing Governance of the Sciences* (Sociology of the Sciences Yearbook; Dordrecht: Springer), 225–42.

Crawford, Elisabeth, Shinn, Terry, and Sorlin, Sverker (eds.) (1993) *Denationalizing Science: The Context of International Scientific Practice* (Sociology of the Sciences Yearbook; Dordrecht: Kluwer).

Croissant, Jennifer, and Sal Restivo (eds.) (2001) *Degrees of Compromise: Industrial Interests and Academic Values* (Albany, NY: State University of New York Press).

Drori, Gili S., Meyer, John W., Ramirez, Francisco O., and Schofer, Evan (2003) *Science in the Modern World Polity: Institutionalization and Globalization* (Stanford, Calif.: Stanford University Press).

Geiger, Roger (1986) *To Advance Knowledge: The Growth of American Research Universities 1900–1940* (Oxford: Oxford University Press).

—— (2004) *Knowledge and Money: Research universities and the Paradox of the Marketplace* (Stanford, Calif.: Stanford University Press).

—— and Sa, Creso (2005) 'Beyond Technology Transfer: US State Policies to Harness University Research for Economic Development', *Minerva*, 43: 1–21.

Geuna, Aldo, and Martin, Ben (2003) 'University Research Evaluation and Funding: An International Comparison', *Minerva*, 41: 277–304.

—— and Nesta, Lionel (2006) 'University Patenting and its Effects on Academic Research: The Emerging European Evidence', *Research Policy*, 35: 790–807.

Gläser, Jochen, and Laudel, Grit (2007) 'Evaluation without Evaluators: The Impact of Funding Formulae on Australian University Research', in Whitley and Gläser (eds.), *The Changing Governance of the Sciences* (Sociology of the Sciences Yearbook; Dordrecht: Springer), 127–52.

Greenberg, Dan (1966) 'Grant Swinger: Reflections on Six Years of Progress', *Science*, 154: 1424–5.

Guston, David H. (2000) *Politics and Science: Assuring the Integrity and Productivity of Research* (Cambridge: Cambridge University Press).

Hagstrom, Warren (1965) *The Scientific Community* (New York: Basic Books).

Homburg, Ernst (1992) 'The Emergence of Research Laboratories in the Dyestuffs Industry, 1870–1900', *British Journal for the History of Science*, 25: 91–111.

Hughes, Sally Smith (2001) 'Making Dollars out of DNA: The First Major Patent in Biotechnology and the Commercialization of Molecular Biology, 1974–1980', *ISIS* 92: 541–75.

Jang, Yong Suk (2003) 'The Global Diffusion of Ministries of Science and Technology', in Drori *et al.* (eds.), *Science in the Modern World Polity* (Stanford, Calif.: Stanford University Press), 120–35.

Jasanoff, Sheila (ed.) (1997) *Comparative Science and Technology Policy* (Cheltenham: Elgar).

Kleinman, Daniel, and Vallas, Steven (2001) 'Science, Capitalism and the Rise of the "Knowledge Worker": The Changing Structure of Knowledge Production in the United States', *Theory and Society*, 30: 451–92.

Kohler, Robert (1979) 'Warren Weaver and the Rockefeller Foundation Program in Molecular Biology', in Nathan Reingold (ed.), *The Sciences in the American Context* (Washington, DC: Smithsonian Institution Press), 249–93.

Kneller, Robert (2007a) *Bridging Islands: Venture Companies and the Future of Japanese and American Industry* (Oxford: Oxford University Press).

—— (2007b) 'Prospective and Retrospective Evaluation Systems in Context: Insights from Japan', in Whitley and Gläser (eds.), *The Changing Governance of the Sciences* (Sociology of the Sciences Yearbook; Dordrecht: Springer), 51–74.

Krimsky, Sheldon (2003) *Science in the Private Interest: Has the Lure of Profits Corrupted Biomedical Research?* (Lanham, Md.: Rowman & Littlefield).

Kruecken, Georg (2003) 'Learning the "New, New Thing": On the Role of Path Dependency in University Structures', *Higher Education*, 46: 315–39.

—— and Meier, Frank (2006) 'Turning the University into an Organizational Actor', in G. S. Drori, J. W. Meyer, and H. Hwang (eds.), *Globalization and Organization: World Society and Organizational Change* (Oxford: Oxford University Press), 241–57.

—— Kosmutzky, Anna, and Torka, Marc (eds.) (2007) *Towards a Multiversity? Universities between Global Trends and National Traditions* (Bielefeld: transcript Verlag).

Laudel, Grit (2006) 'The "Quality Myth": Promoting and Hindering Conditions for Acquiring Research Funds', *Higher Education*, 52: 375–403.

Lenoir, Timothy (1997) *Instituting Science: The Cultural Production of Scientific Disciplines* (Stanford, Calif.: Stanford University Press).

Leslie, Stuart W. (2000) 'The Biggest "Angel" of Them All: The Military and the Making of Silicon Valley', in M. Kenney (ed.), *Understanding Silicon Valley: The Anatomy of an Entrepreneurial Region* (Stanford, Calif.: Stanford University Press), 48–67.

Liefner, Ingo (2003) 'Funding, Resources Allocation, and Performance in Higher Education Systems', *Higher Education*, 46: 469–89.

Lowen, Rebecca S. (1991) 'Transforming the University: Administrators, Physicists and Industrial and Federal Patronage at Stanford, 1935–1949', *History of Education Quarterly*, 31: 365–88.

McClelland, C. E. (1980) *State, Society and University in Germany, 1700–1914* (Cambridge: Cambridge University Press).

Marginson, Simon, and Considine, Mark (2000) *The Enterprise University: Power, Governance and Reinvention in Australia* (Cambridge: Cambridge University Press).

Martin, Ben (2003) 'The Changing Social Contract for Science and the Evolution of the University', in Aldo Geuna, Ammon J. Salter, and W. Edward Steinmuller (eds.), *Science and Innovation: Rethinking the Rationales for Funding and Governance* (Cheltenham: Edward Elgar), 7–29.

Merrien, Francois-Xavier, and Musselin, Christine (1999) 'Are French Universities Finally Emerging? Path Dependency Phenomena and Innovative Reforms in France', in Braun and Merrien (eds.), *Towards a New Model of Governance for Universities?* (London: Jessica Kingsley Publishers), 220–38.

Metlay, Grischa (2006) 'Reconsidering Renormalization: Stability and Change in Twentieth-Century Views on University Patents', *Social Studies of Science*, 36: 565–97.

Meulen, Barend van der (2007) 'Interfering Governance and Emerging Centres of Control: University Research Evaluation in the Netherlands', in Whitley and Gläser (eds.), *The Changing Governance of the Sciences* (Sociology of the Sciences Yearbook; Dordrecht: Springer), 191–203.

—— and Leydesdorff, Loet (1991) 'Has the Study of Philosophy and Dutch Universities Changed under Economic and Political Pressure?', *Science, Technology, and Human Values*, 16: 288–321.

Mowery, David C., Nelson, Richard R., Sampat, Bhaven N., and Ziedonis, Arvids A. (2004) *Ivory Tower and Industrial Innovation: University–Industry Technology Transfer Before and After the Bayh-Dole Act* (Stanford, Calif.: Stanford University Press).

Muller-Camen, Michael, and Salzgeber, Stefan (2005) 'Changes in Academic Work and the Chair Regime: The Case of German Business Administration Academics', *Organization Studies*, 26: 271–90.

Musselin, Christine (1999) 'State/University Relations and How to Change Them: The Case of France and Germany', in M. Henkel and B. Little (eds.), *Changing Relationships between Higher Education and the State* (London: Jessica Kingsley Publishers), 42–68.

—— (2007) 'Are Universities Specific Organizations?', in Kruecken *et al.* (eds.), *Towards a Multiversity?* (Bielefeld: transcript Verlag), 63–84.

Nelson, Richard (2004) 'The Market Economy, and the Scientific Commons', *Research Policy*, 33: 455–71.

Odagiri, Hiroyuki (1999) 'University–Industry Collaboration in Japan: Facts and Interpretations', in Lewis Branscomb, Fumio Kodama, and Richard Florida (eds.)

Industrializing Knowledge: University–Industry Linkages in Japan and the United States (Cambridge, Mass.: MIT Press), 252–65.

Owen-Smith, Jason (2001) 'New Arenas for University Competition: Accumulative Advantage in Academic Patenting', in Croissant and Restivo (eds.), *Degrees of Compromise* (Albany, NY: State University of New York Press), 23–53.

—— (2003) 'From Separate Systems to a Hybrid Order: Accumulative Advantage across Public and Private Science at Research One Universities', *Research Policy*, 32: 1081–1104.

—— Massimo Riccaboni, Fabio Pammoli and Walter Powell (2002) 'A Comparison of U.S. and European University–Industry Relations in the Life Sciences', *Management Science*, 48: 24–43.

Pavitt, Keith (2001) 'Pubic Policies to Support Basic Research: What can the Rest of the World Learn from the US Theory and practice (and What they should Not Learn)', *Industrial and Corporate Change*, 10(3): 761–79.

Powell, Walter, and Owen-Smith, Jason (1998) 'Universities and the Market for Intellectual Property in the Life Sciences', *Journal of Policy Analysis and Management*, 17: 253–77.

—— —— and Jeannette Colyvas (2007) 'Innovation and Emulation: Lessons from American Universities in Selling Private Rights to Public Knowledge', *Minerva*, 45: 121–42.

Rothblatt, Sheldon, and Wittrock, Bjorn (eds.) (1993) *The European and American University since 1800* (Cambridge: Cambridge University Press).

Schiene, Christof, and Schimank, Uwe (2007) 'Research Evaluation as Organisational Development: The Work of the Academic Advisory Council in Lower Saxony', in Whitley and Gläser (eds.), *The Changing Governance of the Sciences* (Sociology of the Sciences Yearbook; Dordrecht: Springer), 171–90.

Schimank, Uwe (2005) '"New Public Management" and the Academic Profession: Reflections on the German Situation', *Minerva*, 43: 361–76.

Shinn, Terry (1979) 'The French Science Faculty System, 1808–1914: Institutional Change and Research Potential in Mathematics and the Physical Sciences', *Historical Studies in the Physical Sciences*, 10: 271–323.

Siegel, Donald, Waldman, David, and Link, Albert (2003) 'Assessing the Impact of Organizational Practices on the Relative Productivity of University Technology Transfer Offices: An Exploratory Study', *Research Policy*, 32: 27–48.

Slaughter, Sheila, and Leslie, Larry (1997) *Academic Capitalism: Politics, Policies and the Entrepreneurial University* (Baltimore: Johns Hopkins University Press).

Stokes, Donald E. (1997) *Pasteur's Quadrant: Basic Science and Technological Innovation* (Washington, DC: Brookings Institution Press).

Tabil, Ameen Ali (2001) 'The Continued Behavioural Modification of Academics since the 1992 Research Assessment Exercise', *Higher Education Review*, 33: 30–46.

Thelin, John R. (2004) *A History of American Higher Education* (Baltimore: Johns Hopkins University Press).

Torstendahl, Rolf (1993) 'The Transformation of Professional Education in the Nineteenth Century', in Rothblatt and Wittrock (eds.), *The European and American University since 1800* (Cambridge: Cambridge University Press), 109–41.

Trow, Martin (1993) 'Comparative Perspective on British and American Higher Education', in Rothblatt and Wittrock (eds.), *The European and American University since 1800* (Cambridge: Cambridge University Press), 280–99.

—— (1999) 'From Mass Higher Education to Universal Access: The American Advantage', *Minerva*, 37: 303–28.

Weingart, Peter (2003) 'Growth, Differentiation, Expansion and Change of Identity: The Future of Science', in Bernward Joerges and Helga Nowotny (eds.), *Social Studies of Science and Technology: Looking Back, Ahead* (Sociology of the Sciences Yearbook; Dordrecht: Kluwer), 183–200.

—— and Maasen, Sabine (2007) 'Elite through Rankings: The Emergence of the Enterprising University', in Whitley and Gläser (eds.), *The Changing Governance of the Sciences* (Sociology of the Sciences Yearbook; Dordrecht: Springer), 75–100.

Whitley, Richard (2000) *The Intellectual and Social Organization of the Sciences*, 2nd edn. (Oxford: Oxford University Press; 1st edn. 1984).

—— (2003) 'Competition and Pluralism in the Public Sciences: The Impact of Institutional Frameworks on the Organisation of Academic Science', *Research Policy*, 32(6): 1015–29.

—— (2006) 'Project-Based Firms: New Organisational Form or Variations on a Theme?', *Industrial and Corporate Change*, 15: 77–99.

—— (2007) 'Changing Governance of the Public Sciences: The Consequences of Establishing Research Evaluation Systems for Knowledge Production in Different Countries and Scientific Fields', in Whitley and Gläser (eds.), *The Changing Governance of the Sciences* (Sociology of the Sciences Yearbook; Dordrecht: Springer), 3–27.

—— (2008) 'Universities as Strategic Actors: Limitations and Variations', in Lars Engwall and Denis Weaire (eds.), *The University in the Market* (London: Portland Press), 23–37.

—— and Gläser, Jochen (eds.) (2007) *The Changing Governance of the Sciences* (Sociology of the Sciences Yearbook; Dordrecht: Springer).

Wilson, Duncan (2008) *Reconfiguring Biological Sciences in the Late Twentieth Century: A Study of the University of Manchester* (Manchester: Faculty of Life Sciences, University of Manchester).

Wittrock, Bjorn (1993) 'The Modern University: The Three Transformations', in Rothblatt and Wittrock (eds.), *The European and American University since 1800* (Cambridge: Cambridge University Press), 303–62.

Woolgar, Lee (2007) 'New Institutional Policies for University–Industry Links in Japan', *Research Policy*, 36: 1261–74.

Yoxen, Edward (1982) 'Giving Life a New Meaning: The Rise of the Molecular Biology Establishment', in Norbert Elias, Herminio Martins, and Richard Whitley (eds.), *Scientific Establishment s and Hierarchies* (Sociology of the Sciences Yearbook; Dordrecht: Reidel), 123–43.

Ziman, John (1994) *Prometheus Bound: Science in a Dynamic Steady State* (Cambridge: Cambridge University Press).

—— (2000) *Real Science: What it is and What it Means* (Cambridge: Cambridge University Press).

I

Reorganizing Academia

Delegating Authority while Increasing Accountability in Universities

2

The UK Research Assessment Exercise

A Case of Regulatory Capture?

Ben Martin and Richard Whitley

1. Introduction

The UK has experienced all six of the major changes to the organization and direction of the public sciences outlined by Whitley in his introductory chapter. In some respects, these changes have been more prominent and influential than elsewhere, especially with regard to establishing a mechanism for regular, systematic, nationwide assessment of university research. Other countries have adopted various approaches to evaluating academic research (Geuna and Martin 2001; Whitley *et al.* 2007) but none have become as central and influential as the UK Research Assessment Exercise (RAE). The evolution of the mechanism for achieving this and the associated effects on authority relations in UK university research form the subject of this chapter.

The UK was one of the first countries both to institutionalize university research assessment over twenty years ago and to link it to financial allocations. Consequently, the effects on authority relations there are likely to be more profound than in many other states. In what follows, we look first at the historical background to the introduction of the RAE, and then examine the evolution of the RAE as carried out on successive occasions. Section 4 analyses its impact on authority relations—in particular those relating to

An earlier version of this paper was presented at the workshop on 'Reconfiguring the Public Sciences', held at the Royal Swedish Academy of Letters, History and Antiquities, 18–20 Feb. 2009. The authors are grateful for the comments provided by those attending that workshop, and in particular to Jochen Gläser for his review and suggested improvements.

the Funding Councils and the government, to users, to disciplinary elites, to universities and university departments, and to individual researchers. The final section summarizes the conclusions emerging from this analysis.

2. Historical background

The UK Higher Education System in the 1960s and 1970s

In the 1960s and 1970s, UK higher education (HE) remained the preserve of a relatively small group, although student participation had risen following the Robbins expansion from 5 per cent of the student-age population in 1960/1 to 15 per cent in 1986/7 at the time of the first RAE (see Mayhew *et al.* 2004: 66, fig. 1).[1] The country possessed a fairly small number of universities (about fifty), all assumed to be broadly equal in standing ('the equity principle'). Universities at that stage were granted considerable freedom, with relatively 'light touch' steering from the University Grants Committee (UGC), which itself operated with a fair degree of autonomy from the Department of Education and Science (DES)[2] (Dearlove 1998*a*, 1998*b*). Internally, they relied on a largely collegial system of governance with a significant level of direct democracy (for example, deans were often elected, the post being rotated among senior faculty). While one must be careful not to exaggerate, many academics had appreciable autonomy with regard to what research they did and how they did it (Clark 1995; Dearlove 1998*a*, 1998*b*). British university departments tended to be less hierarchical than, say, German or Japanese ones, and less managerially controlled than US ones (Trow 1993). University governance in the UK was also characterized by substantial trust between academics, administrators, and policy-makers. In many respects, research in UK universities resembled the state-delegated discretionary type of public science system discussed by Whitley in his introductory chapter.

Like most countries apart from the US, the UK operated a 'dual-support system' for universities, in which the UGC was responsible for the institutional core funding of teaching and research (and in particular for ensuring 'the well found laboratory'), while Research Councils (RCs) supported the additional costs of specific research projects. Decisions on funding allocations by the UGC were supposedly determined by unit size, but in practice

---

[1] Since then, the figure has again almost trebled, standing at 43% at the time of the most recent RAE.

[2] Originally, the UGC had come directly under the HM Treasury, only moving when the Department of Education and Science was created in the 1960s.

they were also strongly influenced by lobbying, with certain universities (in particular, Oxford and Cambridge) being relatively generously funded. Indeed, 'the equity principle' was probably more rhetoric than an accurate representation of the resource distribution outcomes, and the system as a whole was, and remains, highly stratified in terms of resources and social and intellectual prestige.

The final point to stress is that in the UK most publicly funded research is carried out in universities rather than in free-standing laboratories or institutes. Prior to the 1980s, there was a larger non-university research sector (consisting of government laboratories and RC institutes), but following the reforms of the Thatcher government (when many of these were privatized, merged, or closed), this sector became much less important, and is far smaller than in many European countries such as France, Germany, Italy, and Spain. Consequently, the effects of the RAE apply to a large proportion of publicly funded research in the UK.

Changes in the Late 1970s and Early 1980s[3]

Various forces began to bring about change in the late 1970s.[4] The first was the growing costs of research at a time of increasing pressures on public expenditure. Britain had been suffering from major economic difficulties since the mid-1970s. When Mrs Thatcher came to power in 1979, she was determined to reduce public spending, and substantial funding cuts were imposed on universities (Phillimore 1989: 258). The UGC decided that, in order to protect the best universities and departments, the cuts should be allocated very unequally, with some universities being subject to reductions of over 30 per cent. This generated a storm of controversy, not least because the basis of UGC's selectivity was unclear. While it was apparent from funding projections that a policy of selectivity was necessary,[5] what was less obvious was how to determine where to focus the limited funds.

---

[3] See also the analysis of this period in Chapter 8.

[4] There had been an earlier attempt to link research system more closely to the needs of the state. The Rothschild Report (1971) had advocated that more applied research be funded on a 'customer contractor' basis, with government departments acting as a proxy customer for the state. A substantial part of the budgets of the Agricultural, Natural Environment, and Medical Research Councils was therefore transferred to the respective ministries (although in the case of the MRC, this decision was subsequently reversed). However, the largest research council (the Science Research Council) was unaffected, and university research also largely 'escaped' from the effects of the Rothschild changes.

[5] Indeed, the Science Research Council (the largest of the five Research Councils) had been pursuing a policy of selectivity and concentration since 1970 (see SRC, 1970), although the impact of this appears to have been decidedly limited—see Farina and Gibbons (1979, 1981).

A second major driving force consisted of political demands for greater accountability and better 'value for money'. Much attention was devoted to the '3 Es': economy, efficiency, and effectiveness.[6] This rhetoric was associated with a shift to a more managerialist approach to the public sector and to the development of what became known as 'new public management' (Hood 1991), with its emphasis on efficiency, transparency, accountability, quality assurance, and competition. This, in turn, required that publicly funded institutions develop clear objectives, effective decision-making structures, transparent assessment of performance, and a system of allocating resources based on past performance (Henkel 1999: 107).

Universities, with their heavy dependence on public funding, could not expect to stand aside from these growing pressures. In 1985, the Jarratt Report heavily criticized universities for being inadequately managed and for exhibiting little accountability. Particular criticism focused on an over-reliance on university committees, which made for slow decision-making as well as being vulnerable to sectional interests. The report argued that universities needed to develop more efficient internal management—in particular, improved strategic planning, more monitoring to achieve greater efficiency and cost-effectiveness, and clearer resource allocation procedures (Jarratt 1985; Jones 1986). This was to be achieved through a more hierarchical form of management, with Vice Chancellors becoming Chief Executives, and with more top–down decision-making by senior managers (Dearlove 1998b: 115).

In short, Jarratt advocated a shift to what Trow described as 'hard management', with more hierarchy and less trust (Trow 1994). Universities were now expected to act more like businesses, exposed to the disciplines of the 'market' and with explicit strategies. These changes were to be 'encouraged' by the move to a more competitive system of funding. While RC funding was already subject to a relatively intense form of competition, the other stream of the dual-support system, the research funds provided by the UGC, also needed to be subject to competitive forces (Curran 2000; Bence and Oppenheim 2005: 141–2).

The Emergence of the First Research Selectivity Exercise

In 1985, two months after the Jarratt report appeared, the government published a Green Paper on the future of higher education (DES 1985). This stressed the need for the HE sector to become more prudent and

---

[6] In 1983 the Audit Commission was created, concerned with ensuring quality, effectiveness, performance, and value for money (Shore and Wright, 1999: 562).

selective in its use of public funds. Mindful of the furore created by the somewhat arbitrary approach it had adopted in 1981 in its attempts at more selective funding, the UGC (rather hurriedly) introduced the Research Selectivity Exercise as a more formal mechanism to provide accountability and selectivity.

The first Research Selectivity Exercise (RSE), as it was initially termed, was carried out in 1986. It was based on a relatively simple methodology. Each university department or 'unit of assessment'[7] was asked to complete a questionnaire on research income, research planning, and priorities. It also requested units to identify their five best publications from the previous five years. The responses were considered by UGC subject sub-committees, along with a number of assessors,[8] for each of the UGC's thirty-six 'cost centres'. From the information submitted, the subject committees classified units on the basis of a simple four-category ranking.

The effects of this exercise's results on UGC research funding were initially rather limited. However, the RSE undoubtedly came as something of a shock to the university system (Phillimore 1989). While some in the more established universities paid relatively little attention (hoping, no doubt, that the RSE would 'go away'), others took it much more seriously. Even at this stage, it was clear that the introduction of performance-related funding meant an appreciable increase in the authority of the evaluating/funding agency, UGC. Moreover, within many universities, administrators now became more concerned with the level and quality of staff research and publications, and encouraged the development of more explicit strategies for departments. At least formally, the authority of university management *vis-à-vis* its researchers increased.

3. Evolution of the RAE

1989

As the RSE's wider implications began to sink in, there was much criticism of the assessment procedures. In particular, the request for the five best

---

[7] The terminology here is confusing. UGC (and subsequently UFC and HEFCE) used the term 'unit of assessment' for the *subjects* under which universities could make their various submissions, while universities used the term to describe the *groups* being submitted (often departments but sometimes a combination of several departments or a part of a department).

[8] One criticism was that the assessors' identities were never made public (Bence and Oppenheim 2005: 144).

publications from a unit was seen as being unfair to small departments.[9]
These criticisms were to stimulate various improvements in the 1989 exercise—a process repeated after each subsequent RAE. Over time, the RAE
became ever more sophisticated but also increasingly costly and burden-
some.

In 1987, the Advisory Board for the Research Council (ABRC) produced a
report putting forward a radical proposal to separate universities into R, T,
and X categories, with research resources being concentrated on R institutions and selected departments in X institutions. Fierce opposition from
universities quickly led to the proposal being rejected, but the eventual
compromise was to 'beef up' the RSE as a mechanism for providing greater
selectivity in UGC funding. In 1988, the academic-dominated UGC was
replaced by a smaller executive body, the Universities Funding Council
(UFC), which included representatives from the private sector. This marked
a fundamental shift in role from acting as a 'buffer' between government
and universities to operating as a 'coupling' agency responsible for delivering government policy (Phillimore 1989: 258).

A second Research Selectivity Exercise was carried out in 1989—this time
with a more sophisticated methodology. Units were asked to supply publication data including bibliographic details of up to two publications[10] for
each full-time member of faculty—a change that resulted in a major increase in the workload for the assessors. They were also asked to provide data
on research studentships, research grants, and contracts, and total numbers
of publications in relation to the number of full-time staff (UGC 1988).[11]

Recognizing the increased scale of the assessment task, UFC established
approximately seventy panels to evaluate each subject (or 'unit of assessment').[12] Panellists were reportedly chosen to ensure adequate coverage in
terms of 'the range of specialised expertise needed to cover the spread of

---

[9] If the numbers of publications from a department with a certain level of quality are
assumed to fall on a normal distribution curve, a large department is more likely to have its
five best publications appearing in the right-hand 'tail' of the curve than a smaller department.

[10] However, there was no precise definition of a 'publication'. It was no surprise, therefore,
that the 'publication data was found to be unreliable, and where it was reliable, it said nothing
about the quality of the output' (UFC 1989: para. 23). However, already by 1989 lists of 'leading'
journals (generally drawn up by elite bodies) that would be looked upon favourably by subject
panels were beginning to circulate (such as in economics, see Lee and Harley 1998: 24).

[11] However, 'no facility for systematic verification of the accuracy of the submissions was
built into the exercise, and there was some evidence of deliberate "misreporting"' (Bence and
Oppenheim 2005: 145). Only in the 1992 RAE was a formal audit procedure introduced to
check the accuracy of submissions.

[12] Unfortunately, the full list of panels was not sorted out before the RSE (Bence and
Oppenheim 2005: 144).

research in the subject area to be assessed', 'the spread of institutions being assessed', 'age and current active involvement in research', and 'evidence of wide knowledge of the conduct of research in the relevant subject area' (UFC 1989: 15). The panels also sought confidential advice from external experts. An attempt was made to standardize ratings across subjects by asking panels to judge groups on the basis of 'attainable levels of excellence', but in the event there were large variations between the average ratings awarded by different panels, which led to widespread suspicion that some panels had been 'tougher' than others (Johnes and Taylor 1992: 72–6).

The outcome was that units were ranked on a scale of 1 to 5, based on the proportion of their publications judged to be of national or international excellence. This time, the UFC decided there would be a larger, more explicit link between rankings and research funding. Of the 33 per cent of UFC funds devoted to funding research, nearly half was allocated on the basis of the 1989 ratings, with the remainder being determined mainly by student numbers (see Table 1 in Johnes and Taylor 1992: 69). This substantial increase in the proportion of UFC funding distributed on the basis of the assessment, along with the introduction of a common rating scale and the greater emphasis on publications, represented a significant shift in authority from universities to the UFC, although universities retained control over the internal allocation of funds. Moreover, the change from submitting five publications for the entire unit/department to two publications for each member of staff meant that increasing pressure began to be exerted by university managers on all faculties to produce quality publications.

1992

Prior to the next RAE, a government White Paper (DES 1991: 18) re-emphasized that 'funding for research should be allocated selectively to encourage institutions to concentrate on their strengths'. Another important change came with the 1992 Further and Higher Education Act (DES 1992). This abolished 'the binary divide' between universities and polytechnics, elevating the latter (which previously received public funding to conduct teaching but not research) to the status of universities. The resulting inclusion of these 'new universities' (or '1992 universities') in the 1992 RAE added to the competition for limited funds, although only to a small extent since the new universities received just 9 per cent of the total on the basis of their 1992 RAE ratings (Bence and Oppenheim 2005: 146).

By now, the RAE approach had begun to 'settle down'; although 'tweaked' in subsequent RAEs, the approach remained broadly the same from here onwards. In 1992, units were given the opportunity to include only 'research active' staff (those units with 'a long tail' of weaker research- ers had reportedly been penalized with a lower rating in earlier exercises). Despite this, some universities decided to include everyone to show that they were committed to being a research-led institution or to avoid an adverse affect on staff morale. However, a new rule meant units could only include staff in post on the RAE 'census date', which resulted in units 'losing' the publications of staff who had departed before then. Units were also asked for information on the research environment and future plans, and for quantitative data on all publications classified under specified headings.

By this stage, over 90 per cent of research funds (QR) were distributed by the Funding Councils (FCs[13]) on the basis of RAE ratings, with no funds being awarded to units given the lowest grade. However, as in all the RAEs, the exact relationship of QR funding to RAE results was only announced by HEFCE *after* the ratings were announced. Universities were thus forced to take part in each RAE without knowing what the financial consequences would be: 'They were being asked to participate in a game with a blindfold on, because they had not been told the rules which would govern the distribution of money at the outcome stage' (Johnston 1993: 174). They could only guess whether including more staff as 'research active', and hence increasing the unit's 'volume factor', might compensate for a lower grade (or vice versa).[14] Given this intrinsic uncertainty, universities often concentrated exclusively on achieving the highest possible grade for each unit, irrespective of whether this ultimately resulted in the greatest level of QR funding.

In terms of the impact of the 1992 RAE on authority relations, giving universities the right to decide which faculty were submitted as 'research active' provided university managers with another means to exert influence over faculty work patterns. Likewise, the continuing increase in the

---

[13] UFC by then had been replaced by separate HEFCs for England (HEFCE), Wales, Scotland, and Northern Ireland. In what follows, these are collectively referred to as the Funding Councils (FCs) or HEFC.

[14] Even after the link between RAE grades and funding was made public, the calculations involved in retrospectively determining the optimum strategy were far from simple! However, Johnston (1993) showed that omitting the 'tail' of up to a quarter or third of the weakest staff in order to achieve a rating one point higher was an economically rational strategy.

percentage of research funds distributed by Funding Councils on the basis of the RAE results represented a further shift in authority towards the FCs.

1996

In the 1996 RAE, units were requested to submit up to four publications per member of 'research active' staff, further increasing the importance of publications. They could also include 'indications of peer esteem' (such as journal editorships and invited conference presentations). The submissions were assessed by some sixty subject panels. The chairs of these were appointed by the FCs, with the other panellists being selected from nominees put forward by 1,300 learned societies, professional bodies, and subject associations. This provided disciplinary elites with an opportunity to exercise considerable authority over what forms of research were judged to be excellent.

The 1996 RAE saw an extension from five to seven grades, with a splitting of the '3' grade into 3A and 3B, and the introduction of a new 5* grade. However, a shortage of FC funds (there was a 5 per cent reduction in real terms in 1996–7) meant the FCs had to introduce much larger differentials in funding, with no funds at all for the lowest two grades (1 and 2). This further strengthened the authority of the Funding Councils with respect to universities.

2001

By now, there was mounting evidence of universities actively managing RAE submissions, with many excluding staff with few publications as 'research-inactive'. Although such exclusions reduced a unit's 'volume factor', they could significantly increase the probability of a higher RAE rating (Bence and Oppenheim 2005: 147). There was also growing concern about a 'transfer market' in the period before the RAE census date (ibid.) and complaints that previous RAEs had suffered from inconsistent ratings across the panels, that interdisciplinary research had been treated unfairly, and that the personal circumstances of staff (such as career breaks) were not taken into account. These and other problems resulted in further changes in RAE procedures in 2001, as well as a promise to provide more detailed feedback.

Assessment panels were broadened in an attempt to bring in research 'users' in industry and elsewhere, and further shift the balance of authority with respect to decisions on university funding away from academics.

However, this met with only limited success, as few non-academics were willing to put in the effort required of RAE panel members. The FCs also appointed international experts, in particular to check and confirm the top grades proposed by panels. Panels were now expected to treat each publication on its merits (irrespective of the medium of publication such as the journal and its status), imposing a huge reading burden on panellists. Faced with the virtual impossibility of reading all the submitted publications, some panels decided to take a short-cut and assume that work already subjected to rigorous peer review (such as that published in 'top' journals) should be given more weight (Bence and Oppenheim 2005: 150–1).

Previously low-rated departments (1 or 2) were now less likely to bother submitting to the RAE, as there was no money for such grades (or indeed for 3B after 2001).[15] Over time, the proportion of staff in 5-rated departments rose from 23 per cent in 1992 to 31 per cent (in 5 and 5* departments) in 1996 and to 55 per cent in 2001. The positive interpretation of this was that there was a major improvement in research quality during this period. However, the rise also reflected the fact that universities had learnt to play the RAE 'game' more effectively (for example, who and what to submit, and how best to present their submissions). In addition, over time panels tended to become more lenient (there was an element of 'grade inflation') as they realized that panels that had been tough in previous RAEs (such as economics and geography) ended up financially penalizing their fields in comparisons with others.[16]

After the 2001 results were announced, HEFCE realized that funding the new (and much improved) ratings on the existing basis would cost an additional £200 million. However, the HEFCE budget for 2002–3 had already been set in the government's Comprehensive Spending Review. Therefore, HEFCE was only able to maintain unit funding for 5* departments, and that for all lower ratings was cut. This meant HEFCE ended up with much steeper differentials between 3A, 4, 5, and 5* departments than the Scottish and Welsh FCs. For example, in Scotland and Wales a

---

[15] In 1992, 67 per cent of the units submitted had been rated 1, 2, or 3, but this had fallen to 17 per cent by 2001—see Table 2 in Bence and Oppenheim (2005: 149).

[16] A trenchant critique of the assumption that the improvement in RAE grades reflected improved research quality came from the Pro-Vice-Chancellor of Warwick University: 'What that 55 per cent represents [the proportion of researchers in 5 or 5* departments] . . . is a morass of fiddling, finagling and horse trading. Nobody who works in a university in the UK in 2002 seriously believes that research is improving.' Likewise, certain learned societies openly admitted that 'some or all of this improved RAE score is undoubtedly due to increased familiarity with RAE exercises and the ability of university departments to play the RAE game' (see House of Commons Science and Technology Committee 2002: para. 23).

department rated 5* received 3.2 times as much as one rated 3A, while in England the differential was 8.7 (House of Commons 2002: para. 68, table 6). Indeed, HEFCE subsequently decided that even this degree of concentration of funding on the best performers was not enough, and those departments gaining a 5* rating in both 1996 and 2001 were retrospectively awarded a new rating of 6* with additional funds.

2008

After 2001 there was a growing sense that the RAE had become much too cumbersome. A review was carried out by Roberts (2003), who proposed a much simpler assessment system for less research-intensive institutions. However, fierce opposition from these led to this idea being rejected. A few years later came a proposal by the Chancellor of the Exchequer that the peer-review-based RAE should be replaced with a 'cheaper' approach based on metrics. This, too, met with fierce resistance. Eventually, a compromise was reached, whereby the 2008 RAE would remain based on peer review, but with a more metrics-based 'Research Excellence Framework' being phased in for many sciences after that.

The main change in the 2008 RAE was a switch from a single rating on a seven-point scale to a 'profile' for the research of each department, based in large part on what proportion of its publications were judged to be of national or international quality, but also taking account of other data included in the submissions (such as esteem indicators). Each publication was rated on a scale of 1* to 4*. The results announced at the end of 2008 showed that the great majority of the units submitted ended up with an average 'score' of between 2.0 and 3.0. As a result of this 'flatter' distribution, research funding will now be spread slightly more widely, with some middle-ranking universities experiencing the largest increases in 2009/10 (for example, Nottingham, Queen Mary, Loughborough, Brunel, Cranfield) (Attwood and Corbyn 2009; Corbyn 2009).

4. Impact on Authority Relations

Next, we consider the RAE's impact on authority relations governing research priorities and judgements. For ease of discussion, the analysis has been separated into various subsections (government bodies, users, disciplinary elites, universities, and individuals), although many issues cross-cut these categories.

The Funding Councils, the Research Councils, and the Government

Reflecting its broader composition including representatives from the private sector, HEFCE (and, to a lesser extent, its predecessor, UFC) has been more interventionist with regard to universities than the original UGC, and less independent of the state.[17] Since Margaret Thatcher's premiership, the UK government has required accountability and selectivity with respect to the disbursal and expenditure of all public monies, so UFC and later HEFCE have had little option but to deliver. It is ironic, given the claimed intention of the Thatcher Government to reduce the role of the state, that the effect of the RAE has been to increase the authority of government with respect to universities. In particular, the continuing government emphasis on ever greater concentration of research resources has increased the degree of stratification in the HE sector, leaving universities to compete ever more fiercely for government funding in a 'game' in which the state and its agencies (in particular the Funding Councils) set the rules and determine the financial outcome.

The RAE has also strengthened the authority of the Funding Councils, providing them with several sources of power. First, although they consult widely with universities and others, they are ultimately responsible for the choice of the RAE approach and how this should be changed from one RAE to the next. In particular, they determine the subject panels. Since 1989, these sixty to seventy panels have been fairly stable, tending to reinforce traditional disciplinary boundaries. Following each RAE, there have been complaints that some field has been harshly treated, accompanied by calls for a new panel to be established. After consultation, the FCs then decide how the panel structure should be modified.[18] The FCs also select panel members, drawing from a long list of suggestions. Most importantly, they determine how RAE results are to be translated into funds. Over time, the FCs, and in particular HEFCE, have chosen to increase greatly the differential between 'excellent' research and the rest.[19] For all these reasons, the RAE has greatly strengthened the

---

[17] However, UGC was very opaque in the way it operated, as well as being subject to much lobbying; in short, there were no 'good old days', as some academics imply.

[18] For example, after extensive lobbying by environmental scientists, who claimed that their subject had been harshly treated by the Earth Sciences panel, the field eventually acquired its own panel (Warner 1998).

[19] In addition, the FCs determine the differential unit of resource between fields (for example, between laboratory-based and non-laboratory-based research).

authority of the FCs with respect to universities compared with the pre-1986 regime operated by the UGC.[20]

With regard to the Research Councils (RCs), the first point to note is that during the time the RAE has been in operation, the volume of university research funds received from RCs has grown considerably faster than that from FCs. For example, between 1988/9 and 1999/2000 there was little change in the latter while the former grew by 60 per cent in real terms (see, for example, chart 3.1 in HM Treasury 2002: 31). This has strengthened the authority of RCs over university research. In addition, as the competitive pressures associated with the RAE have increased, so RC funds have become ever more important in supporting the production of publications—the primary indicator of quality in the RAE. In principle, funding decisions on individual research projects by RCs are meant to be independent of the RAE rating of the department of authors of proposals. In practice, it is unclear how great that independence actually is, for three main reasons.

First, authors of a proposal from a highly rated department are likely to be better resourced and therefore (other things being equal) better able to produce a strong proposal. Second, peer reviewers or RC committee members are unlikely to completely ignore whether the proposal comes from a department with a strong or weak RAE rating. Third, analysis of RAE results and RC funding shows a very close correlation (of 0.98); indeed, it was this close correlation that led to the government proposal in 2006 that the RAE be replaced by a metric-based approach focusing on departmental research income (Sastry and Bekhradnia 2006).

Users

In one respect, however, the effort by government to use the RAE as part of its wider policies towards science and technology has been notably unsuccessful. This concerns the attempt by government to get Funding Councils to include 'users'[21] in assessing research, so that research contributions to the economy are reflected in RAE ratings along with 'academic' contributions. Introduced in the 2001 RAE, the effort to include users in assessment panels proved largely ineffective. This failure highlights a fundamental contradiction at the heart of UK government science policy in recent years. Since the 1993

---

[20] However, it should be noted that UGC preferences did influence academic strategies before the RAE, as in the reorientation of biological research towards medically related problems and the promotion of molecular biology at Manchester (Wilson 2008).

[21] The term 'users' in the UK is broader than just industry or firms, including NGOs and even government departments.

White Paper on *Realising our Potential* (OST 1993), publicly funded researchers have been expected to identify potential 'users' and to work with them in trying to ensure that the results of their research are effectively exploited. Addressing the needs of such users tends to involve research that does not fall neatly into a single disciplinary 'pigeonhole', but which is multi-, inter-, or transdisciplinary in nature (Gibbons *et al.* 1994). Yet the RAE is more geared up to assessing (or, more cynically, is 'biased' towards) monodisciplinary mainstream research (Alsop 1999). In short, the UK has an essentially discipline-based assessment system for a world in which government policies are trying to encourage more user-focused and often interdisciplinary research. Those who have gone down the user-influenced route frequently conclude that they have ended up being penalized in the RAE process. In the two most recent RAEs, there has been much HEFC rhetoric about treating basic and applied research in an even-handed way, but in practice the heavy reliance on peer review and the composition of RAE panels mean that discipline-focused research invariably tends to be regarded as higher quality.

Disciplinary Elites

The RAE has been based almost entirely on peer review. Peer review is traditionally the preserve of disciplinary elites. Recall that peer review was originally introduced (by organizations such as the Royal Society) for determining which papers to accept in learned journals. In due course, peer review was adopted by universities in decisions on recruitment and promotion. It was then extended to use in decisions on research funding, first by US foundations and later (after 1945) by government research funding bodies. However, in all of these, the unit of analysis is normally the individual researcher or a small group applying for a grant. The RAE was perhaps the first occasion where peer review was extended to assess entire university departments on a systematic nationwide basis.[22] Yet surprisingly little critical consideration seems to have been given to whether this radical extension in the nature and scope of the assessment task is one for which peer review is well suited.[23] Given the generally intellectually conservative

[22] Prior to the RAE, peer review had sometimes been used to assess an entire research institute, but usually on an individual basis (as opposed to assessing and rating *all* institutes in that field). The nearest equivalent precursor to the RAE was perhaps the US National Research Council (NRC) ratings of graduate schools in US universities, but these were based on little more than an opinion poll of academics.

[23] One of the few discussions of this can be found in Martin and Skea (1992), who found that the limitations of peer review become far more significant as the unit of analysis shifts from the individual to the department.

nature of peer-review judgements, their use in such a strong evaluation system (Whitley 2007) seems likely to limit further intellectual innovation, as Braben (2004) has argued.

From the start, the UGC turned to disciplinary elites to provide the bulk of the RAE panellists (the panels are essentially monodisciplinary in composition). Learned societies, subject associations, and professional bodies have all nominated panel members but the decisions are ultimately taken by the FCs.[24] Although the names of the organizations nominating panellists have been made public, along with broad selection criteria, it has never been clear why particular individuals were chosen. This has proved a bone of contention, since comparison of panel membership and RAE results has shown that departments with a member on a panel tended to do significantly better than those without (for example, Doyle *et al.* 1996). With their strong representation on RAE panels, epistemic elites have been able to exert further influence on research in their discipline—in particular on the criteria used for judging 'excellence'.

Given the increased importance of publications, disciplinary elites as represented among journal editors (especially of 'top' journals) have become more influential in determining who (and what type of research) gets published.[25] Although disciplinary elites already exercised significant control over the conditions of academic production (for example, through the funding of research projects via RCs), the RAE has further strengthened their authority.

As noted above, the membership of RAE panels has drawn heavily from disciplinary elites—often those who established their reputation through their contributions to mainstream disciplinary research. This, together with the fact that most 'top journals' tend to focus on research in the disciplinary mainstream, means that the RAE has reinforced the emphasis on conventional mainstream research, discouraging new developments and interdisciplinary research. There is some evidence that the RAE may also have skewed recruitment in departments. In economics, for example, Harley and Lee (1997) showed that the RAE had reinforced the trend towards the hiring of mainstream economists. In a later paper (Lee and Harley 1998) they argued that if these recruitment trends continued, non-mainstream economics was in danger of being eliminated from many UK economics

---

[24] For a detailed description of how panel members were chosen in one field (economics) in successive RAEs, see Lee and Harley (1998: 30–3).
[25] Near each RAE deadline, they also become crucial in deciding which articles are published before that date.

departments—a prediction apparently borne out by subsequent developments.[26]

In their case study of economics, Lee and Harley produced evidence of how the Royal Economic Society, worried about losing its power over the reputational control system in the field, acted promptly in 1989 and later RAEs 'to capture the process by which assessors were appointed to the economics panel', ensuring that predominantly mainstream economists were appointed (ibid. 42). This study provides a convincing illustration of the process of 'regulatory capture' (Lee 2007: 323) of the RAE by the disciplinary elite, who imposed their views on criteria for assessing 'quality' in economics and on which journals are most important, and hence managed to skew resources towards the disciplinary mainstream. Although we lack direct evidence on other fields, there are good reasons for assuming that a similar process of regulatory capture by epistemic elites has gone on in other RAE panels.

However, in considering the effects of the RAE we need to recognize that there have been wide variations across fields. Some disciplines seem to have been quite well suited to the RAE (for example, laboratory-based sciences or economics: see Power 1999: 136; Becher and Trowler 2001: 198), disciplines in which quality was already closely linked to a hierarchy of journals, and where there was standardization of technical entry requirements, a propensity to engage in incremental research ('normal science' linked to an established paradigm), a strong refereeing culture, and often a weaker relationship between research and teaching (ibid.). In such fields, the mainstream elite already exercised considerable control over the 'reputational system' (Whitley 1986, 1991, 2000).

In contrast, for many humanities and social sciences (where research and teaching form more of a unified whole, where research results are published in a wide range of journals and books, and where research may yield a variety of outputs other than academic publications), the cultural and structural consequences of RAE were more traumatic (Harley 2002). In such fields, RAE brought heightened emphasis on research 'output', on journal publications, and on bringing in research income to produce more outputs (see below). Likewise, for professionally related disciplines such as

---

[26] Lee (2007) later found that two-thirds of UK economics departments had no or only one heterodox economists on their faculty, while three-quarters included only mainstream economics on their teaching curriculum (and for departments that had submitted to the RAE the figure was 88 per cent).

engineering, law, accountancy, and nursing, reconciling their work with the demands of the RAE often proved difficult (Robinson *et al.* 2002: 497).

In short, the RAE has reinforced the centrality of peer review and with it the authority of those responsible for peer-review judgements. In many fields, epistemic elites have seemingly taken advantage of the opportunity afforded by the RAE to strengthen their authority as arbiters of 'excellence' and of the reputation of individuals and departments.

Universities and University Departments[27]

Over time, RAE results have been used to justify the progressive withdrawal of funds from lower-rated departments (with ratings of 1, 2, and 3B). At same time, the differential unit of resource between 3A, 4, 5, and 5* departments has sharply increased (particularly in England). The net effect has been to greatly strengthen the concentration of resources on top departments; that is, increasing the stratification of departments and, with it, universities. This increased stratification has consequences not only for the 'vertical' authority relations between universities and the FCs or the state more generally, but also for the 'horizontal' relations between departments within universities, with highly rated departments having a stronger position with regard to the competition for posts and funding as well as greater ability to resist central control. As a result, their authority concerning the definition of research goals and the direction of research is greater than that of weaker departments (see Chapter 9).

However, the greatest impact of the RAE at the university level has been on university management. Since the mid-1980s, UK universities have come under pressure to become better managed and more 'efficient', to develop clearer strategies, and to exhibit a greater willingness and ability to shift resources from declining fields to growing ones. The RAE has provided an opportunity to take a significant step forward in this process. In particular, it has given university administrators a means of comparing the 'quality' of departments, and of legitimising the development of research strategies with differential funding between fields (Meulen, 2007). It has also facilitated the reorganization of academic units around new intellectual strategies, as in the case of the biological sciences at Manchester University (Wilson 2008; Wilson and Lancelot 2008).

---

[27] This subsection should be read in conjunction with Chapter 8.

Once it became clear that the RAE was here to stay, senior management in many universities began to assume responsibility for overseeing the preparation of RAE submissions by departments. Over time, university officials tended to become more interventionist, deciding which departments were to be combined (or split) to form the 'units of assessment' (often a key issue in the 'new universities' with less developed or more dispersed research capacities), which staff should be included as 'research active' and which should be left out or 'hidden', and what incentives to offer faculty for RAE contributions. Elaborate machinery (perhaps in the form of a central university team led by a Pro-Vice-Chancellor, supported by faculty 'directors of research' or departmental 'research coordinators') was developed to oversee preparations for the next RAE. A 'mini-RAE' was often carried out in advance. In the face of all this, staff were increasingly pressured to conform to the dictates of their department or university (for example, to pursue shorter term or more 'mainstream' research), leaving them feeling that their autonomy was being diminished (Broadhead and Howard 1998).

By the mid-1990s,[28] RAE results were also being used increasingly as an input to university strategy, helping to decide which fields and departments to build up, and which departments to merge or close (McNay 1997b: 196; Henkel 1999: 111). In some institutions, departments assumed a stronger and more active role in university management (Morris 2002). Department heads were explicitly tasked with improving RAE performance. Some adopted a more interventionist approach to managing faculty, advising them which research topics to pursue, whom to collaborate with, and in which journals to publish. They might also decide which 'stars' (with outstanding publication records) the university should attempt to recruit, and which current staff should be retained at all costs, even if it meant entering into private 'deals' over resources or offering time off from teaching to concentrate on research (Wilson and Lancelot 2008).

Since 1986, recruitment and promotion have become much more explicit components of institutional strategy, with recruitment and promotion of individuals increasingly taking account of RAE contributions (actual or potential). Early retirement or sideways moves into administrative or 'teaching only' posts have become more common as formal options for less active RAE contributors (for example, Bence and Oppenheim 2005: 151). As the importance of the RAE has grown, so the degree of centralization of personnel decisions has tended to escalate.

---

[28] A survey by McNay (1997a, 1997b) suggests that, prior to 1996, many universities were still rather reluctant to sacrifice weaker departments, but that has since changed.

Despite all the attention paid to the RAE, it is important to emphasize that in larger research-intensive universities the scale of RAE funding is generally relatively small compared with other research income (less than 10 per cent of the total). Nevertheless, the RAE results have considerable 'symbolic power', signalling to funding and other external bodies as well as the wider academic community where the 'best' departments are, and hence the relative standing of the institution in the status hierarchy. The RAE has been one of the factors contributing to the great prominence now accorded to 'league tables' and to the relative position of universities and of the departments within them (Henkel 1999: 110–11). Hence, the emphasis in many universities on getting the best possible RAE ratings, even if, with a reduced 'volume factor', this results in less research income from HEFCE.

Overall, the RAE has reinforced the shift from collegial governance towards 'managerialism' in most universities, with more centralized control and monitoring, greater use of targets and rewards, more bureaucratic procedures, and a more hierarchical structure. In short, it has strengthened the authority of the institution (McNay 1997a; Dearlove 1998a, 1998b) and its control over academic labour (Harley and Lee 1997; Wilson 2008). However, the degree of impact on culture and organization has varied across HEIs (probably being greatest in 'new universities'; see Yokoyama 2006) and fields (generally being greater in social sciences and humanities).

Individual Researchers

Prior to 1986, many faculty (at pre-1992 universities, at least) enjoyed significant freedom in deciding whether to do research, how much (in relation to other academic roles such as teaching, administration, professional practice, public service, and 'scholarship'), in what form, on what topic, whether to raise external funding, and how to disseminate the results (Dearlove 1998a: 71). Some division of labour would often evolve within departments, with certain individuals doing more teaching or administration depending on their respective strengths and inclinations (something for which their more research-focused colleagues were appreciative), but all this was arrived at through largely informal, tacit, and 'unmanaged' processes sustained by shared norms (Dearlove 1998a: 72; Henkel 1999: 120).

Now, located at the 'bottom of the pile' in these revised authority relations, individuals are more constrained in pursuing their own research agendas and subject to growing pressures to 'perform'. They have experienced notable changes in several respects. First, the growing significance of the RAE means that they face more constraints on the type of research they chose to pursue. Evidence from previous RAEs suggests they will be better placed if they focus on: basic rather than applied research (Harley 2002; Bence and Oppenheim 2005: 145); shorter-term rather longer-term research (Henkel 1999: 119; McNay 2003: 52); incremental rather than more ambitious or open-ended 'pioneering' research (Braben 2004: 80–5); mainstream rather than 'alternative' research or research in a highly specialized sub-field (Harley and Lee 1997: 1437; Barnard 1998: 476; Henkel 1999: 109);[29] monodisciplinary rather than inter- or multidisciplinary research (McNay 2003: 52); 'academic' rather than 'professional' research (for instance, in medicine, management, law, planning; e.g. Nadin 1997: 95; Campbell *et al.* 1999; McNay 2003: 51–2); research that yields journal articles rather than books (Johnes 1995: 10; Becher and Trowler 2001: 105–6; Paisey and Paisey 2005: 422); and research where the results can be published in 'top' journals rather than more specialist (and generally lower status) ones (Henkel 1999: 118).

Secondly, individual academics have been subject to growing publication pressures. The absolute priority is to produce four good publications over the RAE cycle. In certain fields there may be additional pressures to concentrate on an 'A list' of top journals (Harley and Lee 1997) or on high-status book publishers. Academics may additionally succumb to a tendency to 'premature publication' before the research is fully ready, particularly as the RAE deadline looms (for example, Gläser and Laudel 2007; Harley and Lee 1997: 1437). Those seeking a permanent academic position[30] or promotion are especially vulnerable to such pressures. Those with poor publishing records may suffer reduced sense of self-worth and status, with the weakest likely to be coerced into 'teaching only' or administrative posts or encouraged to take early retirement.

Thirdly, because RAE incentives are perceived to be much stronger than those for teaching,[31] many departments have witnessed a widening split

---

[29] Becher and Trowler (2001: 105–11) differentiate between 'urban' research fields characterized by a high 'density' of researchers addressing a small number of research problems, and 'rural' research fields where there are relatively few researchers pursuing a large number of problems. Their study suggests that accountability pressures from the RAE may encourage a shift from the latter to the former.

[30] In the UK, in some fields researchers may obtain their first faculty position shortly after completing their doctorate, while in other fields they may work as a postdoctoral fellow for a number of years before moving to a permanent university post. However, in all fields, the effect of the RAE has been greatly to increase the emphasis on publications in the appointment process. This is another illustration of the effect of the RAE on the career system and the labour market for academics.

[31] Although a Teaching Quality Assessment (TQA) system was introduced in 1993, the results were not used by FCs in allocating resources to teaching. Academics faced with decisions at the margin as to whether to allocate additional time to improve their research or their teaching would hence tend to focus on the former. The TQA system was heavily criticized for various

between teaching and research, with deleterious effects on the former (Jenkins 1995: McNay 1997b; HEFCE 1997). The RAE may have caused individual faculty to modify the balance between research and teaching (Broadhead and Howard 1998). Many universities and departments now struggle to persuade faculty to give due attention to teaching or administration. The emphasis on the RAE means that individuals (especially 'leading researchers') tend to devote less time to lecture preparation or to meetings with students (ibid.). Indeed, some 'RAE stars' may be able to negotiate 'research only' contracts, enabling them to concentrate exclusively on research while strengthening the institution's chances in the competitive research 'market'. At the same time, there is less volunteering by faculty for communal tasks such as serving on university committees, and more formal negotiation is frequently needed with individuals to persuade them to undertake such tasks. Likewise, the RAE has reduced the willingness of faculty to engage in other academic activities such as reviewing, editing, translating, contributing to reference works, writing popular books, engaging in clinical medicine or community service, providing policy advice, and so on (Nadin 1997: 95).

A fourth change observable is a shift in the balance between individualism and collegiality (and between competition and cooperation), with a decrease in organizational loyalty. The growing split between research and teaching has contributed to this decline in collegiality—in particular the offering of 'research only' contracts to leading researchers while colleagues with poor RAE records are required to pick up the increased teaching load. Individuals may view increasing their published output and research profile as a way to maximize their value on the academic 'market', making them more attractive to outside offers. In this respect, the RAE has influenced the horizontal authority relations of researchers with others within their university or in their national community, with their authority increasingly depending on the RAE grades of their departments and on their contribution to those grades.

Fifthly, they may face inducements to engage in 'game-playing', allowing the obsession with research ratings to dominate over the generation of novel research. The heightened sense of competition engendered by the RAE has encouraged widespread 'game-playing'—efforts designed to

---

reasons, including '(1) administrative/cost burden; (2) grade inflation/gamesmanship/ organisational learning; (3) elitist bias within the system; (4) system impact of quality review; (5) reliability of the system; and (6) philosophical objections to the system' (Laughton 2003: 309), and it was eventually dropped. Similar criticisms can be levelled at the RAE.

make the department 'look good' while not necessarily improving the research quality. Examples of this include creating 'teaching only' categories of staff to make it appear that a higher proportion of faculty are 'research active', setting up 'paper' units of assessment specifically for RAE purposes (Henkel 1999: 112), and hiring academic 'stars' on a temporary basis during the RAE census date. The efforts involved in such game-playing may not only detract from research; they may also weaken the motivation of academics.

Lastly, many academics have felt it necessary to make sacrifices to their private life. The RAE has been a factor encouraging overwork and adding to levels of stress. It has disadvantaged those (predominantly women) who have take time off for family or other reasons, resulting in a 'gap' in their published output (Barnard 1998: 476). Those who are disabled or sick or forced by circumstances to work part-time may struggle to produce four good publications in the RAE cycle. Moreover, their departmental colleagues are now often less willing to 'carry' them than in pre-RAE times.

Overall, British academics are under increasing pressure to 'produce', the RAE having tended to weaken their control over their work (Wilson 1991; Halsey 1992) and rendered them more susceptible to institutional managerial control (Harley 2002).[32] They have experienced a shift from a system in which academic identity centred around a relatively loose and informal competition for reputation as judged by peer review to a much more managerial system where meeting the expectations of senior university managers (and indeed departmental colleagues) with respect to the RAE has become ever more important (Harley 2002). Failure to publish may now be seen not just as an individual failing but as having 'let down' the department and university. In short, RAE has, along with a number of other factors, brought about significant change in relationship between the individual, the department, the university, and the discipline relative to the situation before 1986 (Henkel 1999). The system as a whole can reasonably be seen as having being transformed from a state-delegated discretionary one to much more of a state-delegated competitive one in the terms discussed by Whitley in his introductory chapter.

---

[32] Such a perceived loss of power and control may also contribute to a growing sense of frustration and disaffection (cf. Perryman 2007: 173).

5. Conclusions

Governments now require accountability mechanisms for almost all areas of public spending. Those operating a dual-support system for university research (that is, most apart from the USA) therefore seek a mechanism to provide accountability with respect to institutional research funding. In the UK, the political pressures for accountability and for selectivity and concentration have arguably been greater than elsewhere—an apparent legacy of the 1980s Thatcher government. Consequently, the UK has constructed a more rigorous and intrusive assessment mechanism in the form of the RAE. In so doing, it has brought about substantial changes in authority relations—both vertical and horizontal—between individuals, disciplinary elites, departments, universities, funding councils, and government to a greater extent than in most other countries (Henkel 1999: 120).

The effects are particularly apparent with respect to vertical authority relations. As we have seen, the RAE has increased the authority of government and the FCs over universities, concentrating research resources and increasing the degree of stratification in HE. Likewise, within universities the RAE has steadily shifted the balance from departments to central university managers as the financial rewards for achieving top grades have grown (or, more accurately, as the penalties for failing to obtain the top grades have escalated).[33] Universities, under pressure since the Jarratt report to become more 'efficient', have seized upon the RAE to justify both monitoring the research performance of departments more systematically and distributing resources across departments in a more unequal manner than formerly. The RAE has often been used by senior university managers to impose a more centralized, hierarchical form of management on departments, and in some cases by departmental heads to do the same within departments.

The RAE has also altered the authority relations between epistemic elites and individual researchers. Those elites were quick to spot the possibilities of regulatory capture with respect to the RAE. With the assistance of universities, they ensured that their favoured evaluation mechanism of (discipline-based) peer review remained at the heart of the assessment exercise

---

[33] The RAE is clearly not the sole influence bringing about these changes. Other government policies such as the shifts from institutional funding to project-based funding, from 'response mode' projects to directed research programmes, and from researcher-driven projects to user-engagement projects, have all contributed.

(Henkel 1999: 105), and then supplied many of the peers needed to operate the RAE. Peer review, which prior to the RAE was already a significant element in the system for allocating reputational rewards in established disciplines, has, through the RAE, become more prominent and much more closely linked to resource allocation, reinforcing the authority of epistemic elites.

In addition, the RAE has also had a significant effect upon horizontal authority relations—for example, between universities, or between departments within a university, or between individual researchers. In particular, faced with a growing concentration of research resources and the increasing degree of stratification in the HE sector, universities have been forced to compete ever more fiercely with one another. Such inter-institutional competition to some extent runs counter to the growing need for collaboration in order to carry out world-leading research, or indeed to meet the government expectation of a larger university role with regard to 'third mission' activities. Likewise, within universities, departments may sense that they are in increasing competition with one another, since universities often distribute FC research funds on the basis of the relative RAE rankings of constituent departments.[34] This, too, may run counter to the need for cross-departmental collaboration—for example to conduct research in newly emerging interdisciplinary areas. Lastly, within departments individuals may well sense that they are in greater competition with one another, especially now that their individual publications are assessed and ranked. Again, this may act as a constraint on collaborative work, as well as encouraging 'game-playing' and reducing the willingness to volunteer for teaching or other 'public good' activities.

As the RAE has evolved over time, so the effects on authority relations have become more profound. With the impending shift from the peer-review-based RAE to the metrics-based Research Excellence Framework (REF), those changes in authority relations seem set to grow ever stronger. The continuing need, even after the switch to the REF, to submit in pre-defined fields (each with their own panel, agreed set of metrics, and 'weighting' system for different metrics or assessment approaches) may reinforce the boundaries between existing disciplines, impeding the development of new research fields. While it is difficult to prove, there are grounds for believing that the RAE has tended to make the boundaries

---

[34] Departments may also be in direct competition with one another for central university funds to hire RAE 'stars' and hence boost their ranking in the next RAE.

between disciplines more 'rigid', hindering the emergence of new fields[35] and hence slowing the overall growth of knowledge. With the REF assessment system more reliant on metrics and less on informed peer review, such a conservative force is likely to become greater and more damaging in its long-term effects on university research (cf. Braben 2004).

In this chapter, we have seen how the current approach to research assessment in UK universities is reductionistic and primitive, and almost certainly counterproductive in terms of generating a wide variety of intellectual innovations in the longer term. What is required is a move towards a more refined conceptualization of 'accountability'—one that goes some way towards rebuilding trust and strengthening autonomy. As Ackroyd and Ackroyd (1999) argue, the effective governance of universities may actually require *less* external control, not more, and with it an inevitable degree of 'creative tension' between the constituent components of the organization. All this will necessitate a wider range of indicators (including qualitative ones) and assessment approaches, which in turn requires more dialogue between assessors and those being assessed (Shore and Wright 1999: 571).

Where does the UK go from here? The government has now decided that the RAE is to be replaced over coming years by the Research Excellence Framework. Yet this represents a move in precisely the opposite direction to the one suggested here, with greater reliance on metrics, and less on peer review. What are the likely consequences? Undoubtedly, there will be new or expanded opportunities for disciplinary elites to exert greater influence on what papers are published in which leading journals and (if citations included among the performance metrics) on referencing conventions and behaviour. The longer term epistemic consequences of this are unclear but they are likely to be considerable. The switch to the metrics-based REF may well reinforce pressures not to stray too far from 'mainstream' disciplinary research. Indeed, one might venture to predict that it will not be too long before UK academics will be bitterly criticizing the adverse consequences of the REF and its metrics, and reminiscing fondly about 'the good old days' of the RAE and peer review!

---

[35] The case of environmental sciences was mentioned earlier as an example of a newer research field that had to struggle before it finally succeeded in obtaining its own RAE panel. One wonders whether fields like computer sciences would have emerged as easily during the 1970s if the RAE had been in operation then.

References

Ackroyd, Pamela and Ackroyd, Stephen (1999) 'Problems of University Governance in Britain: Is More Accountability the Solution?', *International Journal of Public Sector Management*, 12: 171–85.

Advisory Board for the Research Councils (ABRC) (1987) *A Strategy for the Research Base* (London: HMSO).

Alsop, Adrian (1999) 'The RAE and the Production of Knowledge', *History of the Human Sciences*, 12: 116–20.

Attwood, Rebecca, and Corbyn, Zoë (2009) 'Research Elite Shaken by RAE Settlement', *Times Higher Education* (5 Mar.): http://www.timeshighereducation.co.uk/story.asp?storycode=405676, accessed Sept. 2009.

Barnard, Jayne W. (1998) 'Reflections on Britain's Research Assessment Exercise', *Journal of Legal Education*, 48: 467–95.

Becher, Tony, and Trowler, Paul R. (2001) *Academic Tribes and Territories*, 2nd edn. (Buckingham and Philadelphia: Society for Research into Higher Education and Open University Press).

Bence, Valerie, and Oppenheim, Charles (2005) 'The Evolution of the UK's Research Assessment Exercise: Publications, Performance and Perceptions', *Journal of Educational Administration and History*, 37: 137–55.

Braben, Donald W. (2004) *Pioneering Research: A Risk Worth Taking* (Chichester: Wiley).

Broadhead, Lee-Anne, and Howard, Sean (1998) '"The Art of Punishing": The Research Assessment Exercise and the Ritualisation of Power in Higher Education', *Education Policy Analysis Archives*, 6(8): http://epaa.asu.edu/epaa/v6n8.html, accessed Jan. 2009.

Campbell, Kevin, Vick, Douglas W., Murray, Andrew D., and Little, Gavin F. (1999) 'Journal Publishing, Journal Reputation, and the United Kingdom's Research Assessment Exercise', *Journal of Law and Society*, 26: 470–501.

Clark, Burton (1995) *Places of Inquiry: Research and Advanced Education in Modern Universities* (Berkeley, Calif.: University of California Press).

Corbyn, Zoë (2009) 'Reversal of Fortunes', *Times Higher Education* (5 Mar.): http://www.timeshighereducation.co.uk/story.asp?sectioncode=26&storycode=405690&c=1, accessed Sept. 2009.

Curran, Paul J. (2000) 'Competition in UK Higher Education: Competitive Advantage in the Research Assessment Exercise and Porter's Diamond Model', *Higher Education Quarterly*, 54: 386–410.

Dearlove, John (1998a) 'The Deadly Dull Issue of University "Administration"? Good Governance, Managerialism and Organising Academic Work', *Higher Education Policy*, 11: 59–79.

—— (1998b) 'Fundamental Changes in Institutional Governance Structures: The United Kingdom', *Higher Education Policy*, 11: 111–20.

Department of Education and Science (DES) (1985) *The Development of Higher Education into the 1990s* (Cmnd. 9524; London: HMSO).

—— (1991) *Higher Education: A New Framework* (Cmnd. 1541; London: HMSO).

—— (1992) *Further and Higher Education Act and Further Education (Scotland) Act* (London: HMSO).

Doyle, J. R., Arthurs, A. J., Green, R. H., Mcaulay, L., Pitt, M. R., Bottomley, P. A., and Evans, W. (1996) 'The Judge, the Model of the Judge and the Model of the Judged as Judge: Analyses of the UK 1992 Research Assessment Exercise Data for Business and Management Studies', *Omega: International Journal of Management Science*, 24(1): 13–28.

Farina, C., and Gibbons, M. (1979) 'A Quantitative Analysis of the Science Research Council's Policy of "Selectivity and Concentration"', *Research Policy*, 8: 306–38.

—— and —— (1981) 'The Impact of the Science Research Council's Policy of Selectivity and Concentration on Average Levels of Research Support: 1965–1974', *Research Policy*, 10: 202–20.

Geuna, Aldo, and Martin, Ben R. (2001) 'University Research Evaluation and Funding: An International Comparison', *Minerva*, 41: 277–304.

Gibbons, M., Limoges, C., Nowotny, H., Schwartzman, S., Scott, P., and Trow, M. (1994) *The New Production of Knowledge* (London: Sage).

Gläser, Jochen, and Laudel, Grit (2007) 'Evaluation without Evaluators: The Impact of Funding Formulae on Australian University Research', in Whitley and Gläser (eds.), *The Changing Governance of the Sciences* (Sociology of the Sciences Yearbook; Dordrecht: Springer), 127–52.

Halsey, A. (1992) *The Decline of Donnish Dominion* (Oxford: Clarendon Press).

Harley, Sandra (2002) 'The Impact of Research Selectivity on Academic Work and Identity in UK Universities', *Studies in Higher Education*, 27(2): 187–205.

—— and Lee, Frederic S. (1997) 'Research Selectivity, Managerialism, and the Academic Labour Process: The Future of Nonmainstream Economics in U.K. Universities', *Human Relations*, 50: 1425–60.

HEFCE (1997) *The Impact of the 1992 Research Assessment Exercise on Higher Education Institutions in England* (Bristol: Higher Education Funding Council): http://www.hefce.ac.uk/pubs/hefce/1997/m6_97.htm, accessed Jan. 2009.

Henkel, Mary (1999) 'The Modernisation of Research Evaluation: The Case of the UK', *Higher Education*, 38: 105–22.

Hood, Christopher (1991) 'A Public Management for All Seasons?', *Public Administration*, 69(1): 3–19.

House of Commons Science and Technology Committee (2002) *The Research Assessment Exercise* (Second Report of the Science and Technology Select Committee, HC507; Norwich: Stationery Office): http://www.publications.parliament.uk/pa/cm200102/cmselect/cmsctech/507/50710.htm, accessed Feb. 2009.

Jarratt, Sir Alex (chair) (1985) *Report of the Steering Committee for Efficiency Studies in Universities* ('The Jarratt Report'), report to the Committee of Vice-Chancellors and Principals (CVCP) and to the University Grants Committee (London: CVCP).

Jenkins, A. (1995) 'The Research Assessment Exercise, Funding and Teaching Quality', *Quality Assurance in Education*, 3(2): 4–12.

Johnes, Geraint (1992) 'Performance Indicators in Higher Education: A Survey of Recent Work', *Oxford Review of Economic Policy*, 8(2): 34.

—— (1995) 'Scale and Technical Efficiency in the Production of Economic Research', *Applied Economics Letters*, 2: 7–11.

Johnes, Jill, and Taylor, Jim (1992) 'The 1989 Research Selectivity Exercise: A Statistical Analysis of Differences in Research Rating between Universities at the Cost Centre Level', *Higher Education Quarterly*, 46(1): 67–87.

Johnston, Ron J. (1993) 'Removing the Blindfold After the Game is Over: The Financial Outcomes of the 1992 Research Assessment Exercise', *Journal of Geography in Higher Education*, 17(2): 174–80.

Jones, C. S. (1986) 'Universities, on Becoming What they are Not', *Financial Accountability and Management*, 2: 107–19.

Laughton, David (2003) 'Why was the QAA Approach to Teaching Quality Assessment Rejected by Academics in UK HE?', *Assessment and Evaluation in Higher Education*, 28(3): 309–21.

Lee, Frederic S. (2007) 'The Research Assessment Exercise, the State and the Dominance of Mainstream Economics in British Universities', *Cambridge Journal of Economics*, 31: 309–25.

—— and Harley, Sandra (1998) 'Peer Review, the Research Assessment Exercise and the Demise of Non-Mainstream Economics', *Capital and Class*, 66: 23–51.

McNay, Ian (1997a) 'The Impact of the 1992 Research Assessment Exercise in English Universities', *Higher Education Review*, 29(2): 34–43.

—— (1997b) *The Impact of the 1992 Research Assessment Exercise on Individual and Institutional Behaviour in English Higher Education*, Summary Report and Commentary (Chelmsford: Centre for Higher Education Management, Anglia Polytechnic University).

—— (2003) 'Assessing the Assessment: An Analysis of the UK Research Assessment Exercise, 2001, and its Outcomes, with Special Reference to Research in Education', *Science and Public Policy*, 30: 47–54.

Martin, Ben R., and Skea, Jim E. F. (1992) *Academic Research Performance Indicators: An Assessment of the Possibilities* (Report to the Advisory Board for the Research Councils and the Economic and Social Research Council; Swindon: ESRC).

Mayhew, Ken, Deer, Cécile, and Dua, Mehak (2004) 'The Move to Mass Higher Education in the UK: Many Questions and Some Answers', *Oxford Review of Education*, 30: 65–82.

Morris, Norma (2002) 'The Developing Role of Departments', *Research Policy*, 31: 817–33.

Nadin, Vincent (1997) 'Widening the Gap between the Haves and Have Nots in Research Funding: The UK Research Assessment Exercise', *Planning Practice and Research*, 12(2): 93–7.

Office of Science and Technology (OST) (1993) *Realising our Potential: A Strategy for Science, Engineering and Technology* (Cabinet Office; London: HMSO).

Paisey, Catriona, and Paisey, Nicholas J. (2005) 'The Research Assessment Exercise 2001: Insights and Implications for Accounting Education Research in the UK', *Accounting Education*, 14: 411–26.

Perryman, Jane (2007) 'Inspection and Emotion', *Cambridge Journal of Education*, 37(2): 173–90.

Phillimore, John (1989) 'University Research Performance Indicators in Practice: The University Grants Committee's Evaluation of British Universities, 1985–86', *Research Policy*, 18: 255–71.

Power, Michael (1999) 'Research Assessment Exercise: A Fatal Remedy?', *History of the Human Sciences*, 12: 135–7.

Roberts, Sir Gareth (2003) *Review of Research Assessment* (Report to the UK funding bodies; London: Higher Education Funding Council for England).

Robinson, Jane, Watson, Roger, and Webb, Christine (2002) 'The United Kingdom Research Assessment Exercise (RAE) 2001', *Journal of Advanced Nursing*, 37, 497–8.

Rothschild, Lord (1971) 'The Organisation and Management of Government R&D' ('The Rothschild Report'), published as an appendix to the Green Paper, *A Framework for Government Research and Development* (Cmnd 4814, Nov. 1971; London: HMSO).

Sastry, Tom, and Bekhradnia, Bahram (2006) *Using Metrics to Allocate Research Funds* (Oxford: Higher Education Policy Institute).

Science Research Council (SRC) (1970) *Selectivity and Concentration in Support of Research* (London: SRC).

Shore, Chris, and Wright, Susan (1999) 'Audit Culture and Anthropology: Neo-Liberalism in British Higher Education', *Journal of the Royal Anthropological Institute*, ns 5: 557–75.

HM Treasury (2002) *Investing in Innovation: A Strategy for Science, Engineering and Technology*, a report by HM Treasury, the Department of Trade and Industry, and the Department for Education and Skills (London: HM Treasury).

Trow, Martin (1993) 'Comparative Perspectives on British and American Higher Education', in S. Rothblatt and B. Wittrock (eds.), *The European and American University since 1800* (Cambridge: Cambridge University Press), 280–99.

—— (1994) *Managerialism and the Academic Profession: Quality and Control* (London: SRHE and Open University Press).

Universities Funding Council (UFC) (1989) *Report on the 1989 Research Assessment Exercise* (London: UFC).

University Grants Committee (UGC) (1988) *Circular Letter 45/88* (London: UGC).

Van der Meulen, Barend (2007) 'Interfering Governance and Emerging Centres of Control: University Research Evaluation in the Netherlands', in Whitley and Gläser (eds.), *The Changing Governance of the Sciences* (Sociology of the Sciences Yearbook; Dordrecht: Springer), 191–203.

Velody, Irving (ed.) (2009) 'Knowledge for What? The Intellectual Consequences of the Research Assessment Exercise', *History of the Human Sciences*, 12: 111–46.

Warner, Julian (1998) 'The Public Reception of the Research Assessment Exercise 1996', *Information Research*, 3(4) (Apr.): http://informationr.net/ir/3-4/paper45.html#timh, accessed Feb. 2009.

Whitley, Richard (1986) 'The Structure and Context of Economics as a Scientific Field', *Research in the History of Economic Thought and Methodology*, 4: 179–209.

—— (1991) 'The Organisation and Role of Journals in Economics and Other Scientific Fields', *Economic Notes*, 20: 6–32.

—— (2000) *The Intellectual and Social Organization of the Sciences* (Oxford: Oxford University Press).

—— (2007) 'Changing Governance of the Public Sciences: The Consequences of Establishing Research Evaluation Systems for Knowledge Production in Different Countries and Scientific Fields', in Whitley and Gläser (eds.) *The Changing Governance of the Sciences* (Sociology of the Sciences Yearbook; Dordrecht: Springer), 3–27.

—— and Jochen Gläser (eds.) (2007) *The Changing Governance of the Sciences* (Sociology of the Sciences Yearbook; Dordrecht: Springer).

Wilson, Duncan (2008) *Reconfiguring Biological Sciences in the Late Twentieth Century: A Study of the University of Manchester* (Manchester: Faculty of Life Sciences, University of Manchester).

—— and Lancelot, Gael (2008) 'Making Way for Molecular Biology: Institutionalizing and Managing Reform of Biological Sciences in a UK University during the 1980s and 1990s', *Studies in History and Philosophy of Biological and Biomedical Sciences*, 39: 93–108.

Wilson, Tom (1991) 'The Proletarianization of Academic Labour', *Industrial Relations Journal*, 22: 250–62.

Yokoyama, Keiko (2006) 'The Effect of the Research Assessment Exercise on Organisational Culture in English Universities: Collegiality versus Managerialism', *Tertiary Education Management* 12: 311–22.

3

Research Funding, Authority Relations, and Scientific Production in Switzerland

Martin Benninghoff and Dietmar Braun

1. Introduction

The scientific production of a country, its 'rate' of new knowledge production and capability of integrating new topics and research fields, depends partly on its governance structure—the distribution of authority and resources among corporate actors. Such corporate actors are research organizations (laboratories, institutes, and so on), higher-education institutions (faculties, departments, universities), but also the state apparatus (government and parliament, state bureaucracy) and 'intermediary institutions' (funding agencies, science councils), and private industries or foundations. As stressed in several studies (Bourdieu 1976; Latour and Woolgar 1986; Whitley 2000), how authority is distributed among these actors and how money circulates matters for scientific production. Based on a typology of the public science system of governance, Whitley, for example, has suggested that 'state-centralized' systems, in which research organizations and Higher Education Institutions (HEI) stand in a direct authority line with the state bureaucracy and in which funding is organized within the confines of the state bureaucracy, are inclined to strengthen the individual authority of heads of laboratories who, with the help of institutional funding, may pursue long-term research programmes. Rapid change in research topics that are taken up or the establishing of new research fields is unlikely in such systems (Whitley 2008: 17–18). 'State-delegated competitive' systems by contrast, giving HEI and research organizations a large degree of management authority and obliging researchers to find their main resources by applying for grants outside their employment institution, fosters competition and, hence, a stronger diffusion and acceptance of new topics in research or the creation of new scientific fields (ibid. 20).

This assumed linkage between governance structure, effects on the way the scientific elite and individual researchers deal with research, and scientific production forms the starting point of our analysis of the publicly financed research system in Switzerland. We focus on the changes that have taken place in the political organization of research and its possible consequences for research behaviour and research performance by analysing a specific corporate actor, the Swiss National Science Foundation (SNSF) as Switzerland's main funding agency, and its funding instruments. This empirical focus on a funding agency can be justified by the crucial position of funding agencies in the scientific field in general.

Because funding agencies control a part of the distribution of material (money) and ideal (reputation) resources in the sciences, funding agencies can be considered to be an established power, an expression of scientific authority in the sense that they are allowed to express an opinion on the standards, criteria, values, and other categories of classification that govern the scientific field. This authority of funding agencies gives them a central role in public research systems. At the same time, they are important meeting places of the scientific community and the political administration and have to balance both political and scientific logics (Braun 1998; Godin *et al.*, 2000). This makes funding agencies the most important *intermediary* institutions in the public research system between political and scientific interests.

We will focus our analysis on the Swiss National Science Foundation (SNSF) in general and its *use of funding instruments* in particular. Recent studies have argued that the way resources are allocated (for example, block grants vs. project grants) have an effect on scientific production, in particular on the ability of scientists to develop new research themes (see Chapter 1; Laudel 2006). Funding instruments can be facilitating or directing. By concentrating resources on certain research areas or people or by adding conditions to the granting of money to researchers, funding agencies have the capacity to draw the attention of scientists in certain directions and to make them abandon other lines of research. The very choice of funding instruments is therefore a matter of authority in the guidance of scientific production, and a change in the choice of such instruments can reveal changing authority relations in public science systems.

First, we will briefly present the position of the SNSF in the Swiss public science system. Then, we will discuss the establishment of the SNSF in the early 1950s before we analyse the institutionalization of three funding instruments that in our opinion characterize the key changes in governance rationales during the second part of the twentieth century.

2. The Position of SNSF in the Swiss Public Science System

While Switzerland funds a considerable amount of scientific research at roughly 2.8 per cent of GDP, most of this is spent by enterprises (about 70 per cent) and only 30 per cent can be considered as public funding. However, research conducted in Switzerland does well in terms of citation and impact records, which is largely based on the public research system, as an overview of the CEST demonstrates (CEST, 2003): only two enterprises, Novartis (with 2.7 per cent of publications) and Nestlé (with 0.6 per cent) are mentioned among the top twenty-five institutions responsible for research output in Switzerland.

The Swiss public science system itself is rather decentralized, with a historically weak federal authority and a strong role played by the cantons, even if we observe a move towards a more centralized system, or at least to coordinated organization between the federal government and the cantons, over the last decade (Braun and Leresche 2007; Baschung *et al.* 2009). In contrast to the German or French public science system,[1] Switzerland has only few extra-university research institutions and almost no big science research institutes. The few extra-university research institutes are not under the direct authority of the federal government but are managed by the semi-autonomous governing board of one of the two federal technical universities (the Federal Institutes of Technology Board), apart from a number of agricultural research institutes which are under the direct authority of a federal ministry.

Publicly financed research is thus concentrated in ten cantonal universities, two federal institutes of technology (ETHZ, EPFL), and in some of the new universities of applied science.[2] Table 3.1 provides an overview of the size of Swiss universities (with the exception of the universities of applied science, which we will exclude from our analysis) in terms of the number of staff working in these institutions. The differentiation of staff according to the sources of funding reveals the relative importance of institutional funding and external grants (from the SNSF and other sources).

---

[1] On the French system see Chapter 6.
[2] On the research activities of the universities of applied sciences, see Lepori (2007).

Table 3.1. Staff in 2007 by source of funding in higher-education institutions (full-time)

	ETHZ	UZH	GE	EPFL	BE	BS	LS	FR	NE	SG	USI	LU	Total
Institutional funds	5523	4349	2910	2337	2664	1649	1930	1150	542	430	307	174	24065
SNSF funding	419	467	419	390	295	319	189	165	136	14	37	14	2864
Share SNSF in total (%)	6	8	11	11	8	13	8	11	16	2	10	7	9
Other third-party funds	753	712	500	827	579	418	179	209	157	253	44	4	4636
% grants of total	18	21	24	34	25	31	16	25	35	38	21	9	24
Total	6695	5528	3829	3554	3538	2386	2299	1524	835	697	389	191	31504

Sources: Office fédéral de la statistique, 2008; own percentage calculations. ETHZ: Eidgenössische Technische Hochschule Zürich, UZH: Universität Zürich, GE: Université de Genève, EPFL: Ecole Polytechnique Fédérale de Lausanne, BE: Universität Bern, BS: Universität Basel, LA: Université de Lausanne, FR: Université de Fribourg, NE: Université de Neuchâtel, SG: Universität St.-Gallen, USI: Universita della Svizzera italiana, LU: Universität Luzern.

The table demonstrates the large differences in size. The two federal institutes of technology are in the leading group with the universities of Zürich, Geneva, and Berne. At the bottom, we find two new universities: Svizzera italiana (USI) and Lucerne. Compared to other countries, the average size of Swiss higher-education institutes remains quite low, which corresponds to the overall small size of the country.

It is also obvious from the table that institutional funding is the most important funding source for Swiss universities.[3] This has always been the case. From a comparative point of view, this is a relatively high rate of institutional funding, although the share of institutional funding varies considerably between higher education institutions: namely, from 91 per cent for the university of Lucerne to 62 per cent (St. Gallen). Despite this dominance of block grant funding, the SNSF is important in the Swiss public science system for the following reasons:

- The Swiss public science system has very few state-owned national research centres. Therefore, the only way to increase research activities in a research lab, institute or in any university department is to obtain project grants. Here, the SNSF is the most important project-funding agency and represented 50 per cent of the total project funding in 2002 (Lepori 2006). It predominantly funds basic research projects, but its share of funding schemes for 'oriented research' projects has grown.[4]

- The use of SNSF grants has increased during the last thirty years. If one takes only the (most important) grants for basic research (*freie Projektförderung*, 'free project grants'), the number of applications has more than doubled since 1980.[5]

- Finally, the SNSF project grants differ from most institutional funding by the federal and cantonal states by being competitive and being based on national and international peer reviews.[6] Getting grants from the SNSF is not only materially but also symbolically important for university scientists. Pressure on the performance of universities makes this

---

[3] More generally, project funding represented 23 per cent of public research funding in 2000, against 77 per cent of general funding (Lepori 2006).

[4] The share of 'oriented research grants', starting in 1976, went up from 5.4 per cent to (in fluctuations) 29 per cent in 2004 but was back in 2008 to 11 per cent because of important investments made in parliament to support basic research (figures given by the SNSF).

[5] From 807 applications in 1980 to 2,273 applications in 2008 (personal communication from the National Science Foundation).

[6] The federal government gives a part of its institutional funding (about 30 per cent) on the basis of performance criteria. Overall cantonal universities obtained in 2007 15 per cent of their resources from institutional funding out of the hands of the federal government (Office Fédéral de la Statistique 2008; our calculations).

symbolic capital also important in the institutional context of universities.

To sum up, SNSF grants constitute one of the most important ways for university scientists to obtain additional financial resources dedicated to research activities and to increase their scientific reputations. In the context of fiscal strain and the new 'age of performance evaluation', the 'value' of grants from the SNSF is ever increasing, so that one can safely say that the role of the SNSF in scientific production has steadily grown during the last thirty years.

3. Building a Research Foundation (1945–1970)

During the first part of the twentieth century, the Swiss public science system can be characterized—in terms of Whitley's typology—as a '*state-shared*' one (part of the 'state-coordinated' type) with the particularity that the federal structure made two political authorities part of the governance structure: the cantons that were responsible for universities and the federal government that, at this time, only had the right to manage a university for technology, the Federal Institute of Technology in Zürich (Eidgenössische Technische Hochschule Zürich; ETHZ).[7] This contributed, at that time, to a rather clear-cut institutional distinction between fundamental and technological-oriented research and, hence, also to an institutional separation of scientific and technological elites.[8] Extra-university research institutions were—with a few exceptions—practically non-existent (Braun 2001). The federal and cantonal governments had the authority to decide in matters of financing, personnel, and organizational structures, though the ETHZ had more leeway in this respect than the cantonal universities.

We can also consider the Swiss public science system of this time as 'state-shared', because the state actors (federal government and cantons) had to share their authority within cantonal and federal universities with academic elites who were running everyday activities and had exclusive and substantive rights concerning the determination of content and procedures of

---

[7] In 1969, a second federal institute of technology was 'created': the École Polytechnique Fédérale de Lausanne (EPFL).

[8] Like Whitley (2003: 1016), we see technological research as part of public sciences which is directed to 'investigating the behaviour of artefacts and artificial materials', done by scientists who are primarily interested in gaining scientific reputation. It does not include research for commercial or military purposes.

research. Scientists were employees of the federal or cantonal states. Both political authorities were solely responsible for the financing of their institutions. Such a structure tends to make heads of research units or chairholders largely independent in their choice of topics. During the first part of the twentieth century, competition for external grants remained relatively low. Structurally, reputation in such systems is, according to Whitley, more locally than nationally based, although some scientists had international reputations.

The Creation of the Swiss National Science Foundation: A Funding Agency for Basic Research

After the Second World War, a number of pressures led to a debate on the need for a nationwide funding organization. These pressures arose from the 'brain drain' in the aftermath of the war, and difficulties experienced by universities in financing research on a larger scale and supporting their young scientists. Another impetus was the perception that more financial support was necessary to stay internationally competitive.

The establishment of a national research foundation proceeded in two stages.[9] The first attempt revealed a lasting tension between the technological and scientific elites in the public science system, which resulted in the failure of a top–down initiative by the president of the ETHZ during the Second World War because of opposition from representatives of cantonal universities. They claimed that the project was too strongly oriented towards the interests of industry, that the ETHZ took up too much space in the project, and that the social sciences and humanities were disadvantaged. The cantonal universities already felt disadvantaged by a more generous funding of the ETHZ in comparison to their revenues and saw the creation of a funding organization as a means of overcoming their disadvantages.

After the war, the president of the Swiss Society of Natural Sciences established a Commission to consider this project in which all disciplines were represented. At this time, it was obvious that the public science system needed a stronger influx of money for research and a national effort to raise its attractiveness in general, but any initiative was limited by the existing distribution of power relations between the Swiss Confederation and the cantons in different policy domains that could not be altered. Any solution

[9] Data related to the creation of the Swiss National Science Foundation mostly come from the study conducted by Fleury and Joye (2002).

had to respect the autonomy of the cantons and should not increase federal authority in the guidance of universities.

The Commission proposed the following structure for the new Foundation: a foundation board (*Stiftungsrat*), a research board (*Forschungsrat*), and an authority for financial control. The make-up of the foundation board was to be very diverse, with representatives from the federal government, universities, scientific associations, and various stakeholders of society and linguistic regions. According to Fleury and Joye (2002: 148), the board was meant to be a representation of the whole political system and society and to establish 'a showcase of the national foundation'. The foundation board served both as an arena for the exchange of interests between political administration, societal stakeholders, and scientific elites,[10] and as a device to sensitize citizens to research activities and to secure the political and financial support of influential actors. The research board was to be composed exclusively of scientists and would allocate funding support for research.

In 1950 the cantonal universities, the ETHZ, and scientific academies submitted a report to the federal government in which they asked for the establishment of a national science foundation. They justified the creation of this foundation by: (1) a 'delay' in Switzerland compared with other countries; (2) insufficient resources for research which were dispersed among many institutions; (3) insufficient resources for training young researchers; and (4) the 'brain drain'. The purpose of the National Foundation should be to encourage scientific research in Switzerland—above all with subsidies to researchers who lacked sufficient institutional financial support. This could entail the payment of a salary, research grants, participation in the costs of publication, and scholarships for young and advanced researchers.

One year later the federal government proposed to parliament the means of financing the Swiss National Science Foundation (SNSF). It covered all the elements the commission had asked for. It should be a private law institution to manifest its chief role as a service institution of science and give it some freedom from direct political interference. In this way it could, when it was created, be seen as a scientific institution located within the governance structure of research. Its organizational structure allowed, however, the influence of politics and society, and its dependence on federal

---

[10] The technological elite of the ETHZ were under-represented, as the main focus of the foundation was on the support of basic research. In this sense, the ETHZ had lost the struggle for governance power against cantonal universities.

government money made it subject to annual budget discussions in parliament and therefore vulnerable. The project was well received by parliamentarians. They saw it as a turning point in relations between the federal government and the cantons in the academic field. The proposal was accepted without objections.

The SNSF in a State-Shared System of Research

What did the creation of the SNSF signify in the context of the state-shared system of public financed research? Several points can be stressed. The SNSF was a new authoritative agency in the Swiss public science system and practically had a monopoly position in the external funding of research.[11] Though it was created as a scientific institution for basic research, the federal government had an influence on resources, structures, and even on discussions about funding instruments within the SNSF foundation board. In addition, via SNSF funding, the federal government could for the first time contribute to the financing of research throughout the country.

Without any doubt the foundation of the SNSF enabled the 'creation' of a national scientific elite that organized the distribution of symbolic and material resources. With the creation of the SNSF, scientists had to accept in this sense a stratification of authority in the scientific field. At the same time, the scientific elite in all disciplines became part of the governance structure of research. Scientists from cantonal universities dominated the organization.[12] The establishment of the SNSF therefore strengthened cantonal universities.

Concerning scientific activities, the SNSF was an important step in securing more resources for research and to address particular problems, such as the support of young scientists. Due to the division of labour between the foundation and research boards, the selection of 'quality' in research became an exclusive matter for the scientific elite.

There were no changes in the relationships between the managers of universities and their scientists. Scientists' research remained independent. Additional resources were beneficial to both university management and scientists. The additional resources coming from the SNSF also constituted

---

[11] Beside the SNSF there were only some small funding councils for applied research or nuclear energy at the federal level, which had been founded after the Second World War (see Joye-Cagnard 2010).

[12] In 1952 the research council was composed of eleven professors: nine from cantonal universities and two from ETHZ.

a way of altering the authority structure in departments and their scientific orientation, and allowed the development of new themes in a discipline.[13] Though resources were limited at first, the elements of competition between researchers were introduced.

The state-shared system was in no way challenged by the introduction of the SNSF. Its main element—the introduction of federal money for public research activities—did not compete with the main role of the cantonal states in general or the authority of the academic profession in research. The academic community was even strengthened by the establishment of the SNSF as scientific elites became, through the SNSF, responsible for the distribution of federal government money to research. The sums of money invested in the beginning were too small to compete with institutional resources given to universities or the ETHZ. However, the distribution of grants ('project funding' and 'individual and career development funding') by competitive selection became the nucleus of a more competitive orientation in science.

4. Increasing Demands for 'Useful' Research: The Creation of National Research Programmes (1970–1985)

The political context changed during the 1970s. The oil crisis, stagflation, and rising budget deficits created a climate in which government expenditures increasingly came under pressure in the intense political debates about revenues and spending needs. At the same time, OECD-countries had started institutionalizing research and technology policies by creating ministries of research and science councils. The changing financial context led to a questioning of expenditures that seemed to have no overt and immediate utility, like the funding of basic research. The institutionalization of research policy and discussions about the 'return' of funding of research provoked increasing demands for 'useful research' (Elzinga and Jamison 1995; Joye-Cagnard 2010). These tendencies occurred in most OECD-countries, and Switzerland was no exception.[14]

As in other countries, a (Swiss) Science Council (SSC), composed of scientists and politicians, was created in the mid-1960s to advise the federal

---

[13] For example, the institutionalization of molecular biology at the University of Geneva during the 1950s would have been impossible without the financial and scientific support of the SNSF (see Strasser 2006).

[14] On the Swiss–OECD relations during this period, see Gees (2006).

government in the design of a science policy.[15] A first administrative unit for education and research was created within the federal administration. This was not only a response to the overall climate of a more active policy stance in matters of education and research but also necessary because the federal subsidies for universities that had been paid since 1968 through a federal law needed implementation structures. The cantons finally accepted the involvement of the federal government in the financing of universities because of rising numbers of students and a lack of resources (Benninghoff and Leresche 2003). The federal subsidies did not imply any co-decision rights in higher-education policy, but it nevertheless made the federal government a partner with a voice in matters concerning the cantonal universities.

The National Research Programmes

The changing climate had implications for the SNSF. The first move in this regard came from the Swiss Science Council, which proposed, in 1973, in its report on Swiss research, the creation of 'large interconnected research centres' and suggested the federal government shift its funding support to oriented research. The SSC went as far as launching the idea of a second national science foundation for applied research.

The SNSF had to react to this proposal, as it jeopardized its legitimacy among policy-makers and the public, and there was a fear that parliament would redistribute resources in favour of such a new organization. The agency used the discussion about its budget for the period 1975–79 to make its counterproposal. It was aware that the political climate was not favourable, and that any call for an increase in funding for basic research would be in vain. Instead, it suggested taking the funding of oriented research into its own hands and creating a new instrument, the National Research Programmes (NRP). This proposal was heavily contested within the SNSF itself, but the advantages of internalizing the promotion of application-oriented research were obvious. It avoided the creation of a competitor at the intermediary level, it would help to re-establish the SNSF's

---

[15] Even though its role and authority has varied between the 1960s and today, the SSC has never been a dominant actor in matters of research policy. Its main domain was higher education policy, which—because of the important research position of universities—had linkages to research. The lack of influence of this council was caused in the first period (until 1999 when a reorganization took place) by its 'corporatist' style of decision-making. It was more a debate club than a planning body. After 1999, when only scientists were active within the council, it was the lack of direct links to policy-makers that made the council powerless (Braun and Leresche 2007).

legitimacy within parliament, and it authorized the SNSF—and with it the scientific elite—to debate the terms of setting up such an instrument, as well as entitling it to implement the instrument on its own terms (OFES 1994).

The federal government agreed with the introduction of NRP but claimed an important role for political authorities in the management of such an instrument, which would have been a qualitative change because of the rather passive stance of policy-makers which prevailed until then. At the same time, the federal government rejected the proposal to substantially increase the budget of the SNSF or to allocate a special budget to NRPs. For policy-makers it was clear that the introduction of the NRP must be a 'zero-sum game': it did not allow an increase in the budget of the SNSF. The federal government therefore proposed that 10 per cent of the SNSF budget (which was entirely dedicated to basic research) should be attributed to the application-oriented programmes. The parliament went even further than the federal government and demanded that 12 per cent of the SNSF should be allocated to the NRP while the total budget should be reduced by 10 per cent. As a result of these parliamentary debates, the SNSF suffered a loss of 5 per cent of its total budget but was able to implement the new NRP. Though the SNSF could therefore avoid the emergence of a 'new competitor' and strengthen its legitimacy, it had to pay in terms of financial resources. The funding of basic research had to be cut not only by 5 per cent but in addition by the 10 per cent that now was reserved for the new instrument.[16]

Though the idea of establishing a funding instrument to promote applied research was widely accepted, the goals of such a programme remained vague, and so it was able to meet all expectations: it could result in more funds for universities; it could be used as an incentive programme for the Science Council or for the development of research within the federal administration; and it allowed for a better concentration of research. This indeterminate nature of the programme gave them political credit, since 'every pressure group could hope to achieve, one time or another, to have its own national program' (OFES 1994).

---

[16] The debate on the introduction of the new instrument reveals one distinctive feature in Swiss science politics: parliament has an influential voice in budget discussions, and as these discussions cover all substantial elements of the funding policy it can deliberate on such strategic elements as the introduction of a new instrument. This puts strong pressure on the funding agency to be concerned about its 'public appeal' and societal concerns represented by political representatives.

During the parliamentary debates, parliamentarians had already made a series of proposals for research themes. Discussions continued between the SNSF, the Science Council, and the Federal Office for Education and Science, which was responsible for education and research. Each agency wanted authority over future research topics. Finally, the Ministry of Home Affairs (responsible for education matters) decided that scientists should propose a number of research themes, which were then selected by the SNSF according to scientific standards. But it was the federal administration which made the final choice.

The internalization of oriented research within the SNSF resulted therefore in a stronger co-decision role of political forces in the funding of public research. The advantage for the scientific community was obviously that scientists still had an important weight in choosing topics relevant to them and society, even if at the beginning the federal administration was a key actor in the choice of the research theme. The first four NRPs were dedicated to cardiovascular disease, water pollution, social integration, and energy.

The introduction of NRPs had other consequences that diminished the authority of the SNSF: to implement these national research programmes, the SNSF created a fourth division in the SNSF Research Council.[17] Of the twelve members of that division, six were directly appointed by the federal government, while the other six were nominated by the foundation board. These appointments were not trouble-free. Researchers were reluctant to join for two reasons. First, they considered this division as too dependent on political power and supporting a kind of 'second-class' research. Second, appointments were not only based on scientific merit but also on political criteria such as language, disciplines, and universities. Furthermore, the division was made up of representatives of the scientific community, the federal government, industry, and cantonal authority (Latzel 1979: 145). This organizational set-up went even further than the assumed authority of the federal government to select members of the Research Council: for the first time in the history of the SNSF, stakeholders and political representatives at the parliamentary level had the right to manage such a programme along with the scientists. This meant that even the implementation of the programme now had to be shared with 'outsiders', and a balance of diverse interests had to be struck.

---

[17] Divisions I (social science and humanities), II (mathematic, natural sciences, and engineering), III (biology and medicine) are dedicated to free and fundamental research.

The Establishment of a Tool for Oriented Research

The implications of the new grant scheme for scientific production were not as important, as the NRP took only about 15 per cent of the resources of the SNSF while the rest of funding was distributed according to usual scientific standards for basic research. However, application-oriented research became an option in academia. The more scarce the resources, the more interesting the NRP became. For those who participated in the new funding scheme, discretion over research topics was reduced but not dramatically so: topics were co-decided by scientists and by political bodies (SSC, federal administration, and federal government) even though policy-makers made the final choice.[18] In this way, research remained anchored within scientific interests even though the aspect of applicability and usefulness had to be taken into account. Since the programmes functioned as umbrellas for a number of projects, they also encouraged more contacts between scientists and generated a stronger national orientation of research which further reduced the 'localism' that had been strong before the creation of the SNSF. The monopoly of the scientific elite in selection of research projects was, of course, lost. The voice of non-scientific actors added other criteria into the handling of such projects, though this could not question the priority of scientific quality in the selection and evaluation of programmes.

The changes for universities were not immediately visible. The NRP did not change anything in the existing 'state-shared' structure, but universities were of course confronted by the use of their infrastructural resources for purposes defined by federal actors. In this way, the federal government enlarged its impact on the use of scientific resources, although its support for research did not lead to more authority over research strategies. The NRPs, however, could render visible what scientists in universities did and focus their attention and activities on domains of interest to the federal government.

Finally, individual researchers did not immediately feel the implications of the new funding instrument. As mentioned above, the money allocated to the NRP was limited. The 15 per cent reduction in basic research created, however, a more competitive climate as fewer projects could be funded. Success rates remained high but diminished compared to the end of the 1960s. Thus, the implications for the functioning of the SNSF and its role in

---

[18] The way the themes of NRP were decided changed considerably after the 1970s, with scientists gaining more and more weight in the selection of topics.

the governance structure were stronger than those for scientific production as such. The scarcer resources became, the more competition for project money increased. Diminishing revenues of universities at the end of the 1980s and in the beginning of the 1990s led to more competitive bidding for SNSF grants.

5. The Struggle for Priority Programmes (1985–1995)

In the context of an economic crisis and a further deterioration of the public finances of both political authorities (federal and cantonal), the 1990s were characterized by a reorganization of the federal administration based on new public management doctrines. The reforms were supposed to increase the efficiency and effectiveness of the public administration and policies. In the research policy domain, new administrative structures were set up in order to increase coordination and efficiency at the national level and to be more active at the international one. This reorganization of the federal administration set up a stronger 'dual system', with on the one hand a Ministry of Home Affairs responsible for education and research and, on the other, a Ministry of Economy in charge of technology and innovation policy (for more details, see Baschung *et al.* 2009). This dual system was reinforced by the strengthening of the Commission for Technology and Innovation (CTI)[19] located in the newly created Federal Office for Professional Education and Technology—an administrative agency of the Ministry of Economy. The landscape of research institutions also changed during this period because seven universities of applied science were established in 1997 and considerably enlarged the institutional base of the technological elite.

Beside this push towards technological development, the federal parliament approved a proposal of the government to create the post of state secretary for science and education. The newly nominated state secretary—

---

[19] We deliberately omit discussion of this commission. The sums of money distributed to technology research are relatively small in comparison with the SNSF (one fifth of the SNSF budget). The SNSF and the CTI work on different funding domains. The CTI actually focuses only on technology research while oriented research within the domain of the SNSF links fundamental and technology research. Funding instruments are addressed to the scientific and technological elites that are still interested in scientific reputation and not in commercial products like technological research. There have been discussions, at different times, to unite both agencies in one agency, but the different juridical status—the CTI was until recently part of the Ministry of Economy, while the SNSF is a foundation—and political interests—two different departments responsible for the management of funding lines—have so far prevented such a move (Braun *et al.* 2007).

he had formerly been president of the Federal Institutes of Technology in Zurich—revived an old policy concept of the 1970s called *Hochschule Schweiz* (University Switzerland). The 'University Switzerland' was intended to be a platform for promoting the coordination of the cantonal HEI and research organizations. It should also help to define research priorities at the national level. In this context a new funding instrument was introduced: the so-called 'priority research programmes' (PRP). This instrument was supposed to promote innovation and technological development and at the same time to make the support of research activities more efficient.

The Rise and Fall of the Priority Research Programmes

In the beginning of the 1990s, in a climate of increasing budget deficits and economic stagnation, a more far-reaching 'attack' on the self-determination of the scientific elite in the SNSF took place when the newly nominated state secretary for science and research (who was also president of the board of the two federal technical institutes, the so-called ETH-Board) launched a new nationwide programme. It was initially to be administrated by the ETH-Board itself. The state secretary mandated professors of the federal technical institutes to propose an 'implementation structure' for future priority research programmes in their institute. But the Swiss University Conference (SUC)—a coordination body of the federal government and cantons in matters of higher education—and the Swiss Science Council were against the proposal as they saw such priority programmes as an additional funding venue for the federal technical institutes from which cantonal universities could not benefit. Hence, after a political debate on the leadership of this programme, the federal government decided to give half of the research project to the SNSF and half to the ETH-Board in order to not overemphasize technology policy and the federal technical institutes. This decision was supported by the cantons as the funding institutions of cantonal universities. In parliament the lack of integration of the social sciences and the exclusive focus on technology was criticized.[20] Despite these criticisms, the programme was approved by parliament and received almost 400 million

---

[20] The priority research programme focused exclusively on technological innovation. Six research domains were identified: electronics, optical technologies, environmental sciences, biotechnology, material science, and informatics. Three of them were supposed to be managed by the ETH-Board and three by the SNSF in coordination with the federal administration. The programme was to be financed for six to ten years and receive financial support from industry (ABB, Ascom, Sulzer AG, and so on).

francs for the period 1992–5 as part of a 2-billion-franc federal budget dedicated to research and education (Benninghoff and Leresche 2003: 87–93). A social science programme was later added to the existing technology programmes.

The programme expressed two new priorities of the federal government: to target funding resources at specific research domains (opting against a policy of indiscriminately pouring out money), and to encourage more strongly technological research and transfer activities in order to help small and medium enterprises. The prerogatives of scientists to define areas of interest within the SNSF had left the administration with few options in the end, and the link between fundamental research and technological applications remained a problem. This led to the more proactive stance by the federal government. The priority research programmes (PRP) were conceived top–down and not bottom–up. It was the administration, parliament, and stakeholders who had the right to define areas of interest. But it was decided not to have such a programme organized by the administration itself. Again, 'outsourcing' was the dominant strategy of the federal administration.

The new programme was abandoned after eight years, mostly because of a mismatch between the topics chosen by stakeholders and the few scientists in universities who were able and/or willing to take up these topics. Another reason to abandon the programme was the difficulty of establishing the research topics in universities after the financing of the programme ended, which was one of the goals of the programme.

A number of interesting points can be raised on the basis of this experience:

- The mismatch demonstrated not only that topics that are not genuinely anchored within the scientific community run the risk of finding no researchers, but also that apparently the financial situation of researchers in Switzerland was still not desperate enough to grasp all opportunities to fund their research, especially in the domains of natural and technical science.

- The creation of the PRP was an attempt by policy-makers to get a firmer grip on research funding. Its primary aim was to foster a stronger linkage of fundamental and technological research, which was not taken into account sufficiently by existing institutions. Instead of launching a new attempt to draw the SNSF into this direction, it was decided to create a new organizational structure. The stronger role of the ETH-Board started, however, a struggle for dominance in the research field between the SNSF and the ETH-Board.

- The conflict between the Federal Institutes of Technology Board (ETH-Board) and the SNSF concerning the direction of this research programme reveals larger and older tensions between scientific and technological elites, which had existed since the Second World War. The SNSF had become an organization that was mostly based on university scientists. Priority research programmes were an attempt to institutionalize support for technological elites and industrial interests.

- Again, the introduction of such a programme for technology research did not change the basic authority structures, even if we observe a struggle between the ETH-Board and the SNSF for the leadership of research policy. The federal government, as provider of financial resources for both actors, took the position of a 'referee' in such conflicts.

- The HEIs were again faced with the integration of research topics that were exogenously defined, but the sums of money invested in the programmes were not so significant as to make this a qualitative leap in the organization of research in universities. Individual researchers still had sufficient options to finance fundamental research projects, and the NRP remained the strong and more accepted programme among scientists.

6. The National Centres of Research: Towards a Science Policy for a 'Knowledge-Based Economy' (1995–2005)

At the end of the 1990s there was a qualitative change in the way science in general and HEIs in particular were managed by federal and cantonal governments. This period was not one of public deficit reduction; in fact the government began to increase its spending on scientific research. The key phrase of the federal government became: a 'strong economy needs a strong science'. That meant, in terms of science policy, the concentration of financial resources on specific disciplines (such as 'life sciences'), the definition of priorities (for HEI), and an efficiency approach to the organization of research organizations (based on competition, coordination rules, networking, and so on).

The Creation of National Centres of Competence in Research

The creation of a new instrument of oriented research at the end of the 1990s was, on the one hand, embedded in this political and economical context of a 'knowledge-based economy', and on the other hand it was partly the answer of the SNSF to the launching of the PRP. But the trigger for the new funding instrument was an evaluation of the SNSF by an international panel that had insisted on a stronger and more proactive role by the SNSF in science policy in general.

For these experts, the SNSF should no longer be the target of the federal administration in setting up programmes of applied research but must become a player in its own right. In addition, it was seen as pertinent to claim back the loss of domain power, which had taken place with the creation of the PRP. Any new programme should emerge within the confines of the SNSF, which meant that it should be controlled by the scientific elite. It was for these purposes that the SNSF suggested replacing the PRP with a new and better instrument. Based on experiences with similar instruments in Germany and the USA, the SNSF proposed the creation of National Centres of Competence in Research (NCCRs, Braun and Benninghoff 2003; Benninghoff *et al.* 2009).

These NCCRs were an answer to the failure in linking fundamental research and technological application. It was planned to create long-term (four to twelve years) interdisciplinary centres within universities that were composed of large national networks of researchers working together around a core theme that would allow for 'strategic research'; that is, research at the interface of fundamental and applied research. The social sciences and humanities were explicitly included, though it proved difficult to get them funded.[21] Elements of the PRP were integrated within this proposal as, for example, topics that were of 'strategic importance' for Switzerland. However, selection procedures were analogous to the NRP; that is, bottom–up with the final selection authority of the secretary of state for education.

This proposal re-established the *status quo ante* PRPs in funding habits in Switzerland. The proposition to leave the development of research topics for funding to the scientific community again was just the logical conclusion to the failure of the PRPs to find sufficiently qualified researchers. The ultimate authority of the secretary of state in selecting topics upheld the role of politics within applied research, and was important for gaining support in parliament.

In the beginning, the SNSF project met some resistance, particularly from the SSC and the technological elites in the federal technical institutes who had initiated and supported the priority programmes of research. Critical

---

[21] In fact, it was necessary to launch a second call reserved only for the social sciences and humanities in order to have a substantial number of NCCR in this domain.

points were raised about the need to reform the PRP, the meaning of 'centres of competence in research', how to transfer results, the political criteria in the selection of research centres, the duration and the amount of funding, and the required involvement of universities. Despite such criticism, the SNSF refined its project and decided to enter into direct discussions with stakeholders, the administration, and parliament.

In spring 1999 a new secretary of state for science and research presented in a White Paper his vision of what should be the new 'university landscape', launching the idea of a 'Swiss network of universities'. The implementation of such an idea would be, as the secretary of state said, by creating 'centres of competence'. This instrument should enable the entire system of Swiss higher education to be reorganized (Kleiber 1999). The SNSF project fitted perfectly within these ideas and began a closer relationship between the federal decision-makers and the SNSF.

After a consultation procedure, the federal government presented the project to the federal parliament as part of its budget proposal concerning the promotion of education, research, and technology for the years 2000 to 2003. Parliament accepted this new instrument for funding research managed by the SNSF. The objective of this programme, as presented at the call, was as follows: promoting high-level research, maintaining and consolidating Switzerland's position in strategically important areas for policy, linking fundamental research, transfer of knowledge and technology, and training. Similarly, the programme was intended to optimize the distribution of tasks between universities, to promote collaboration between academic institutions and public and private sectors, to encourage interdisciplinary and innovative approaches within disciplines, and to promote young scientists and in particular the place of women. Each Centre should be composed of a 'leading house' (research institute responsible for the Centre), which directed and coordinated a network of research institutions (public or private) working over a maximum of twelve years around the same theme.

The development of this new instrument also changed the structure within the SNSF research council. The division (IV) for oriented research was reorganized into two sections: one for national research programmes[22] and a second one for the NCCR. In the latter section we find scientists designated by the SNSF[23] as well as one representative of the federal

---

[22] Today, three members come from Swiss federal institutes of technology, and six from cantonal universities.

[23] Today, five members come from federal technical institutes, and five from cantonal universities.

government with the status of an 'observer', and a second one from the innovation promotion agency of the Ministry of Economics.

Meanwhile, the overall structure of the SNSF was changed (in 2005) after an external evaluation launched by the federal government: the research board with its four divisions was now completely given into the hands of the scientific elite and partly into the hands of technological elites. Stakeholders no longer had a voice in selection procedures. This is why we speak of 'observers' in the division for applied research. The influence of the foundation board has been strengthened and streamlined. It must approve the main strategic decisions and the budget and decide on the organizational structure of the SNSF. Stakeholders dominate the foundation board. It is apparently an important arena for gaining public support for the SNSF in exchange for public control and influence on structure and policies of the SNSF. The stronger anchoring of the research board within the scientific elite was a sign that policy-makers did not want to intervene in procedural management any more and were happy to leave this to the scientific community. In other words, the new structure respected in a certain sense the basic ideas of the 'new public management', which recommends clearly separating strategic and operational authority. The strategic authority of stakeholders in the foundation board is, of course, relative, as agenda-setting is done by the SNSF itself in the form of proposals. But consensus has to be searched for with stakeholders, and this gives stakeholders an important influential role in funding policies. Nevertheless, it is not going too far to say that the reorganization of the SNSF has reinforced its position as a 'buffer' between politics and science. An important factor has been the reinternalization of the priority programmes within the confines of the SNSF, which made the SNSF again the only authority in matters of oriented research, except for applied research realized through the innovation promotion agency.

The NCCR: A New Instrument for Strategic Research

The institutionalization of the NCCR allowed the SNSF to increase its authority in science policy. But does that mean greater autonomy? It is clear that the autonomy of the SNSF is relative: the choice to launch the NCCR was a constrained choice. The SNSF urgently needed the support of policy-makers and the parliament. The adoption of the PRP demonstrated that the monopoly position of the SNSF was not guaranteed. The setting up of another instrument for oriented research within the SNSF re-established the *status quo ante* in terms of power, but it meant another concession to political logics. Thus, even if the SNSF appears to have increased its room for manœuvre in defining its policy, it is nevertheless obliged to take political expectations into account, sometimes even before they are formed. The establishment of the NCCR illustrates the tensions in terms of the mode of legitimacy which the SNSF must face and which is linked to its social and political context of the 1990s.

The NCCR funding scheme presupposes a strong involvement of universities in the funding of an NCCR, above all in terms of financial and infrastructural support. This is not without effects on the configuration of authority relations in science policy. Scientists had to win the consent of the university management for their NCCRs, and often found themselves competing for such support with their colleagues within the university. The dependence of university scientists on the university management therefore increased during this period.[24] For universities, NCCR were partly an advantage and partly a disadvantage. The advantage was that management could— by selecting scientific projects for support—insist on priority programmes within universities that became more and more relevant with the stronger strategic orientation of universities. The disadvantage was that it was difficult to refuse to supply matching funds in order to finance such NCCR. This demanded considerable sums of money that had to be invested even if the management would have preferred to spend it otherwise. The discretion of action of university management was therefore reduced to some extent.

Beside these aspects of 'power' configuration in funding policies and struggles to get more authority, the new programme had also important consequences for scientific production.

The NCCRs are based on the ideas of networks that are of importance especially for new 'search regimes'[25] (Bonaccorsi 2008) that rely on 'complementarities' in terms of collaboration among scientists within and between disciplines. The long-term structure made it possible to build up knowledge in research fields over a certain period of time without being subject to the uncertainties of refunding that are produced by funding instruments of more limited scope of time and money. This again is

---

[24] With the new century universities became more independent of policy-makers thanks to the introduction of new university laws based on ideas of new public management (contract, autonomy, delegation, *ex post* evaluation, and so on). See Braun and Merrien (1999).

[25] Search regimes are based on resources needed to do adequate research ('complementarities'), on the degree of productivity (research output in time) and the expansion of research domains in terms of the formation of sub-fields of research. 'New' search regimes like biomedical or computer sciences are characterized by very high productivity, strong differentiation in sub-fields, and a strong need for complementarities above all in terms of collaboration among disciplines.

favourable to research on more complex and interdisciplinary topics. Finally, it was possible to ask for sums of money that allowed the launch of large investments in the new areas. Competition among scientists became stronger; not only because the sums of money invested were attractive and universities had started to demand more research output as a qualifying criterion for their employed scientists, but also because the nationwide level allowed for comparison of research output. The NCCR contributed therefore—and this was of course the intention—to a stronger competitive orientation within the scientific community. The integration of the new programme within the SNSF did not, however, change the dominant position of a small scientific elite within the SNSF that has almost a monopoly position in the overall funding system. To a certain extent such a danger is reduced with the growing number of international experts that are summoned to evaluate project proposals, including NCCR proposals.

The introduction of the NCCR as a new funding instrument held considerable financial attraction and the high number of submitted projects demonstrate this. The balance between basic and oriented research did not, however, dramatically change. All instruments for oriented research together made up, as mentioned in the beginning, about 11 per cent of the budget of the SNSF in 2008, which is low by international comparison. And even oriented research itself remains to a large degree rooted in scientific interests as proposals, though they have to take into account stakeholders' interests, and include fundamental interests of scientists in knowledge production for reputation purposes. Switzerland has not developed a culture of top–down funding, where scientific and technological elites or stakeholders define research frameworks. Propositions for NCCR are made by individual researchers in the scientific community, and the first selection within the SNSF is based on scientific quality. Concerning the NRP, the topics may be suggested by anybody.

7. Changes in Authority Relations in the Swiss Public Science System

Compared to the 'state-shared system' with a strong localism of the academic community of the first period of the SNSF history, the Swiss science system has seen several changes that justify classifying the country at the beginning of the twenty-first century into the group of 'state-delegated systems' with a combination of 'stable' and 'competitive' elements in terms of allocations procedures (Benninghoff *et al.* 2005; Lepori 2003).

1. A significant change was the gradual introduction of new public management based universities. All cantonal universities and the federal technical universities have experienced new university laws that in general have followed general tendencies in other countries: university management was strengthened within the organization and received more autonomy in terms of budget use, curricula, and even personnel policy. The state— cantons and the federal government—remains the main financing organization, and still has authority over the overall budget, structures, and strategic objectives, which are negotiated within contracts. This has turned university management into a more proactive player in university research policies. Researchers must look for a consensus with strategic intentions of the management, at least if significant university resources are involved. That is what happened in case of the NCCR funding scheme that required substantial contributions from universities. This new dependence reduces researchers' discretion to pursue certain research topics; for example, if the management does not support the demands submitted to the SNSF.

Another outcome of this change is an increased attention of the university to the performance of its employees, which is measured mainly by numbers of publications but also by the acquisition of external funding. This puts pressure on researchers to compete for reputation and resources. The federal government adds to this pressure on both researchers and the management by binding a part of its subsidies to performance criteria in teaching and research. These developments strengthen the search for SNSF funds. As indicated in the beginning, the number of submitted proposals has risen considerably since 1980, both in basic and oriented research.

2. Since its foundation in 1952 the SNSF has gained authority. It has been crucial to the emergence of a competitive funding system, whose evolution was accompanied by a relative decrease of institutional funding and therefore of opportunities for long-term research based on institutional resources. The size of the SNSF has grown, the number of funding instruments has been considerably diversified, and the number of projects evaluated has increased exponentially. This demonstrates that Switzerland, like other OECD-countries, has made competition among researchers a strong element in the research system.[26] This should in principle contribute to a rapid expansion of topics, new research themes, and diffusion of research results. However, the problem is that the SNSF has

---

[26] If one measures competition in 'success rates' of submitted projects, the situation for Swiss researchers is still favourable as it varies between 35 and 45 per cent. There is competition, but in comparison to other countries competition may be less harsh and demanding.

remained—with the exception of technologically and commercialized research—an almost monopolistic institution in basic and oriented research.

According to Whitley, the existence of a relatively small scientific elite located within one funding organization may have conservative effects on the distribution of new topics and the degree of change in research contents. Discussions about the transparency of decision-making within the SNSF have taken place, and fears of 'conservative effects' (the 'Matthew-effect'; cf. Merton 1968) have been taken into account by revising a number of internal evaluation procedures. The role of international reviewers as a base for evaluation has in addition increased substantially. This only demonstrates that the monopoly position of a relatively small scientific elite has indeed raised concerns in the academic community and that such effects of 'conservatism' cannot be excluded for scientific research in Switzerland.

3. Another tendency that has not been mentioned before but is important in this context is the rise of European Union and European Science Foundation funding of Swiss research. By way of bilateral agreements Swiss researchers are treated as researchers from member countries of the EU. One sees again a substantial rise of submitted projects at this level. This contributes at least to some extent to a diversification of funding structures and hence to the increase of opportunities to launch ideas that possibly does not fit into mainstream research in Switzerland (especially in the technological domain and applied science). However, the financing of the EU and ESF remains limited compared to SNSF funding.[27]

4. Funding for oriented research has become part of SNSF funding. This had several effects. The NRP and especially the NCCR have without a doubt contributed to stronger national networks and therefore the national integration of the scientific community. This is important for the strengthening of Swiss research in the international competition for reputation and resources. The reduction of discretion for individual researchers by obliging researchers to embark on the road to application is, in comparison to other countries, most notably the Anglo-Saxon ones or the Netherlands, limited. This has been demonstrated on several occasions. The resources that are spent by the SNSF on oriented research (11 per cent of its budget) still allow the financing of a substantial number of purely investigator-driven basic science projects. The choice of topics is done in

---

[27] Lepori (2006) estimates the share of EU framework programmes in Swiss project funding at about 15 per cent in 2002, while the SNF has a share of 44 per cent.

the first stage by scientists, and the first selection is only based on criteria of scientific quality. Only at the last stage can policy-makers select on the base of their priorities. It is nevertheless true that propositions of topics are constrained to some extent because the component of oriented research looms large among the reference points for developing a project. Researchers, however, have every possibility to anchor these topics strongly within questions of fundamental research.

5. The role of the SNSF in the governance structure of research has been and still is crucial. The SNSF succeeded in internalizing oriented research into its own confines. This made it possible on the one hand to get a grip on contents and procedures of oriented research funding, but signified on the other hand that the SNSF became the arena for balancing political and scientific interests. This makes it vulnerable, as it must accommodate interests of stakeholders to a certain extent. On the other hand, discussions and negotiations contribute to its legitimacy within parliament, governments, and the public. In this way the scientific elite has maintained a strong control on the spending of most non-university funding for research in Switzerland.

6. If we summarize these findings by sketching the image of the Swiss individual researcher one would state that today he or she has lost some discretion in defining their own research topics compared to institutionalized university funding in state-shared systems: the researcher is more constrained by judgements from peers in the SNSF, by intra-organizational discussions with the management, and by application for projects in oriented research. He or she must accept more competition and risk of failure when applying for funds, and is obliged to take up the latest developments in research to increase his or her chances of obtaining non-institutional support. 'Localism' has been reduced. The search for networks and other partners in research has become an important element in doing research. European funding has helped to find other sources for launching ideas and above all to build up international cooperation. Application-oriented research is an option but not an obligation.

What does all this mean for scientific production in Switzerland today? Though conservative tendencies remain in the development of new research ideas and diffusion of new research results, scientific production has become more intense: the increasing competition and the growing role of the European level has helped to diversify the scientific community and allowed for a more rapid diffusion of new topics and results. The NCCR and European projects have an important function in delivering the 'complementarities' necessary above all in new 'search regimes'. The linking of researchers within network structures has helped to increase the quality of the research output and helped the embarking on more interdisciplinary and comprehensive and complex topics that also characterize new search regimes. One obvious advantage of the Swiss system is that it avoids putting its researchers under pressure to think in terms of application and technologies. The existing funding instruments help to support such a view, but they do not force most funding activities in this direction. There are still large sums of money available for exploring new ideas. This helps innovation. One point remains a problem though: the apparent institutional fragmentation of scientific and technological elites. Strategic research (Martin and Irving 1989) remains difficult to set up in this way, though the NCCR are conceived to solve this problem to a certain extent. But especially research that can result in technological application needs a stronger linking of both communities.

References

Baschung, L., Benninghoff, M., Goastellec, G., and Perellon J. (2009) 'Switzerland: between Cooperation and Competition', in C. Paradeise, E. Reale, I. Bleiklie, and E. Ferlie (eds.), *University Governance: Western European Comparative Perspectives* (Dordrecht: Springer), 153–75.

Benninghoff, M., and Leresche, J.-P. (2003) *La Recherche affaire d'État* (Lausanne: PPUR).

—— Perellon, J., and Leresche, J.-P. (2005) 'L'Efficacité des mesures de financement dans le domaine de la formation, de la recherche et de la technologie: Perspectives européennes comparées et leçons pour la Suisse', *Cahiers de l'Observatoire*, 12 (Lausanne: OSPS-University of Lausanne).

—— Goastellec, G., and Leresche, J.-P. (2009) 'L'International comme ressource cognitive et symbolique: Changements dans l'instrumentation de la recherche et de l'enseignement supérieur en Suisse', in J.-P. Leresche, P. Laredo, and K. Weber (eds.), *L'Internationalisation des systèmes scientifiques* (Lausanne: PPUR), 235–55.

Bonaccorsi, A. (2008) 'Search Regime and the Industrial Dynamics of Science', *Minerva*, 46(3): 285–315.

Bourdieu, P. (1976) 'Le Champ scientifique', *Actes de la Recherche en Sciences Sociales* (June 2/3): 88–104.

Braun, D. (1998) 'The Role of Funding Agencies in the Cognitive Development of Science', *Research Policy*, 24: 807–21.

—— (2001) *Staatliche Förderung ausseruniversitärer Forschungseinrichtungen am Beispiel der Niederlande und Deutschlands: Kritische Begutachtung eines Förderinstruments* (Berne: Centre d'études de la science et de la technologie; CEST 2001/10).

—— and Merrien, F. (eds.) (1999) *Towards a New Model of Governance for Universities? A Comparative View* (London: Jessica Kingsley).

—— and Leresche, J.-P. (2007) 'Research and Technology Policy', *Handbook of Swiss Politics*, 2nd edn. (Zürich: Verlag Neue Zürcher Zeitung), 765–90.

—— Griessen, T., Baschung, L., Benninghoff, M., and Leresche J.-P. (2007) 'Zusammenlegung aller Bundeskompetenzen für Bildung, Forschung und Innovation in einem Departement', *Les Cahiers de l'Observatoire* (Lausanne: University of Lausanne).

CEST (2003) *Place scientifique Suisse 2001: Développement de la recherche en comparaison internationale sur la base d'indicateurs bibliométriques 1981–2001* (Berne: CEST).

Elzinga, A., and Jamison A. (1995) 'Changing Policy Agenda in Science and Technology', in S. Jasanoff, G. E. Markle, J. C. Peterson, and T. Pinch (eds.), *Handbook of Science and Technology Studies* (Beverly Hills, Calif.: Sage), 572–97.

Fleury, A., and Joye F. (2002) *Les Débuts de la politique de la recherche en Suisse: Histoire de la création du Fonds National Suisse de la Recherche Scientifique 1934–1952* (Geneva: Droz).

Gees, T. (2006) *Die Schweiz im Europäisierungsprozess: Wirtschafts- und Gesellschaftspolitische Konzepte am Beispiel der Arbeitsmigrations-, Agrar- und Wissenschaftspolitik, 1947–1974* (Zürich: Chronos).

Godin, B., Trépanier, M., and Albert M. (2000) 'Des Organismes sous tension: Les Conseils subventionnaires et la politique scientifique', *Sociologie et Société*, 32(1): 17–42.

Joye-Cagnard, F. (2010) *La Construction de la politique de la science en Suisse: Enjeux scientifiques, stratégiques et politiques (1944–1974)* (Neuchâtel: Alphil).

Kleiber, C. (1999) *Pour l'université* (Berne: Swiss Confederation).

Latour, B., and Woolgar, S. (1986) *Laboratory Life* (Princeton: Princeton University Press; [1979]).

Latzel, G. (1979) *Prioritäten der Schweizerischen Forschungspolitik im Internationalen Vergleich: Die Nationalen Forschungsprogramme* (Berne and Stuttgart: Haupt).

Laudel, G. (2006) 'The "Quality Myth": Promoting and Hindering Conditions for Acquiring Research Funds', *Higher Education*, 53: 375–403.

Lepori, B. (2003) 'Understanding the Dynamics of Research Policies: The Case of Switzerland', *Studies in Communication Sciences*, 2(1): 77–111.

—— (2006) 'Public Research Funding and Research Policy: A Long-Term Analysis of the Swiss Case', *Science and Public Policy*, 22(3): 205–16.

—— (2007) *La Politique de la Recherche en Suisse* (Berne: Haupt).

—— (2008) 'Research in Non-University Higher Education Institutions: The Case of the Swiss Universities of Applied Sciences', *Higher Education*, 56(1): 45–58.

—— Benninghoff, M., Jongbloed, B., and Salerno, C. (2007) 'Changing Models and Patterns of Higher Education Funding: Some Empirical Evidence', in A. Bonaccorsi and C. Daraio (eds.), *Universities and Strategic Knowledge Creation: Specialization and Performance in Europe* (Cheltenham: Edward Elgar), 85–111.

Martin, B. R., and Irvine, J. (1989) *Research Foresight: Priority-Setting in Science* (London: Pinter).

Merton, R. K. (1968) 'The Matthew Effect in Science', *Science*, 159(3810): 55–63.

OFES (1994) *Expertise des Programmes Nationaux de Recherche (PNR): Rapport du Groupe d'Experts pour l'Office Fédéral de l'Education et de la Science (OFES)* (Berne: OFES).

Office Fédéral de la Statistique (2008) *Personnel des hautes écoles universitaires 2007* (Neuchâtel: Office Fédéral de la Statistique).

Perellon, J. (2003) 'The Creation of a Vocational Sector in Swiss Higher Education: Balancing Trends of System Differentiation and Integration', *European Journal of Education*, 38(4): 357–70.

Strasser, B. J. (2006) *La Fabrique d'une nouvelle science: La Biologie moléculaire à l'âge atomique, 1945–1964* (Florence: Olschki).

Whitley, R. (2000) *The Intellectual and Social Organization of the Sciences, 2nd edn.* (Oxford: Oxford University Press).

—— (2003) 'Competition and Pluralism in the Public Sciences: The Impact of Institutional Frameworks on the Organisation of Academic Science', *Research Policy*, 32: 1015–29.

—— (2008) 'Changing Authority Relations in Public Science Systems and their Consequences for the Direction and Organisation of Research', *Manchester Business School Working Paper*, 556.

4

The Changing Governance of Japanese Public Science

Robert Kneller

1. Overview

This chapter provides an overview of Japan's public science system (PSS) and the major changes under way concerning its governance. From the end of the Second World War until 2004, Japan's national universities, which account for the great majority of basic science research, were under the direct control of the Ministry of Education, Culture, Sports, Science and Technology (known as Monbusho before 2001 and Monbu-kagaku-sho, or MEXT, thereafter). Government research institutes (GRIs) similarly were under the direct control of their responsible ministries until about the same time. Since 2000–4, various policies have been initiated to change the governance and administrative system of national universities and GRIs across a wide front. Financial and administrative independence and mobility of researchers have been encouraged and comprehensive evaluation procedures implemented. The push towards greater financial independence is indicated by a continuing decline in general operational and administration subsidies (block grants) to universities, which now must compete for a growing supply of competitive research funds. However, some of these competitive programmes are aimed explicitly at creating a limited number of centres of excellence. Some are under the direct control of ministries and their academic advisers to the extent that these groups strongly influence research themes and award decisions.

This chapter adopts a system-wide analytical perspective. It is based on twelve years of experience in an interdisciplinary graduate-level education and research centre of the University of Tokyo, and frequent contacts with scientists and students. It also draws upon a large number of interviews over the past decade with companies that deal with universities. However, most first-hand observations are from Japan's best endowed and most prestigious university. Therefore, the perspectives of researchers in some of Japan's GRIs and good but lesser known universities may not be adequately represented in this chapter.

2. Japanese Public Science Institutions

Japan's public science system consists of (1) universities, (2) government research institutes (GRIs), and (3) consortia and collaborative centres.

Universities

In 2008 Japan had 86 national universities that since 2004 have been incorporated under the jurisdiction of the Ministry of Education, Culture, Sports, Science, and Technology. In addition there are 90 universities under the jurisdiction of local governments and 589 private universities offering at least a four-year basic (bachelor's) degree. National universities account for the majority of university science and engineering research, as well as the bulk of graduate education conducted in Japan.[1] The most prestigious of the national universities are those designated as Imperial Universities prior to the Second World War. These are the University of Tokyo and Kyoto, Osaka, Tohoku, Nagoya, Hokkaido, and Kyushu Universities. Among these, the *big four* (Tokyo, Kyoto, Osaka, and Tohoku) receive the largest amount of support under almost any government programme. For example, in the three years 2006–8, they accounted for 44 per cent of all MEXT grants-in-aid for scientific research—hereinafter grants-in-aid or GIA—to national universities and 35 per cent to all types of universities. The University of Tokyo alone accounted for 16 per cent of all MEXT GIA to national universities in 2007 (http://www.jsps.go.jp/j-grantsinaid/index.

---

[1] Thus, although national universities accounted for only 22 per cent of total student enrolment in 2008, they enrolled 59 per cent of Japan's 263,000 total graduate students, including 65 per cent of masters students and 76 per cent of doctoral students in science and engineering. In 2008 they received 68 per cent of R&D funding under MEXT's grants-in-aid programme—the largest single source of government funding for project specific university R&D: http://www.mext.go.jp/b_menu/toukei/001/08121201/index.htm.

html). This concentration of funding among Japan's top four or five universities is at least twice that among the UK's top four or five universities (http://www.hesa.ac.uk/index.php/content/view/807/251/).

The best-known private universities are Keio and Waseda Universities in Tokyo, but their research funding is less than any of the previously named national universities. They rank 12th and 16th respectively among 2008 recipients of MEXT GIA. But this amounted to only 12 and 10 per cent, respectively, of the University of Tokyo's share. A few other private universities also attract excellent science and engineering researchers. Among local government universities, the leading recipients of 2007 GIA were Osaka City and Osaka Prefectural Universities, ranked 25th and 28th respectively (http://www.jsps.go.jp/j-grantsinaid/index.html).

Overall, the national government is by far the largest funder of university R&D. Even private universities receive substantial subsidies from the national government. The contribution of prefectural and other local governments to research in national universities is negligible so far. However, very recently there have been cases of local governments donating land and infrastructure for the expansion of national universities. This may mark the beginning of local governments engaging with national universities as a means to promote regional development.

Government Research Institutes (GRIs)

The GRIs are research institutes linked to various science-related ministries.[2] Their overall funding for natural science and engineering research and development (R&D) is about 60 per cent that in universities. Many GRIs with large budgets are laboratories within mission-specific agencies such as the Defence Ministry, the space agency (JAXA), and the atomic energy agency (JAEA). The most important GRIs with a multidisciplinary scope and basic science orientation are Riken (in English: the Institute for Physical and Chemical Research) and the National Institute for Advanced Industrial Science and Technology (AIST). Riken is an independent administrative institution under MEXT, while AIST is an independent administrative institution under the Ministry of Economy, Trade, and Industry (METI). Among all GRIs, Riken probably hosts more graduate

---

[2] Often GRI researchers are also eligible to apply for project funding from other ministries, including MEXT grants-in-aid.

students and postdoctoral researchers than any other. Riken's 2008 budget was M¥ 111,500 (M¥ 100 ≈ 1 MUSD), while AIST's 2009 budget was M¥ 88,700.[3] In comparison, the University of Tokyo's total 2007 budget was nearly twice Riken's, and external support for research in the University of Tokyo (excluding salaries of permanent staff and general research allowances of professors) was two-thirds of Riken's total budget including salaries. In other words, universities as a whole (particularly national universities) are the dominant actors in Japanese public R&D. Moreover, the largest national universities conduct more research than the largest multidisciplinary, fundamental-science-oriented GRIs.

Consortia and Collaborative Centres

Other major venues for public science include *consortia* of various companies organized by the government (usually METI) to conduct R&D on a particular theme in a free-standing laboratory. A recent example is the Extreme Ultraviolet Association to develop techniques of ultraviolet lithography. However, the trend is probably towards consortium research simply being dispersed among the various partner laboratories rather than being centred in a new free-standing laboratory.

Another public science programme involving dedicated research facilities (sometimes rented from universities) is the Japan Science and Technology Agency's (JST's) Exploratory Research for Advanced Technology (ERATO) programme. ERATO projects are well funded and are formulated by JST (since 2001 part of MEXT) and its advisers to address important, cutting-edge scientific issues. Only about four new projects are initiated each year. Usually they involve researchers drawn from several university or GRI laboratories. Participation by overseas academics is not uncommon. Nor is participation by industry researchers, but unlike the 1980s, the vast majority of ERATO projects are now directed by academic scientists. Of all Japan's public science programmes, ERATO has probably received the most praise in Japan and abroad, in terms of both quantifiable metrics (Hayashi 2003) and recognized scientific achievements (JTEC 1996).

[3] http://www.riken.jp/engn/r-world/riken/outline/index.html
http://www.riken.go.jp/r-world/riken/info/pdf/keikaku2008.pdf
http://www.aist.go.jp/aist_j/outline/h21_plan/h21_plan_8.html

3. Changes in the Governance of Japanese Public Science

Analysis of the governance of the Japanese public science system focuses first on the legal status of national universities, which provides the basis for their limited autonomy. This leads naturally to an examination of recruitment of academic staff, resource allocation, evaluation procedures, and technology transfer.

Legal Status of National Universities: Incorporation

In 2004 the national universities were incorporated—a step that provided the basis for them to develop as autonomous institutions. Nevertheless their autonomy remains limited by dependence upon the MEXT for salaries of full-time staff as well as most infrastructure costs. Salaries of full-time faculty and administrative staff are paid from an *operational and administration (O&A) subsidy* (in Japanese: *unei koufu kin*) from MEXT to each national university. These subsidies are determined by a formula, but the important variables, such as number of faculty and numbers of expected graduate students, are based upon precedent, with any changes negotiated between MEXT and the individual university. MEXT also provides *operating expense subsidies* to most private universities which amount to about 15 per cent of their total budgets. This is less than the 45–60 per cent of national university budgets covered by O&A subsidies, which implies that private universities have to cover a greater proportion of their costs through tuition charges. Nevertheless MEXT operating expense subsidies are important for private universities. All these subsidies come from MEXT's *general university accounts* budget, which MEXT must negotiate with the Ministry of Finance (MOF). MOF has placed steady pressure on MEXT to reduce this budget item, and as a result these subsidies have been reduced by about 1 per cent annually since 2005, although reductions have varied among universities.[4]

The number of full-time equivalent (FTE) faculty and administrative positions that MEXT will fund in each major department of each university is generally still a matter of negotiation between the university and MEXT. Universities can borrow FTEs from various departments to form a new department.[5] However, this freedom is limited both by the existing

[4] See http://www.mof.go.jp/singikai/zaiseseido/siryou/zaiseib190521/02_a.pdf, and also http://www.zendaikyo.or.jp/siryou/2009/04-09-koufukin-itiran.pdf for O&A subsidies for individual universities.

[5] For example, a university can borrow FTEs from engineering and economics to form a subdepartment of technology management within the engineering or economics department.

departmental structures and the need to obtain MEXT's approval for major changes. This need for MEXT approval may be more a result of universities' strategies to obtain funding from MEXT for new FTEs, than of MEXT's wanting to have control over their allocation. A university desiring new FTEs usually submits to MEXT a plan to fund a new centre or subdepartment, and these requests specify the number and level of various associated faculty positions, and also the anticipated number of graduate students. Thus, it may seem inappropriate for a university to say a few years later that it does not need FTEs in particular departments and to start shifting FTE allocations on its own. A member of several high-level policy committees indicated that MEXT may nevertheless be willing to give universities substantial say over allocation of FTEs among disciplines. However, currently the main issue is how to absorb cuts in O&A subsidies rather than distribution of that funding (see below).

Universities have flexibility in how they manage funding from competitive MEXT programmes such as *Centres of Excellence (COE)* and *Special Coordination Funds*. The former provide mainly programmatic funding, some of which is used to upgrade research facilities and to hire non-permanent researchers, especially post-docs and junior faculty on time-limited appointments. The latter cover a wide range of activities ranging from support for training and non-permanent researchers to co-funding of cutting-edge research with private companies.

Major universities have compiled 'action plans' setting forth visions of how they want to develop, and outlining major steps to reach these goals. For example the University of Tokyo's Action Plan seeks to

- create an educational system for interdisciplinary fields;
- secure university-wide common spaces (including research facilities);
- form links with local communities around its various campuses.

It also lists goals indicating that it is beginning to establish a framework for autonomous financial management, a prerequisite for it to act autonomously, strategically, or entrepreneurially. Such goals include:[6]

- easing institutional restrictions, in particular deregulation of funds management, asset utilization, long-term borrowing, issuance of bonds, investments, and tax provisions governing donations;
- establishing a budget system that can support autonomous and decentralized basic education and research;

---

[6] See http://www.u-tokyo.ac.jp/gen03/pdf/ActionPlan2005–2008.pdf.

• establishing rules for the effective use of the president's and department heads' discretionary funds.

As required under the terms of the incorporation law, national universities have also drawn up specific *mid-term plans* setting forth the goals they hoped to achieve in the first four years following incorporation (April 2004 to March 2008) and their progress with respect to those goals. These mid-term plans are a major focus of the evaluation exercises (see further below). GRIs have had to submit similar mid-term plans as part of their evaluation process.

The University of Tokyo's Mid-Term Plan submitted to MEXT in June 2008 is 179 pages long. It covers a wide range of educational, research, administrative, and infrastructure programmes in some detail. It indicates that most financial support will have to come from MEXT. However, it also celebrates the raising of M¥ 13,000 from 2005–8 for the university's endowment fund. This represents about 6 per cent of the University's total FY 2008 budget of M¥ 220,000. While this amount is small in comparison to many UK and Canadian universities, and much smaller in comparison to many US universities, it represents a credible start.[7]

Recruitment of Academic Staff: Increasing Openness and Mobility

Coleman (1999) and Kneller (2007*b*) describe the traditional *kouza* system (modelled on the traditional German professor *chair system*) that constituted the organizational basis for Japanese university teaching and research. The full professors were simultaneously the heads of laboratories and had considerable freedom to choose their research directions—within the limitations of available funding and often limited human resources. Each head was lord of his castle, even though it might be small (Bartholomew 1989; Coleman 1999; and personal observations beginning 1997). This independence extended to recruitment and promotions, with the professor choosing who would fill vacancies at the laboratory's assistant level, often from among his graduating students. These assistants would gradually advance to fill vacancies emerging at the associate and finally full professor levels—a process well described by Coleman. A typical *kouza* consisted of one professor, one associate professor, two assistants (often

---

[7] See http://www.u-tokyo.ac.jp/gen01/pdf/keikaku1904.pdf
http://www.u-tokyo.ac.jp/gen02/pdf/chuki_keikaku21.pdf, and *University of Tokyo Data Book* (2008).

euphemistically called *assistant professors* in English), and sometimes one *koushi* (instructor) whose primary responsibility would be teaching.

However, open recruitment (wide and open solicitation of applications and selection on merit) may finally be gaining traction, though it is probably not yet the norm in major universities. Data from the Japan Science and Technology Agency (JST) website, where academic positions are most frequently advertised, shows a steady year on year increase. In 2008 about 1,500 vacancies in national universities were advertised at the associate professor level.[8] Since there were in this year about 17,600 associate professors in national universities nationwide,[9] and the average duration of an associate professorship can be assumed to be six years, about 3,000 vacancies would arise each year, so applications to fill these vacancies appear to have been openly solicited in about half the cases. Even when applications are widely solicited by open advertisement, sometimes a lead candidate has already been designated by influential professors and advertisement of the position serves mainly to legitimize a pre-determined selection. Called 'empty open recruitment' (in Japanese, *kara koubo*), this probably applied to a significant proportion of advertised positions prior to 2000 (Coleman 1999). However, recent discussions with faculty who have taken part in open recruitments suggest that, more often than not, such solicitations are genuine, at least in the sense that the selection committees are willing to consider seriously strong candidates from the outside.

A review of the national universities' open advertising on the mentioned site for associate professor positions in the eight months from November 2008 to June 2009 indicates that most vacancies are in lesser known national universities. Only about 10 per cent are in the big four, an additional 15 per cent in other former Imperial Universities, and yet another 10 per cent in other major national universities; in particular, Tokyo Institute of Technology and Tsukuba, Hiroshima, Kobe, and Okayama Universities. This suggests that the trend towards increased reliance on open recruitment is proceeding more slowly in elite universities.

In the University of Tokyo and probably other elite universities, recruitment for faculty vacancies typically begins with formation of a search committee of five or six professors, usually from the same department or

---

[8] http://jrecin.jst.go.jp/seek/SeekDescription?id=006
http://jrecin.jst.go.jp/seek/html/h20/jobinfo2.pdf. In terms of changing one's institutional affiliation the transition from assistant to associate professorship usually involves the most significant change, one that is often mandatory (see below).
[9] MEXT (2008): *Basic School Survey*, table 29.

117

centre. This committee will recommend a name to the entire department which then has the opportunity to discuss the nominee. Usually discussion is minimal, and it is extremely rare for an alternative candidate to be proposed 'from the floor'. This system means that one or two influential professors on the search committee usually play the main role in filling the vacancy.

But even without open advertisement, capable candidates hear about vacancies in elite universities and make sure they are on the radar screens of professors likely to serve on recruitment committees. How they do so probably depends more upon their actively seeking out contacts or trying to build their own professional reputations than upon introductions by their professors, which probably was more common in the past (author's inquiries and Coleman 1999). One important way to build a professional reputation is publication in English in respected international journals, although such publications are rarer (and thus their importance for promotion also less) in the humanities and social sciences (author's inquiries). In any case, it has become easier for good science and engineering graduates of second-tier universities to compete for vacancies in the elite universities, as described below.

There is a Ph.D. glut in Japan, and many graduates from elite universities are looking for teaching positions, even in second-tier universities (Ledford 2007; Shodo 2007). Open recruitment enables these universities to compete for such talent. Also, open recruitment seems to be well entrenched in the GRIs. Between 2004 and 2007 all 1,059 full-time research positions at Riken were filled by open advertisements widely soliciting applications, as were 94 per cent of 511 full-time research positions at AIST.[10]

Engineering and natural science professors have explained that, until even as late as 2000, the traditional model described by Coleman (1999) applied. However expectations have altered, and now researchers aspiring to academic careers should change institutions, either just after receiving their doctoral degree or when moving from an assistant to associate professorship. These professors also say that they are less obligated now than in the past to find jobs for their students. 'It is now up to them to find jobs' is a common refrain. These changes in attitudes were encouraged, perhaps even initiated by, the Government's Second and Third Science and Technology (S&T) Basic Plans issued in 2001 and 2006, respectively, which made the independence and mobility of young researchers national goals.

---

[10] http://www8.cao.go.jp/cstp/siryo/haihu77/siryo4-1-3-5.pdf.

A review of biographical information on over seventy researchers in over thirty laboratories in the big four national universities, plus Nagoya, Hiroshima, and Okayama Universities indicates these changes are indeed occurring, although it also shows that even among older cohorts, some researchers did move between institutions early in their careers. The review considered laboratories that were mostly (about 80 per cent) working in various fields of mechanical engineering (including biomedical engineering, nano-materials, and robotics) with the remainder mostly in chemistry.[11]

Most full professors were born before 1970. Their career paths can be classified as matching one of three patterns.

PATTERN A

About a quarter spent their entire academic career from undergraduate studies to their current professorship in the same institution, with the possible exception of a few years in a non-permanent position in an overseas university.

PATTERN B

About 30 per cent have earlier in their careers worked at another institution, usually a private company or government research institute, but are now professors in the same university where they did all their undergraduate and graduate studies. The sojourn usually occurred after a few years as a research assistant (since 2007: assistant professor). Pattern B is probably more common among engineering than natural science faculties, mainly because outside of engineering it is rare for academics to spend time in industry early in their careers.

PATTERN C

About 45 per cent had more significant career shifts, involving in almost all cases education as well as some faculty experience in a different university. Consistent with what has already been said with respect to pattern B, the most frequent time for a career change came after serving several years as a research assistant. Although numbers are small, pattern C is less common in the case of University of Tokyo and Kyoto University faculty. It seems

[11] A less systematic analysis suggests that the findings described below apply also to biology and physics. However, in medicine it is more common for promotions to occur within the same institution. Career patterns for the social sciences and humanities were not analysed. Note 13 and the accompanying text suggests that the same trends are likely to be found for economics, but not law.

that, in the past, these most elite universities often trained persons who became faculty in other prestigious universities, but the opposite happened less frequently.

The effects of policies that encourage greater mobility would only be seen in the career paths of associate professors, typically born in the latter 1960s or early 1970s, or assistant professors (before 2006: research assistants), who mostly were born after 1970.

In the case of associate professors, about 60 per cent have followed pattern C. Most of these completed undergraduate to doctoral work plus several years as research assistant (assistant professor) in another university before assuming the associate professorship in a new university. Patterns A and B are about equally common, approximately 20 per cent each. These percentages appear to apply even to the University of Tokyo.

In the case of assistant professors pattern C is the most common, but at about 45 per cent, less so than in the case of associate professors. An almost equal number of assistant professors have followed pattern A; that is, they have no experience in any other academic institution, except in a few cases in an overseas university.

The above indicates that the policy of encouraging young academic researchers to change institutions is resulting mainly in transfers between the assistant and associate professor stages in their careers. Also, it seems that opportunities are expanding for assistant professors from highly regarded universities such as Tohoku and Waseda to obtain permanent associate professorships in the most prestigious universities—Tokyo and Kyoto. Whether assistant professors who leave Tokyo and Kyoto for other universities will return to their Alma Maters at a higher rate than occurred in the past remains to be seen.[12] However, this would require the government to establish more FTE positions at the associate and full professor level at a time when it is trying to reduce such positions.

The above findings are supported by annual surveys ranking various departments according to percentage of so-called 'pure blood' faculty (percentage of full-time permanent instructors (*koushi*)) associate professors, and full professors who graduated from the same university. These show declines in inbreeding between 2002 and 2007 in major universities. For example, in engineering, University of Tokyo topped the pure blood rankings in 2002 at 87 per cent, but by 2007 this had declined to 72 per cent. Kyoto (second in 2002) declined from 81 to 72 per cent. Declines also

---

[12] About half of current full professors in Tokyo and Kyoto followed pattern B, in that they left, usually to work several years in companies, and then returned.

occurred in major private universities such as Keio. Among the disciplines analysed, law faculties tend to have the highest rates of internal recruitment and economics faculties the lowest.[13] In all disciplines, rates of internal recruitment and promotion vary greatly between universities. Tokyo and Kyoto Universities are always near the top, tending to confirm the above findings that, until quite recently, these two universities often trained faculty for other universities but rarely recruited from other universities (Ikeuchi 2004; Imatani 2008).

In the past, laboratories were basically inherited by the associate professor when the full professor retired. Under such circumstances opportunities for young researchers to strike out to explore their own interests were limited (Normile 2004). However, at least in the large universities, opportunities for young faculty to pursue their own projects and sometimes to head their own laboratories have increased. The following are among the major underlying factors:

1. New laboratories that are headed by young associate professors are being established. This is often the result of negotiations between universities and MEXT to establish new FTEs, as discussed above. At the same time, more grants are being made available specifically for young researchers. Some of the larger of these, such as JST's PRESTO grants, have enabled young researchers to equip independent labs. In other words, these funds are often combined with salary support from MEXT to provide a complete package that allows associate professors in their late thirties or early forties to pursue independent research.

2. The influence of the traditional *kouza* system has diminished, so that now the filling of vacancies becomes the concern of the department, or more commonly a subdepartment of three to six laboratories. This does not diminish the importance of patronage by established professors in recruitment and promotions, but it does mean that promotions are no longer seen primarily as continuing a particular professor's line of research, but rather as promoting the (sub)department and its particular scientific field.[14]

---

[13] In law, the University of Tokyo topped the rankings in both years (97 to 91 per cent from 2004 to 2007) while Kyoto University was second (80 to 76 per cent). In economics Kyoto University topped the rankings in both years, but the decline was more pronounced (73 to 57 per cent), while in the University of Tokyo the proportion of 'pure blood' faculty declined from 58 to 42 per cent.

[14] For example, in the author's own centre in the University of Tokyo, when professors in artificial organs, informatics, economics, and history of science retired or moved, their particular field of research was not continued by a new hire or new promotion. Rather influential search committees were formed to identify respected researchers who were seen as

3. Since the reforms in 2000 that enabled researchers to be hired on time-limited appointments using competitive research funds, a large number of time-limited faculty positions have been created. In 2006 these amounted to 15 per cent of all national university faculty positions, and 26 per cent of assistant professors (Cabinet Office data). Often these non-permanent faculty are called *project* assistant (or associate, and so on) professors, indicating they are employed with funds for a large research project. As these projects tend to be awarded to elite universities (see below), these project faculty tend to be similarly concentrated. Most are young. Fewer than 20 per cent are near the end of their careers. Although they usually do not have their own laboratory, because it is understood that they will have to compete for positions three or five years hence, they often are given freedom to pursue their own research interests so long as they fit within the scope of the project.

As in the case of increasing mobility, probably the main impetus for these changes came from the Second and Third Science and Technology Basic Plans. In other words, they were centrally initiated.

In summary, although faculty patronage is still necessary for hiring and promotions in elite universities, open recruitment and mobility are increasing, along with opportunities for young researchers to pursue their own interests. However, rather than talent becoming more evenly distributed the trend seems to be the opposite. Centripetal forces are stronger.

Resource Allocation: Elite Universities vs. Others

Beginning in 2005, one year after incorporation of the national universities, the Ministry of Finance and MEXT started to cut O&A subsidies by about 1 per cent annually for all national universities. O&A subsidies account for 43 per cent of the total overall budget of national universities,

---

likely to increase the centre's visibility, make significant contributions, and so on. Within the centre there is consensus that a balance should be maintained between the centre's main fields of research: engineering, technology for disabled persons, and life, environmental, material, and information science. But as professors retire or move, fields of research do shift. Similarly, the author has heard detailed descriptions of recruitment procedures in the mathematics and synthetic biochemistry depts in Tokyo and Kyoto Universities. When vacancies arise, the department members meet to consider potential candidates and also the types of expertise most desired in terms of current scientific frontiers and the possibility of synergy with other department members. Little weight is given to continuing the specific line of research of the departing faculty member. Less systematic evidence suggests these considerations have become the norm in first-tier universities, at least in the natural sciences and engineering. While this system may promote quality research and synergies at the subdepartment level (sometimes a group of only five faculty researchers), it does not help build wider cross-disciplinary synergies.

and 59 per cent if patient fees for university-affiliated hospitals are excluded. The subsidies to private universities are also being reduced.[15]

Aside from lowering the overall government budget deficit, part of the rationale for these reductions dates from the so called Toyama plan set forth in 2001 by the then Education Minister Ms Atsuko Toyama. This plan envisaged a shift to competitive funding for university research as part of an overall effort to make universities more competitive and efficient. It also envisaged about thirty universities rising to world-class educational and research levels. The remainder would be classified as 'education oriented' universities and their research funding limited accordingly. The plan did not go into details about how universities would be allocated among these two tiers or how a university might boot-strap itself from the lower to higher tier (Cyranoski 2002).

The Toyama Plan gave rise to the 21st Century Centers of Excellence (COE) Programme which made funds available competitively, usually for a fairly broad department (or subdepartment) level research and educational programme or for a theme that crosses department lines. Probably from its inception, COE funding was considered to be a partial substitute for O&A subsidies—but a substitute that only a relatively small number of universities would receive, allowing some to expand advanced level education and research while others would have to scale back such programmes (Cyranoski 2002; Shinohara 2002).

A commonly mentioned (but perhaps never officially set forth) corollary of the Toyama plan was that O&A subsidies would be reduced across the board for all universities by 1 per cent annually for the first five years beginning in 2005. Subsequently, the education oriented (second-tier) universities would face steeper reductions of 2 per cent annually, while the approximately thirty top-tier research-oriented universities would continue to face only 1 per cent annual cuts. However, current policy documents are less precise. They mention growing projected shortfalls in funding for permanent staff and infrastructure as well as the need to take into consideration each university's unique situation and the results of the *mid-term evaluations*, discussed below.[16] Although speculation is rife that beginning in 2010, O&A subsidies will be cut even by 3 per cent annually, university

---

[15] For FY 2008 national university and private university budgets, respectively:
http://www.mext.go.jp/b_menu/shingi/gijyutu/gijyutu4/010/siryo/08020812/003.pdf
http://www.shidairen.or.jp/blog/files/doc/h191227seifuyosan.pdf. For other years similar URLs can be found using same headings as in 2008 documents. See also n. 4 above.
[16] See http://www.zendaikyo.or.jp/katudou/kenkai/daigaku/08-7-30seimei.pdf and
http://www.gyoukaku.go.jp/genryoukourituka/dai67/shiryou2.pdf.

officials suggest that it might be more likely that 1 per cent cuts will continue for the foreseeable future, especially considering that the present evaluation reports do not provide a clear basis to determine which universities should be subjected to more severe cuts (see below).

In the case of the University of Tokyo, O&A subsidy cuts from FY 2005 through FY 2009 averaged about M¥ 950 (slightly over 1 per cent) annually. However, this has been more than offset by increases in competitive or industry funding. In particular since 2005, University of Tokyo awards of government commissioned research and corporate joint research have increased substantially—the former by about M¥ 2,800 annually and the latter by M¥ 400 annually.

The same phenomenon applies to the other former Imperial Universities and indeed to national universities as a whole. Overall O&A subsidies decreased by M¥ 37,200 from between FY 2004 and FY 2007. However, funding from all the other principal sources increased by more than double this amount. 65 per cent of this increase was accounted for by government-commissioned research (mainly from JST and NEDO), 15 per cent by industry-sponsored joint research, and most of the remainder by donations, primarily to the University of Tokyo.[17]

The Appendix (p.140) contains figures showing trends for all the major funding and brief descriptions of the programmes. Figure A1 shows that MEXT O&A subsidies remain the most important source of funding by far for national universities, although both absolutely and as a proportion of total funding the share is decreasing. Figure A2 shows that most other major funding sources support project-specific research. The exceptions are COE funding which aims to enhance research infrastructure—but by covering personnel costs it indirectly supports specific projects—and donations which, in addition to often supporting specific research, sometimes support ancillary activities such as travel, holding of conferences, and even building construction. MEXT GIA have traditionally been the largest source other than O&A subsidies. But GIAs have plateaued, while COE, joint research, donation, and especially commissioned research continue to increase—to the point where the latter almost equals GIA. Commissioned research is usually funded by government agencies such as JST and NEDO—an issue discussed

---

[17] Indeed among the 19 national universities that receive the most overall funding (those listed in Figs. 4.1 and 4.3), only one seems to have suffered a clear decline in total funding since 2005. This is Hiroshima University, which generally is regarded as among the top ten national universities. Kanazawa University's total funding also declined slightly comparing 2004–5 and 2006–7 total funding. So the tipping point at which most universities experience overall losses is probably below the twentieth university ranked in terms of overall funding.

below. Figures A3 and A4 show that these trends are magnified in the case of the University of Tokyo, where O&A subsidies are proportionately less than for national universities as a whole and commissioned research is the second largest source of funding.[18]

The non-O&A subsidy programmes, in particular government-commissioned research and the new Global COE, are highly skewed in favour of elite universities. The big four accounted for 47 per cent of commissioned research paid to national universities in 2007 and 2008, which is even higher than their 44 per cent share of GIA. Their share of the new Global COE programme over the same two years (51 per cent) is likewise higher. As for Special Coordination Funds and joint research with industry, the big four accounted for 41 and 39 per cent respectively in 2007. In contrast, they accounted for only 21 per cent of O&A subsidies in 2007–8.

In summary, the most equitable programme is being scaled back, and although the loss is more than made up for by competitive programmes, these tend to be awarded to elite universities. Moreover, the fastest growing and probably soon-to-be largest category of competitive programmes—government-commissioned research—is not only one of the most skewed towards elite universities, but also one of the most dominated by the ministries and small elite groups of academic advisers in terms of selection of themes and award decisions (see below).

Thus, as cuts in O&A subsidies continue or even accelerate in second-tier universities, severe costs reductions will probably be necessary. Criticisms are heard.[19] However, to an outside observer these seem more muted than similar protests in Europe. In addition to standard cultural explanations, a possible explanation is widespread acknowledgement that consolidation among universities, especially the nearly 600 private universities, is necessary. Another is lack of a unified response from universities, with the elite universities feeling less threatened than second-tier national universities, and both these groups feeling they ought not to be subject to the same consolidation pressures as most private universities. A third explanation might be the different role the research assessment exercise is playing in these reductions (see below).

But even the elite universities are facing severe challenges, because so far all rely on O&A subsidies to pay salaries of permanent faculty and administrative

---

[18] In some other major national universities, specifically Osaka, Kyushu, Hokkaido, and Tokyo Institute of Technology, commissioned research funding in 2007 also surpassed MEXT GIA (Cabinet Office data).

[19] See, for example, the analysis from Miyazaki National University, http://meg.cube.kyushu-u.ac.jp/~miyoshi/jsa_symposium/4-4.pdf, and the position paper by the All Japan Association of University Faculties: http://www.zendaikyo.or.jp/katudou/kenkai/daigaku/08-7-30seimei.pdf.

and support staff. None has adopted a soft money system that would tap the increasing funding for project-specific research to pay such salaries. However, according to senior advisory committee members and university officials, MEXT would probably permit this. Indeed, the ultimate goal of the MEXT and the Ministry of Finance may be to force at least a partial adoption of such a system: for example, the pooling of some of the project-specific funds to pay the salaries of assistant professors who have permanent employee status. Holding back this step is not so much the government—although some officials in MEXT do oppose soft funding for any permanent faculty positions—but senior professors who would probably lose control over assistant professorship positions allocated to their laboratories by their universities. Another reason concerns pension regulations, it being harder to accrue pension benefits if one's salary is not guaranteed. However, many familiar with university policy debates believe that some movement in this direction will inevitably occur over the next few years.

Evaluation: A System Loosely Coupled to Funding

The 2004 university incorporation law mandated recurring evaluation of each university's performance. Similar provisions are included in the laws transforming many GRIs into 'special administrative entities' with a similar degree of autonomy as the national university corporations. Beginning in 2005, as part of the evaluation process, each faculty member must report annually the numbers of publications (noting separately those in international journals), collaborative research projects, awards, patents, and so on for that year. These data can be used for promotion decisions. They are also aggregated by department as part of the evaluation process.

Another key input into the evaluation process are the mid-term plans mentioned above. All national universities submitted these plans in late 2007. These were reviewed by the National Institute for Advanced Degrees and University Evaluation (NIAD-UE) and National University Evaluation Committee (NUEC), with the latter playing a supervisory role and the former doing much of the actual evaluation. A former member of the NIAD-UE described this as a complex, time-consuming process. Sometimes the evaluations included site visits, but more often they involved teams from the universities making presentations to NIAD-UE and NUEC staff.[20]

---

[20] These reports are now publicly available at http://www.mext.go.jp/a_menu/koutou/houjin/1260234.htm. Major GRIs also submitted similar mid-term plans around the same time, and their evaluation reports are also publicly available.

The evaluation report on the University of Tokyo is nearly 300 pages long, with separate analyses for education and research in each major department. Reading through it, and the reports on a few other institutions, gives a sense of the report writers bending over backwards to offer understanding and constructive guidance. There is no ranking or overall score, either by departments or universities. According to a high-level advisory committee member, this reflects the philosophy of the recently retired head of the NIAD-UE, Takeshi Kimura, that the main goal of the evaluation process should be constructive guidance, not ranking of universities. Such an approach resembles that taken by some European states, such as the Netherlands (Meulen 2007).

Nevertheless, as noted in the previous subsection, some recent MEXT documents indicate that the mid-term evaluations probably will influence the allocation O&A subsidies.[21] However, even Japanese university officials and senior members of advisory committees feel it is not easy to use the mid-term evaluations to decide which universities ought to suffer more severe reductions in O&A subsidies. One official indicated that it was unlikely that the evaluations would play much role at all in any near-term decisions to reduce O&A subsidies. The nature of these evaluations may even be causing the government to rethink any plans for two-tiered reductions in O&A subsidies, and to opt instead for continuing uniform reductions.

Japan may be taking a unique approach of eschewing use of research evaluations to determine funding, but instead forcing all universities to rely more on competitive funding. However, for this strategy to be effective, universities will have to solve the salary dilemma and competitive funds will have to be allocated fairly, so that the researchers and projects that receive funding really are likely to be the ones that deserve it most. Whether this is the case is explored in the section on peer review below. Also the section on government influence notes that evaluation of individual large projects has become strict.

Technology Transfer: Joint Research with Large Companies

A series of reforms between 1998 and 2004 were intended to facilitate cooperation between universities and industry and to give universities incentives to commercialize their discoveries. However, growth in licensing

---

[21] http://www.gyoukaku.go.jp/genryoukourituka/dai67/shiryou2.pdf.

of independently created university inventions has been feeble. The same is true for formation of startups with strong business prospects, except for some startups in software or life sciences. The dominant mechanism of university–industry cooperation and transfer of university discoveries has become collaborative (joint) research. In major national universities, about half of all university discoveries on which patent applications are filed are attributed to joint research with private companies. About three-quarters of all inventions that are transferred to industry (either by joint patent applications or licenses) are attributed to joint research.

Except in the life sciences, joint research partners are overwhelmingly large established companies. However, joint research accounts for less than 7 per cent of these universities' research funding, *not including* salaries for permanent staff. This means that publicly funded commercially relevant discoveries are being leveraged by the companies that provide joint research funding. This has recreated the situation that existed before the 1998–2004 reforms, when the vast majority of university discoveries were passed by university inventors directly to companies that gave donations to their laboratories. This method of transfer is usually quick and transaction costs are fairly low. However, large companies exclusively control the intellectual property, and their obligations to develop the discoveries are weak (Kneller 2008, 2007*a*, 2006).

Among industrialized countries, Japan has by far the highest proportion of issued patents covering university discoveries that are co-owned by private companies (approximately 60 per cent). Almost all of the co-owning companies are large domestic companies that, by virtue of co-ownership, have the exclusive right to develop and use the inventions.

4. Factors Underlying Skewed Resource Allocation

Peer review

In most natural science laboratories O&A subsidies (block grants) are insufficient to obtain the instruments and hire technicians and post-docs to conduct ground-breaking research. The same is true for social scientists who carry out large-scale surveys or who need expensive databases, and researchers in any discipline who need more than just occasional international travel funds. Nevertheless, it should be noted that graduate students have traditionally been self-funded. They must pay tuition (although in national universities this is relatively modest) and bear all their own living expenses. Thus for Japanese faculty, graduate students have generally been a free resource (observations and inquiries by author).

Standard MEXT GIA are the largest source of project-specific funding, providing the bread and butter of research support throughout Japanese universities. Even in prestigious universities these are the main source of support for researchers in the humanities and social sciences, although in some elite universities COE funding is supporting a significant amount of teaching and research in these fields (author's observations).

The peer-review committees that evaluate funding applications to these programmes are rarely dominated by faculty in prestigious universities. In the case of MEXT grants-in-aid, reviewers are recruited from members of academic societies and other university faculty who have published or previously received GIAs in the field. Each reviewer serves for two years, so turnover is frequent and many academics from across Japan serve as reviewers during their careers. Each application is usually assigned to six reviewers. An analysis of the academic positions of each of the reviewers on eight committees (two each from chemistry, biomedicine, engineering, and social science) in both 2002 and 2007 shows that in 2002, on average 16 per cent were from the big four. In 2007 this had climbed to 25 per cent. In each of these periods, at least a quarter of reviewers were from universities that are not well known.[22] Yet in this most democratic of Japanese funding programmes, awards are still highly skewed, although they are so numerous that much funding does filter down to less prestigious universities.

Rather than dominance by epistemic elites, a more pertinent criticism is that the review process for GIA applications is rushed, with few procedures to ensure quality. Except in cases of extremely negative or positive evaluations, there is no need for reviewers to explain any of their scores. Sometimes reviewers are not familiar with proposed research fields. Often they must squeeze reviews of a large number of applications into already tight schedules. There is no exchange of opinions among reviewers that might afford some proposals a second, more careful, look.[23]

---

[22] http://www.jsps.go.jp/j-grantsinaid/14_kouho/meibo.html.

[23] Compare the summary of NSF and NIH peer review in Kneller (2007b). These and other shortcomings were noted by Coleman (1999). The author has observed that junior faculty tend to write the majority of GIA applications although they usually add the name of the laboratory head as one of the applicants, often as the principal investigator. In such cases, the laboratory head's publications are included in the application document, and as Coleman (1999) has noted, laboratory heads are usually named as co-authors on all publications from their laboratories. Thus the combination of capable young researchers and laboratory heads with impressive publication lists may give laboratories in elite universities an advantage in competing for GIA funds. Compared to second-tier universities they attract more capable young researchers, and their senior faculty tend to have more impressive publication lists.

Similarly it would be difficult to argue that the COE peer review committees are dominated by elite academics, although perhaps there is a trend in this direction. In the 2002 review process, two of the five proposal review committees—interdisciplinary studies and medicine/life science—lacked members from the big four, while for the humanities and social science review committee, eight of twenty-three members (35 per cent) came from this group. In 2007, big four representation on the interdisciplinary studies committee had climbed to 18 per cent and on the medicine/life science committee to 32 per cent, although it had fallen to 9 per cent on the humanities/social science committee.

Nevertheless, what seems most striking about the membership of these committees is their professional diversity, with members drawn from companies, GRIs, journalism as well as academia.[24] Herein may lie one of the shortcomings of the COE evaluation process. While the reviewers can scrutinize the programmatic aspects of proposals, they probably do not have sufficient expertise, time, or energy to probe deeply into what sorts of new research and training programmes will be conducted, or their implications for science and the national economy. Also the pattern of awards suggests that proposals from prestigious institutions that already have considerable programmatic experience and infrastructure tend to receive higher scores than proposals from innovative researchers in smaller institutions where improved programmatic capabilities would help innovative research to take off and have an impact.

Similar to the perspectives among natural science academic researchers in the UK discussed in Chapter 8, those in Japan seem to agree that forefront research in their field requires expensive infrastructure and large numbers of researchers. For many forefront natural science projects, most types of GIA are insufficient (exceptions being two categories that provide awards over 500,000 USD per year). Aside from these limited forms of GIA and COE programmatic funding, mechanisms for funding large projects generally are limited to MEXT Special Coordination Funds (SCF), commissioned research by government agencies such as JST and NEDO, and industry-sponsored commissioned or collaborative research.

As in the case of GIA and COE, it would be difficult to argue that the allocation of SCF is dominated by academic elites. The programme has only seven peer-review committees. Each has typically ten members, although the largest, training in advanced interdisciplinary fields, has approximately

---

[24] http://www.jsps.go.jp/j-globalcoe/03_iinkai_meibo.html.

twenty reviewers representing a wide spectrum of universities and GRIs. Among the interdisciplinary fields panel 8 per cent were members of the big four in 2003, and 16 per cent in 2007. Among the panel dealing with biomedicine and bioethics, 36 per cent were from the big four in 2003, 30 per cent in 2007.[25]

Government

On the other hand, themes for the major government-commissioned research programmes such as those of JST and NEDO (often called *national projects*) are determined in a top–down manner, and the chief scientists that lead proposal review teams tend to be well-known, elite scientists (Kneller 2007b). Moreover, government-commissioned research is the fastest growing category of competitive research funding with the aggregate total approaching that of GIA (see Figure A2). Thus, while most funding programmes are not dominated by academic elites either in terms of setting research goals or selecting among competing applicants, in one of the most important categories of programmes, they probably are. This raises the prospect of the government playing an increasingly influential role in determining university research themes and in shaping the scientific careers of young scientists.

In addition, compared at least to the USA, a larger proportion of government university research funding (including the mechanisms mentioned above, except for some of the main types of GIA awards) is awarded to groups of laboratories (Kneller 2007a: 61). Typically the senior professors leading these multi-laboratory projects distribute funding among junior researchers in a trickle-down fashion. It seems likely that this creates another barrier to encouraging young and mid-career researchers to formulate and pursue their own creative projects. Most of these large projects undergo mid-project and end of project reviews by panels of ministry officials and outside scientists (*ad hoc* peer-review panels) that are strict in the sense that projects are rated and suspension in mid-term is not uncommon. Assessment is partly on the basis of whether the projects met their stated goals, and identifiable near-term achievements are often important to obtain a high score.[26] This type of immediate results-oriented evaluation

---

[25] http://www.mext.go.jp/b_menu/houdou/19/05/07051420/002/005.htm, http://www.mext.go.jp/a_menu/kagaku/chousei/1284635.htm, and similar pages.

[26] For example, mid-term reviews over the past two years of about twenty-five large-scale university–industry interdisciplinary projects under the Special Coordination Funds Programme, resulted in cancellation or suspension of over a third of the projects. Those that

may divert researchers from challenging, risky projects whose break-throughs are not likely to be recognized for several years.

The main R&D ministries are staffed predominantly by graduates from elite universities: 57 and 85 per cent of career employees joining MEXT and METI, respectively, in 2005 were graduates of the University of Tokyo or Kyoto University. If graduates from Tohoku and Osaka Universities are included, the proportion for MEXT would increase to 64 per cent (that for METI would remain unchanged).[27] For a variety of reasons, the ministries may believe that concentration of resources is beneficial, or a traditional practice that need not be altered (especially if the practice favours their Alma Maters).

But some of bureaucracy's support for resource concentration may be based on misperceptions. For example, conversations with Cabinet Office staff members suggest that NIH's comprehensive Cancer Centers Programme was probably construed by the Cabinet Office as a programme to establish a small number of elite cancer research and clinical care centres and thus a model for Japanese centres of excellence programmes. In fact these centres are fairly widely distributed and there is probably no underlying assumption within NIH that concentrating funding in a limited number of national centres of excellence will bring economies of scale in research discoveries or patient care.

Finally, unlike in North America, Germany, or Switzerland, there is scant scope in Japan for competition between regions to build outstanding universities. They do not offer an alternative source of funding for university R&D. The appropriateness of Tokyo and Kyoto Universities remaining at the apex of higher education has never been seriously questioned. Universities were not supposed to compete for funding until the Toyama plan. Now as the proportion of competitive funding increases, there is scope for competition. But pay scales remain regulated and central ministries determine the allocation of the bulk of funding; that is, of O&A subsidies plus commissioned research. University of Tokyo presidents and department heads may rotate every two or three years, but their positions guarantee them close ties with the ministries and this probably ensures a sympathetic hearing to proposals that ensure the University continues to receive the lion's share of funding. Those who hold key positions in the other big four

received particularly high marks had early stage concrete achievements such as working prototypes of diagnostic instruments (discussions with ministry officials and US NSF personnel who have compared NSF and Japanese government project evaluation procedures).

[27] Data from various ministries between 2002 and 2005 compiled annually at http://www.geocities.jp/plus10101/the-todai.html.

universities probably have similar ties with the ministries. (See for example Shodo 2007, confirmed by personal observations.)

Academic Advisory Committees

The influence of elite academics can also be felt through advisory committees on S&T policy. The highest-level committee for coordinating S&T policy is the Council for Science and Technology Policy (CSTP) whose members consists of seven cabinet members with S&T responsibilities, and eight 'experts' from academia and industry. Among the eight experts, two are directors of major corporations and six are academics. Half of the experts either received most of their education in the big four universities or spent most of their research and teaching careers there. However, three of the academics, including the chair among the experts, had little or no affiliation with former Imperial Universities, and at least two of these three followed untraditional and difficult career paths.[28]

The thirty-member Council for Science and Technology (CST) is MEXT's main advisory council for S&T affairs.[29] Just over half its members either were educated primarily in big four universities or spent most of their professional careers in these universities. This proportion was somewhat smaller in most of the CST's eleven subcommittees.

The Centre for Research Development Strategy (CRDS) is an advisory body within JST whose mission is to identify priority research topics for government support.[30] Thus it influences the direction of government-commissioned research. Its eleven members include four well-known academics, three government officials, a former vice president of AIST, and two representatives from industry. All but two or three of the eleven either graduated from or spent much of their professional careers in the big four universities.

Finally the Science Council of Japan (SCJ) advises the Prime Minister and the country at large on a broad range of science-related issues. Somewhat akin to the US National Academy of Science, its 210 council members consist mainly of heads of various universities, GRIs, and major departments, as well as professors (many from the humanities and social sciences) from a cross-section of universities.

---

[28] http://www8.cao.go.jp/cstp/yushikisyahoka.html.
[29] http://www.mext.go.jp/english/org/struct/049.ht.
[30] http://crds.jst.go.jp/en/activity/Our_approach.html.

As shown by Tanaka and Hirasawa (1996) the advisory councils can have an important influence on S&T policies. However, they also found that council recommendations have been mediated by the ministries, and that the ministries often shape the agenda for the advisory councils. Moreover, because members are appointed by the ministries, they 'tend to respect the ministries' intentions'. Thus in Japan's case the influence of bureaucrats (often mid-level section heads) may be considerable, and in some cases greater than that of the elite academics on the advisory committees.

In summary, graduates or faculty of the most elite universities are highly represented in senior advisory committees. However, the influence they have relative to bureaucrats may not be great. Thus it is unlikely that academic elites are the main force in perpetuating the concentration of research resources in a few universities. Rather, a symbiotic relationship between ministry personnel (middle as well as high-level bureaucrats) and respected scientists and administrator in the elite universities ensures these universities have privileged access to resources. The ministries rely on these universities for the nation's most important basic scientific output, while the universities rely on the ministries for funding. Countervailing, centrifugal forces that might nurture competing institutions are weak. Indeed, the following factors probably reinforce the trend towards concentration: the perception among young researchers that facilities are best in the elite universities, increasing opportunities for capable researchers from the outside to compete for junior faculty positions in the elite universities, and a seemingly widely held perception in academia and the bureaucracies that concentration of resources is necessary to produce world-class science.

5. Conclusions

Salient Features of the Governance of the Japanese PSS

A review of the other chapters in this volume suggests some areas in which Japan's system of public science governance is unique. Foremost among these is the high degree of concentration of resources in a small number of universities. Another is close cooperation between universities and companies—particularly the large proportion of patented university discoveries that are exclusively controlled by large collaborative research partners. This relationship and its implications for science and Japan's economy have been discussed elsewhere (Kneller 2007a). Finally, Japan's soft approach to research evaluation, coupled with a steady march towards an American-style system of soft-money funding even for permanent staff salaries, seems unique. More speculatively, there is the possibility of a tilt towards applied research in the major government-funded commissioned research pro-grammes. However, the extent to which these programmes really are applica-tion oriented (or designed to promote national industrial competitiveness), and whether Japan's funding priorities in this regard are any different from those of most other countries, are beyond the scope of this chapter.

Implications

Although a complete analysis of the implications of resource concentration is not possible here, insights can be gained from citation data (Negishi 2009; Thomson Reuters 2009). These data attribute citations among Japa-nese universities in proportion to co-authorship. In other words, a publica-tion co-authored by two University of Tokyo, one Nagoya, and two Cambridge researchers would be attributed 40 per cent to Tokyo, 20 per cent to Nagoya, and 40 per cent to Cambridge, and any citations would be allocated proportionately (Negishi 2009).

The following analysis matches numbers of apportioned 2006 and 2007 citations to 2003–7 publications to various measures of inputs into novel scientific research. Efforts were made to measure inputs at times when they would be likely to have had the greatest impact on publications cited in 2006 and 2007. However, limitations on some of the input data con-strained the possibilities for matching the time periods of inputs with the 2006–7 citations. The nineteen universities selected for this analysis are the top-ranked Japanese universities for the inputs under consideration. GRIs were excluded because their funding and personnel structures are different from those of universities. However, in terms of crude numbers of citations between 1998 and 2008, Riken, AIST, and the MEXT's National Institute of Natural Science, would rank 7th, 10th, and 14th respectively.[31]

Numbers of full-time faculty (assistants to full professors, including persons on time-limited contracts) plus numbers of graduating doctoral students

---

[31] http://science.thomsonreuters.com/press/2009/top_japan_research_institutions/. The ratio of science/engineering/agriculture/health to arts/humanities/law/social faculty and students is roughly 3:1 in most of these universities. Therefore, different mixes of disciplines (with some disciplines more likely to produce more highly cited papers than others) is probably not a factor in explaining the trends in Figs. 4.1–4.3, except in the case of Tokyo Institute of Technology (TIT) and Tokyo Medical and Dental University, where almost all students and faculty are in science/engineering and health sciences, respectively. Also all universities have medical schools except TIT and Waseda, so a significant proportion (roughly one-third) of research in all universities except these two is biomedical related.

was used as a metric for personnel inputs into creative science, since this would include university researchers who would be involved in formulating research, analysing results, and writing the resulting papers. Figure 4.1 lists the nineteen universities in order of decreasing number of full-time faculty, and shows a declining trend in citations per scientist as university size decreases. Especially after removing Tokyo Institute of Technology (TIT) and Tokyo Medical and Dental University (TMDU) which have distinctive personnel structures, the trend is so clear that, even if all doctoral students were included in this metric, the trend would still be apparent.[32] In other words, on their face these data do not suggest that concentration of resources is harming scientific output. Moreover, the top private universities seem to have a weaker output compared to similar-size national universities.

These findings are probably explained in part by the fact that faculty in the smaller universities concentrate more on teaching and less on research, and by higher percentages of young researchers on the faculties of large, compared to small, universities. In 2007, 28 per cent of faculty in the big four were younger than 37 years compared to only 19 per cent of faculty

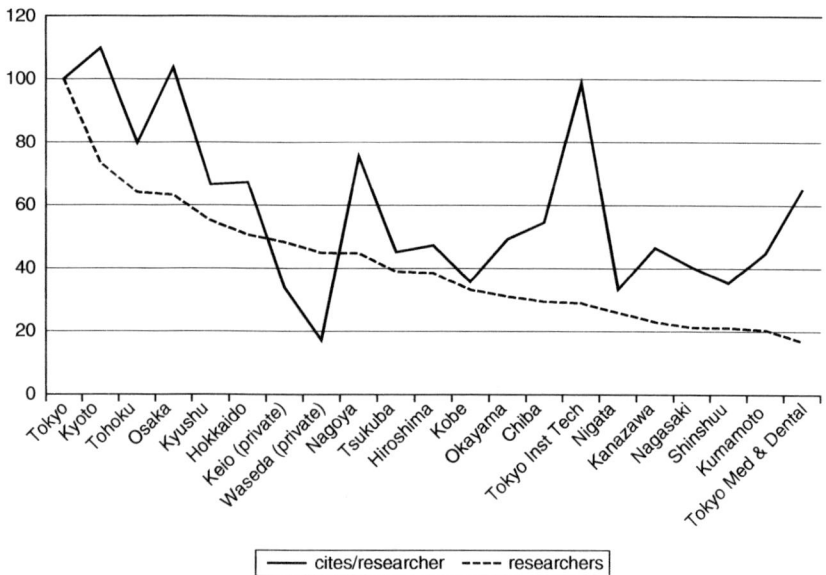

Figure 4.1. 2006–07 cites to 2003–07 pubs (apportioned by university)/full-time 2005–06 reseacher (faculty + doctoral graduates/year) (University of Tokyo = 100)

---

[32] A doctoral programme in science or engineering usually takes three years, with the first year devoted largely to course work and the last two years to research.

in Kanazawa, Nagasaki, Shinshuu, and Kumamoto Universities (Cabinet Office data). Direct observations confirm that, even in the best private universities, faculty tend to have higher teaching loads than in national universities of roughly equal or larger size.

When using total funding as a metric for total inputs into creative science and ordering the universities accordingly, citations to publications per unit of funding shows a modest declining trend as funding decreases (Figure 4.2).[33] Again, this might seem to suggest that it would not make sense to try to redistribute resources to smaller universities. However, the declining trend in citations per unit of funding is not as sharp as the decline in citations per scientist (Figure 4.1). Also, compared to the University of Tokyo, productivity in terms of highly cited research is higher in Kyoto, Osaka, and even Chiba Universities (not to mention TIT, which might have an advantage because almost all its research is in science and engineering). Productivity in Nagoya and Okayama Universities appears approximately equal to that of the University of Tokyo.

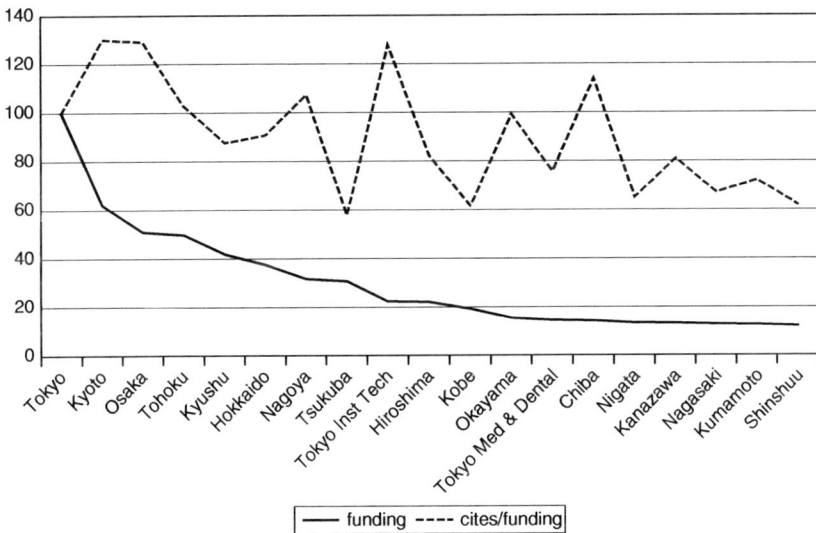

Figure 4.2. 2006–07 cites to 2003–07 pubs (apportioned by university)/$1M total 2004–05 funding (excluding tuition payments and patient hospital charges) (University of Tokyo = 100, national universities only)

---

[33] Strictly speaking Fig. 4.2 does not include all funding, because it excludes student tuition payments, and patient hospital charges. However, tuition accounts for less than 10 per cent of the University of Tokyo's total income, net hospital patient charges and tuition rates are basically the same in all national universities.

Recall from Figure A1 in the Appendix that total funding constitutes about 75 per cent of O&A subsidies that cover mainly personnel costs and about 25 per cent of competitive funding which covers project-specific (mainly non-personnel) costs. Indeed, the curve showing citations per unit of O&A subsidies (not shown) is similar to the curve for citations per scientist (Figure 4.1). Thus the difference between the trends in Figures 4.1 and 4.2 probably is mainly accounted for by the impact of project-specific funding that does not include salaries for permanent staff. This effect seems to moderate the phenomenon shown in Figure 4.1, which shows sharply declining productivity per researcher as university size decreases.

Figure 4.3 is consistent with the moderating effect of competitive, project-specific funding proposed above. It shows that citations per unit of GIA have a clear inverse association with total GIA received by each university. Curves showing citations per unit of commissioned research or COE funding show the same trend, but it is even more pronounced. Of course, casual inferences can only be drawn with extreme care. Taken together, Figures 4.1–4.3 simply show that, along the continuum from large to small universities, competitive funding declines more quickly than citations, which decline more quickly than total funding, which declines more quickly than numbers of scientists. At most, Figure 4.3 tentatively suggests that capable

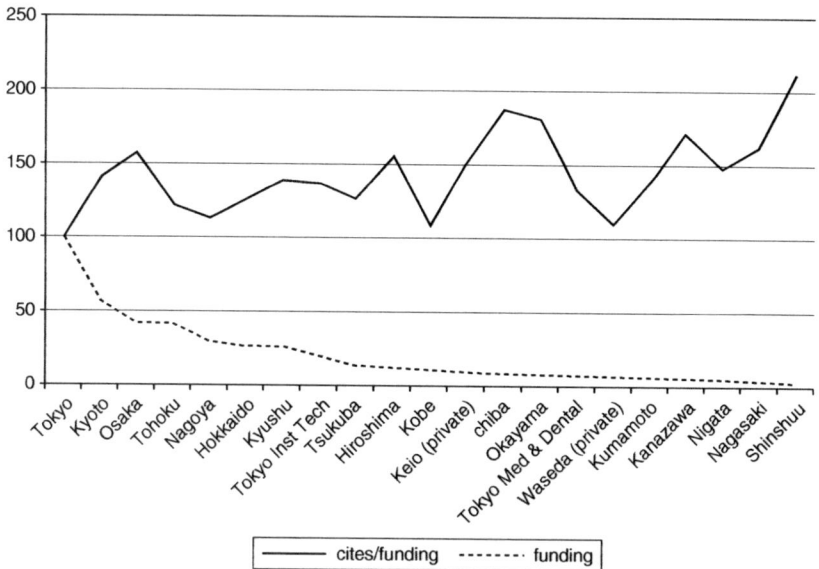

Figure 4.3. 2006–07 cites to 2002–07 pubs (apportioned by university)/$1M MEXT Grants-in-aid (2001–2006) (University of Tokyo = 100)

researchers in the smaller universities are relatively starved of non-salary research funding, and that redirecting some of these funds to smaller universities will result in a net gain of citations for Japanese university research.

Besides, if the peer-reviewers evaluate GIA applications properly, they ought to be aware of highly cited articles in their field from the universities submitting applications and, if these articles come from the same laboratories as the applicants, this ought to have a positive bearing on their evaluations. However, the large disparity between GIA awards and citations shown in Figure 4.3 (about 70 per cent more citations per unit of funding attributable to small universities compared to the University of Tokyo) suggests that peer-review panellists sometimes are not aware of such publications.

In addition, these trends probably suggest that the relative strength of the big universities lies more in attracting skilled scientists than in economies of scale associated with their having lots of equipment, data access, travel opportunities, and so on. The latter are usually purchased with competitive funds. If these, not brains, are what gives elite universities greater scientific productivity, we would probably not see the moderating effect of competitive funding (Figure 4.2 compared with Figure 4.1) nor the trend in favour of non-elite universities shown in Figure 4.3.

Of course, this may vary according to discipline. It may be that trends in Figures 4.1–4.3 would be found for separate analyses of citations to biology, chemistry, and some engineering publications, but not for citations to high-energy physics research, where access to very expensive equipment may be essential for ground-breaking research. In fields such as experimental physics, the advantage of elite universities with respect to citation productivity may be just as strong with respect to competitive funding as it is with respect to scientific personnel. However, data for such field-specific analyses are not available.

In addition, it is not clear if the advantage of elite universities with respect to having, on average, more capable scientists is additive or multiplicative—whether bringing many bright researchers together increases creative output beyond what they could achieve as individuals. Collaboration between laboratories is not common in Japanese universities (Cyranoski 2002; Kneller 2007a; and recent interviews with companies collaborating with universities). If the effects are mainly additive, then providing bright energetic researchers with incentives to work in universities other than the big four probably would not hurt Japan's total output of high-quality science. Figures 4.2 and 4.3 even suggest (albeit tentatively) that if this is accompanied by access to competitive funding, overall scientific productivity might increase.

Conversely, as O&A subsidies diminish, if competitive funding is concentrated in a few elite universities, bright scientists will be reluctant to work anywhere else and the process of concentration will continue. Also, the increasing mobility and meritocracy of the recruitment process may accelerate the concentration of the best scientists in a small number of universities. The largest and most favoured universities probably will then have little need to be entrepreneurial or more independent of government influence.

Appendix

Figure A1 shows trends from 2004–7 (the years for which data for all categories are available) for all sources of funding for all national universities. Figure A2 shows trends for the main categories of project-specific research funding in national universities over a longer time period. Figures A3 and A4 show the same trends for the University of Tokyo alone. Figures A3 and

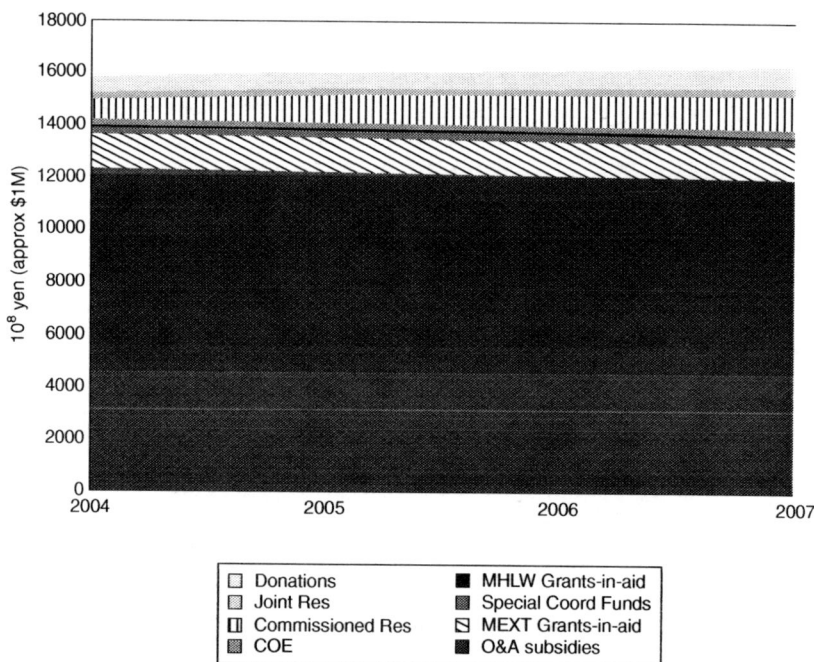

Figure A1. National university funding, all principal sources including O&A subsidies (but excluding tuition and hospital patient charges)

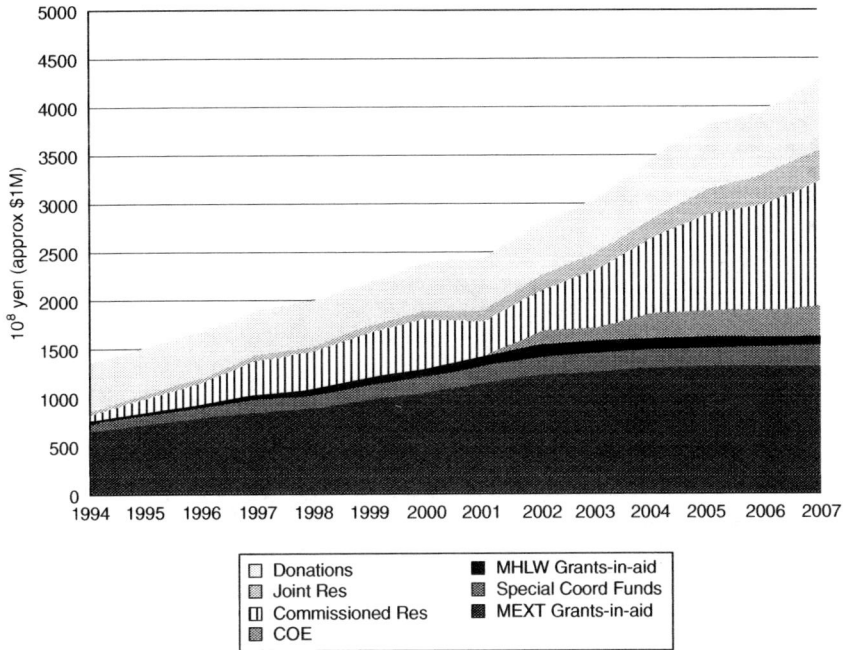

Figure A2. National university funding, all principal sources excluding O&A subsidies, tuition, and hospital patient charges

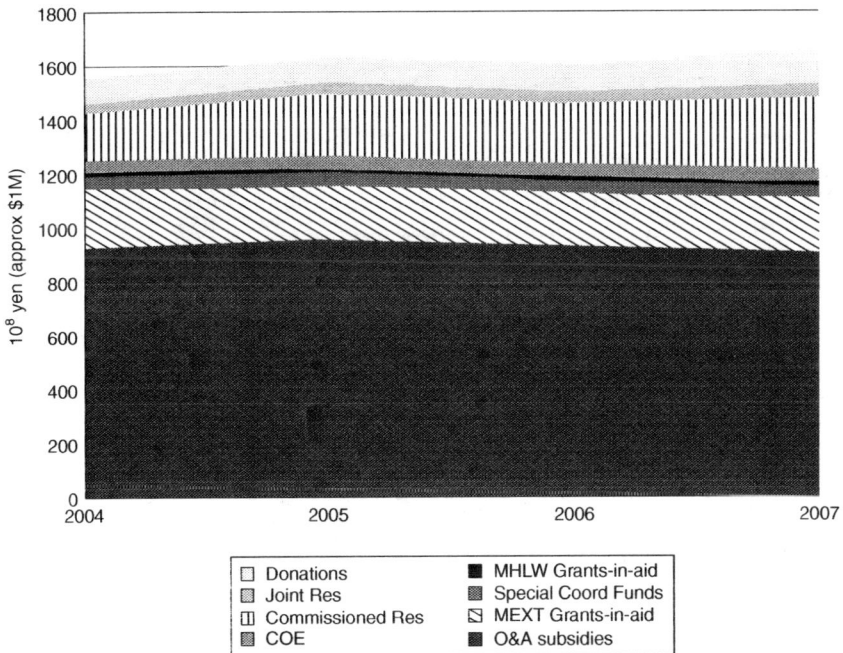

Figure A3. University of Tokyo funding, all principal sources including O&A subsidies (but excluding tuition and hospital patient charges)

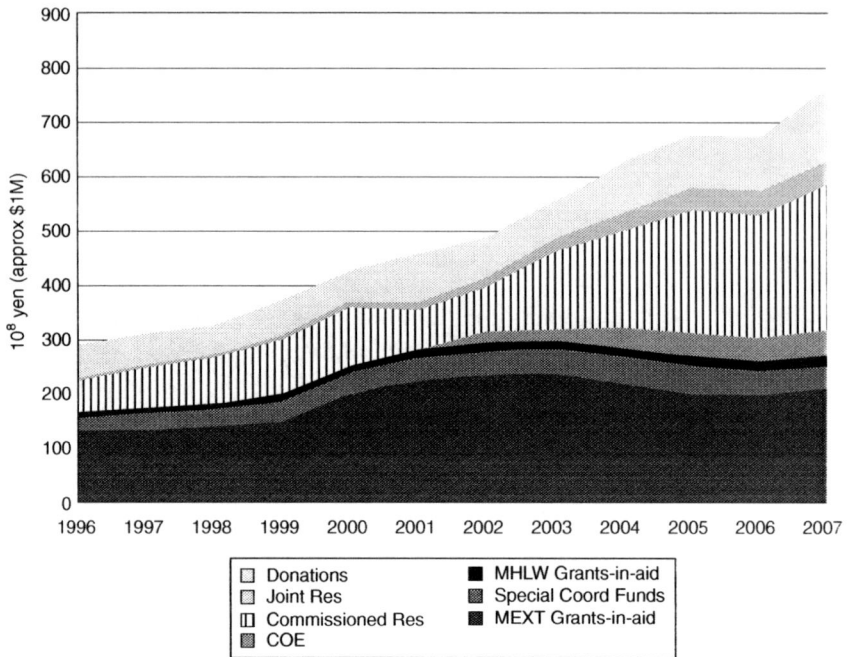

Figure A4. University of Tokyo funding, all principal sources excluding O&A subsidies, tuition, and hospital patient charges

A4 illustrate the more pronounced and growing impact of large, government-funded commissioned research projects in the most prestigious national universities.

The main data sources for these charts are Cabinet Office, MEXT, and the *University of Tokyo Data Book* for various years.

As noted in the text, the increase in university R&D funding is due mainly to an increase in commissioned research. These usually are large projects funded by government agencies, such as METI's New Energy Development Organization (NEDO) or the Japan Science and Technology Agency (JST) under MEXT. A small percentage of commissioned research (barely 4 per cent in national universities in 2005) is commissioned by private companies (Kneller 2003; JST 2007, 2009). As noted in Kneller 2007*b*, these are top–down programmes where research topics and awardees are determined by agency officials in consultation with a small number of well-known academics. JST and NEDO are the largest sources of university-commissioned research (JST: 503×10^8 yen, NEDO: approximately 170×10^8 yen in 2008) (NSF Tokyo Office, Rept. Memo. #09-01) although substantial research is also commissioned by the Ministry of

Health, Labour, and Welfare, and the Ministry of Internal Affairs and Communications (Kneller 2007b; Tanaka 2006).

GIA contain subcategories that run the gamut from small grants for young researchers of about 10,000 USD to Specially Promoted Research Projects that receive about 1 million USD annually. However, the largest portion of funding consists of grants to individual investigators of about 20,000–50,000 USD per year. For researchers in second- and third-tier universities, this is the main source of research support.

Joint research funding is primarily from private companies.[34] It has been increasing steadily since reforms in 2000 greatly reduced administrative hurdles associated with joint research. Somewhat surprisingly, donations from private companies have continued to grow slowly, even though in terms of the IP rights companies obtain, and their ability to specify the research that is to be done, joint research contracts offer significant advantages over donations. By far the largest donations in 2007 were to the University of Tokyo to meet its goal of establishing an endowment of 130×10^8 yen before its 130th anniversary in 2008.

References

Bartholomew, James (1989) *Formation of Science in Japan: Building a Research Tradition* (New Haven: Yale University Press).

Cabinet Office (various years) *Understanding the Scientific and Technical Activities of GRIs and National University Corporations* (Dokuritsu gyousei houjin, kokuritsu daigaku houjin nado no kagku gijutsu katsudou no haaku, shoken torimatome): at http://www8.cao.go.jp/cstp/budget/trimatome.html.

Coleman, Samuel (1999) *Japanese Science: From the Inside* (London: Routledge).

Cyranoski, David (2002) 'Independence Days', *Nature*, 419 (Oct.): 875–6.

Hayashi, Takayuki (2003), 'Effect of R&D Programmes on the Formation of University–Industry Government Networks: Comparative Analysis of Japanese R&D Programmes', *Research Policy*, 32: 1421–42.

Ikeuchi, Satoru (2004) 'Universities Ought to Let in Outside Wind, Faculty Also Need to be Exposed to Outside Currents (Daigaku wa gaibu kara kaze wo ireru beki,

---

[34] From 2005 through 2007, national or local government agencies, or government corporations such as JST and NEDO, accounted for 20 per cent of the total 1092×10^8 yen joint research funding to Japanese universities (JST 2009). The small proportion of public-sector funding for joint research and the small proportion of private-sector funding for commissioned research tend to balance each other out. For a 'big picture' understanding of proportionate sources of funding, it is appropriate to consider that joint research funding is from the private sector, while commissioned research is from government affiliated entities.

kyouin mo soto no kaze in ataru hitsuyou ga aru)', in *University Rankings (Daigaku Rankingu) 2004* (Tokyo: Asahi Newspaper Publishers), 114–17.

Imatani, Akiru (2008) 'Too Many Barriers to Movement (Idousuru ni wa houheki ga oosugiru)', in *University Rankings (Daigaku Rankingu) 2009* (Tokyo: Asahi Newspaper Publishers), 148–55.

JST (2007) *Industry–University Databook 2007* [in Japanese] (Tokyo: JST).

JST (2009) *Industry–University Databook 2008–2009* [in Japanese] (Tokyo: JST).

JTEC (Japan Technology Evaluation Center) (1996) *Japan's ERATO and PRESTO Basic Research Programs*: http://www.wtec.org/loyola/erato/toc.htm (accessed May 2007).

Kneller, R. W. (2003) 'University–Industry Cooperation and Technology Transfer in Japan Compared with the US: Another Reason for Japan's Economic Malaise?' *University of Pennsylvania Journal of International Economic Law*, 24(2): 329–449.

—— (2006) 'Japan's New Technology Transfer System and the Preemption of University Discoveries by Sponsored Research and Co-inventorship', *Journal of the Association of University Technology Managers*, 18(1): 15–35. Reprinted in *Industry and Higher Education*, 21(3).

—— (2007a) *Bridging Islands: Venture Companies and the Future of Japanese and American Industry* (Oxford: Oxford University Press).

—— (2007b) 'Prospective and Retrospective Evaluation Systems in Context: Insights from Japan', in Richard Whitley and Jochen Gläser (eds.), *Changing Governance of the Sciences: the Advent of Research Evaluation Systems* (Sociology of the Sciences Yearbook; Dordrecht: Springer), 51–74.

—— (2008) 'Invention Management in a Major Japanese University and its Implications for Innovation', in *AUTM Technology Transfer Practice Manual*, 3rd edn. (Deerfield, Ill.: AUTM), iii/2.

Ledford, Heidi (2007) 'Scientists to Spare', *Nature*, 449: 1084–5.

Meulen, Barend van der (2007) 'Interfering Governance and Emerging Centres of Control: University Research and Evaluation in the Netherlands', in Richard Whitley and Jochen Gläser (eds.), *Changing Governance of the Sciences: The Advent of Research Evaluation Systems* (Sociology of the Sciences Yearbook; Dordrecht: Springer), 191–202.

MEXT (2008) *Basic School Survey* (Gakkou kihon chousa hokoku sho) (Tokyo: Kokuritsu insatsu kyoku): available at http://www.mext.go.jp/b_menu/toukei/001/08121201/002.pdf.

MEXT (various years) 'Industry–University–Government Cooperation Achievements (San-gaku-kan renkei no jisseki)': data at http://www.mext.go.jp/a_menu/shinkou/sangaku/sangakub.htm.

National Science Board (NSB) (2008) *Science and Engineering Indicators 2008* (Arlington, Va.: US National Science Foundation).

Normile, Dennis (2004) 'Older Scientists Win Majority of Funding', *Science*, 303 (19 Mar.): 1746.

Negishi, Masamitsu (2009) 'Trend Analysis on the Disparities among Japanese Universities Observed with the Number of Publications and Citation Counts (Ronbun su, inyou su kara mita wagakuni no daigaku aida kakusa no doukou bunseki)', *Journal of the Japan Society of Information and Knowledge* (Jouhou chishiki gakukai shi), 19: 158–69.

Preparation for the Civil Service Examination (Koumuin shaken juuken ja-naru) (2004–5): data summary at http://www15.atwiki.jp/gakureking-mirror/pages/16.html.

Shinohara, Kazuko (2002) *Toyama Plan: Center of Excellence Program for the 21st Century* (Tokyo Regional Office Report Memorandum #02-05 (21 June); Tokyo: National Science Foundation): http://www.nsftokyo.org/rm02-05.html.

Shodo, Mizuki (2007) *Ko Gakureki Wakingu Puua* (Highly Educated Working Poor) (Tokyo: Kobunsha Shinsho).

Tanaka, Hisanori (2006) 'The Competitive Research Funding System: Expanding the Review System and Measures to Prevent Wrongdoing (Kyousou teki kenkyuu shikin seidou: fusei boushi taisaku to shinsa seido no kakujuu wo chuushin ni)', Issue Brief No. 55 prepared for the National Diet Library (6 Dec. 2006) at http://www.ndl.go.jp/jp/data/publication/issue/0555.pdf.

Tanaka, Yoichi, and Hiraswawa, Ryo (1996) 'Features of Policy-Making Processes in Japan's Council of Science and Technology', *Research Policy*, 25: 999–1011.

Thomson Reuters (2009) Statistical Analysis by M. Negishi on 'National Citation Report (NCR) for Japan'.

University of Tokyo Data Book (various years) (Tokyo: University of Tokyo Press; issued annually in English and Japanese).

II

Reorganizing Research Organizations

Shifting Authority Relations between Teams, Departments, and Employers

5

Informed Authority?

*The Limited Use of Research Evaluation Systems
for Managerial Control in Universities*

Jochen Gläser, Stefan Lange, Grit Laudel, and Uwe Schimank

1. The New Information Needs of Universities

The university is a key authoritative agency in the public sciences that is facing increasing political demands to improve research performance. In many countries, national research evaluation systems (RES) have been institutionalized that hold universities accountable for their research by assessing its quality and feed back this information to universities through public comparative assessments of research quality and/or by differential funding based on quality assessments. In all cases, universities receive both a signal that a major authoritative agency—the state—expects high-quality research and information about the extent to which they fulfil this expectation. From the perspective of the universities, this amounts to increasing pressure to improve their research (Whitley and Gläser 2007).

At the same time, governance reforms in many countries have increased the autonomy of universities from the state and strengthened the position of their central administrators *vis-à-vis* academics conducting research. Comparative studies have shown that, although governance reforms occur at different rates in different countries, there is a distinctive trend towards stronger hierarchical management in universities, which is accompanied by a weakening of academic self-governance (Schimank 2005). In many countries this amounts to a paradigm shift in authority relationships, with authority concerning decisions on education and research shifting from the state to universities (Lange and Schimank 2007; Paradeise *et al.* 2009).

As a result of these two developments, university managers experience both pressure to improve research performance and increasing formal opportunities to 'manage' research activities. However, effective management of research requires accurate information about research. This is not to say that no management decision can be made without such information. The sociology of organizations tells that rational management based on accurate information is by no means commonplace (Isenberg 1984). Universities are no exception here—after all, it was a university that provided the inspiration for the garbage-can model of organizational decision-making (Cohen *et al.* 1972).

Having said this, we must also note that acquiring and processing information is a crucial process for organizational control. In order to maintain and improve their operations, the managers of organizations constantly need information about both their internal activities and their environment. Universities are increasingly expected to improve their research, and their managers' authority to do so is growing—but do they have the information needed?

This information is partly produced by the RES themselves. By assessing research quality, RES not only create incentives to improve research but also provide information that can be used by universities in that very process. This function of RES has so far been insufficiently appreciated, although it has surfaced in the distinction between summative and formative evaluations, in which the latter are described as being aimed at assisting units to improve their research performance (Geuna and Martin 2003: 278).[1] In political discussions of RES, their use for university managers has been mentioned in a critique of the British plans to replace peer review by an indicator-based assessment procedure. According to Bekhradnia (2008), the indicator-based procedure would deprive universities of the information they need to improve their research.

The informational yield of different RES for university management varies, as do the uses to which this information can be put. The aim of this chapter is to suggest how the different evaluation techniques used in RES affect their informational yield for the exercise of authority by

---

[1] Unfortunately, the distinction between formative and summative evaluations (those 'making judgments about the performance of a unit by comparison with similar units' (Geuna and Martin 2003: 278) suggests that summative evaluations neither have the purpose to support units' improvement of their research nor do actually contribute to such an improvement. Our chapter will demonstrate that this suggestion is misleading.

university management. Our analysis is based on the comparison of five RES, which between them cover the major variations in the autonomy of universities, purposes of RES, evaluation procedures, and informational yield.

2. Analytical Framework

The Contingent Authority of University Management

As managers of organizations we might expect university managers to set goals, to create conditions for task performance, to monitor goal attainment, and to respond to the outcomes of this monitoring by changing conditions of work. However, in practice the authority of university management in most public science systems varies between 'almost non-existent' to 'limited', for three main reasons. First, even though research is one of the core activities of the research university, it is only loosely coupled with the university's authoritative structure. Universities provide researchers with some material resources and an organizational frame that serves as interface between research and society. However, they are not the actor that produces knowledge even though they provide an important social context for research. Goal setting, choice of methods, collaboration, the acquisition of critical resources (information), integration of results, and quality control are all primarily governed by the scientific communities to which academics contribute. Most of these communities transgress organizational and, usually, national boundaries.

Owing to their specific decision structure, scientific communities can produce new knowledge regardless of the uncertainty inherent to that process, which includes uncertainty about the problem, the existence of a solution, the approach to the solution, the resources required for solving the problem, and the meaning of the solution (Gläser 2006; 2007). For the same reason, only senior academics working in the same field as their colleagues can exercise effective authority over research activities. The impossibility for university management in general of 'managing' problem definition and selection, evaluating research strategies and techniques, and assessing the intellectual value of results limits the extent to which universities can turn into strategic actors (Whitley 2008: 24–6)—by which they can move beyond the rhetorical construction of 'actorhood' that has been discussed by Weingart and Maasen (2007).

Beyond this principal limitation for university management, its authority and action capabilities depend on the way in which universities are formally institutionalized (Whitley 2008: 26–31). The authority of universities depends on their autonomy *vis-à-vis* the state and *vis-à-vis* their academics, and thus varies considerably between countries. Simplifying the distinction introduced by Whitley in this volume, the poles of the spectrum are the 'hollow organization' (the traditional German model) and the 'employment organization' (mostly established in the Anglophone countries). The former was in place in Germany, Austria, several Nordic countries, and Japan in its pure form until a decade or so ago. It is characterized by largely powerless universities, whose personnel matters, budgets, and resources are controlled by the state. The high autonomy of university professors, who were appointed and equipped by the state, prevented universities from interventions in matters of teaching and research. Although most countries featuring the traditional German university model have initiated reforms in order to increase the autonomy of universities, many of these limitations still exist. In contrast, the 'employment organization' model features much more formally autonomous universities which have considerable control of budgets, capital investment, personnel, and salaries, as well as their internal structure and the courses taught. In many such university systems, internal hierarchies are replacing more collegial forms of governance (Clark 1998).

A third limitation faced by all universities is a consequence of universities' limited access to 'their' research. In order to 'manage' research effectively, universities would need information about the organization and direction of research. Obtaining such information is difficult because work processes are embedded in scientific communities rather than being governed by formal organizations, and are opaque to university management. Most information about the research process, the quality of conditions of work, and the quality of results is generated in the course of conducting research. However, this information is largely internal to the work process, idiosyncratic, and understandable only to other academics—in most cases only to those from the same field.

Thus, the increasing pressure for universities to produce high-quality and societal-relevant research highlights a major problem for them: they need to be able to affect research processes but cannot generate the information to do so. This problem is aggravated in some countries by the limited formal authority and capabilities of university management, but exists in all countries because of the specific way in which the production of scientific knowledge is embedded in universities.

Research Evaluation Systems as Sources of Information

The emerging RES not only create incentives for universities, and thus their demand for information about research, but are also an important source of such information. RES are created for three purposes: (1) to inform universities about the quality of research, (2) to provide incentives for improving research, and (in many but not all cases) (3) to improve research by redistributing resources from weak to strong performers. The information they generate can be used for university management decisions. However, for most RES this is a side effect rather than their main purpose.

RES differ widely in the information they use as input, assessment procedures, and their output of information about research quality (see Gläser 2008 for an overview). The most important distinction is that between RES based on peer review and indicator-based RES. The former employ an assessment of research that is conducted by colleagues working in the same field as the academics whose research is evaluated, while the latter use quantitative data about research and its outputs. The two kinds of RES are easy to tell apart because even though peer-review processes often also use quantitative information on research performance, indicator-based systems are characterized by the absence of peer judgements, as will become clear from our case studies.

RES that are based on peer review take the view that a researcher's peers are the only people who can judge the quality of research by analysing its content.[2] Assessment of content is achieved by requesting the submission of research outputs (mostly publications) which are read by the assessors. In most cases, contextual information including statistical information on staff, resources, external funding, and outputs is used as additional input to the peer-review process. This information is provided for 'units of assessment', which might be constituted by all research of a university in a certain field, an organizational sub-unit of the university such as an institute or research group, or an individual academic. Apart from reading submitted research outputs, peer review procedures may include interactions with the evaluated units such as interviews of scientists or site visits. The outcomes of peer review-based evaluations are commonly provided as formalized ratings of the 'units of assessment' in one or more dimensions,

---

[2] The first RES ever—those of the Netherlands (introduced in 1983) and the UK (1986)—utilized peer review as the assessment method. The British RES—the Research Assessment Exercise (RAE)—has been copied (with varying degrees of match) by Hong Kong (first review in 1993), New Zealand (2003), Italy (2004), and Australia (planned for 2008 but abandoned by the new government that was elected late in 2007).

which make the assessments comparable across universities and disciplines. The ratings are usually accompanied by a short text that explains the judgement.

In contrast, indicator-based RES rely on quantifiable properties of the input, process, or output of research as proxies for research quality. Countries currently using indicator-based RES include Australia, Germany (at the level of federal states), Ireland, Norway, and Belgium (the Flemish region). The most frequently used indicators are

- the amount of external funding, which is seen as an indicator of quality because winning external grants usually depends on the project passing a peer review;
- numbers of publications, which are sometimes weighted by the quality of publication types (such as 'international, peer-reviewed journals'); and
- numbers of Ph.D. graduations.

These indicators obviously measure the amount rather than the quality of research. Their use can be explained by two of their properties: the opportunity to easily collect the information about them, and their applicability to most fields of research. The former property makes indicator-based RES a cheap alternative to the expensive peer reviews, while the latter is essential for applying homogeneous RES to universities covering all fields of learning. More sophisticated and more valid indicators such as those based on citations cannot be currently applied to all fields, and require expensive methods of data collection and data cleansing (van Raan 1996: 403, 405).

Variables

Our analysis compares the informational yield of each of the RES and the use universities make of this information. Since university management is by and large unable to prescribe inputs, procedures, or outcomes of the research process, its opportunities to influence research are limited to shaping some of its basic conditions and to managing some of the behaviour of its actors—the researchers. These opportunities arise mainly in three kinds of decision. *Internal resource distribution* allocates the university's block funding to sub-units of the university (faculties, schools, and individual academics) in order to maintain the infrastructure for teaching and research tasks. A second major task of the university management is internal *(re)structuring of the university*. Much of this restructuring occurs because of changes in student demand (which still is the major basis of funding for universities). However, RES and in particular the performance-based allocations of block grants create strong incentives for universities to change their structures of research as well. A third kind of decisions aimed at improving research performance concerns the *management of individual performance*. This includes hiring decisions, the use of probationary periods, recommendations to academics about their conduct of research, decisions on tenure and promotion, and other uses of incentives.[3]

While these decisions mostly utilize different information, they share the requirement for information about research performance and about favourable conditions for increasing performance. This kind of information is—to varying degrees—provided by RES. In order to compare the informational yield of RES, we use five dimensions: (1) richness, (2) timeliness, (3) validity, (4) legitimacy, and (5) comparability. These are, as will be evident below, partly derived from the organizational sociology literature on management information processing. The evaluation techniques can be ranked according to their yield in each dimension (Figure 5.1). With the exception of timeliness, the rankings must be based on relative rather than absolute scales because no objective 'yardstick' exists against which richness, validity, legitimacy, and comparability of information could be assessed.

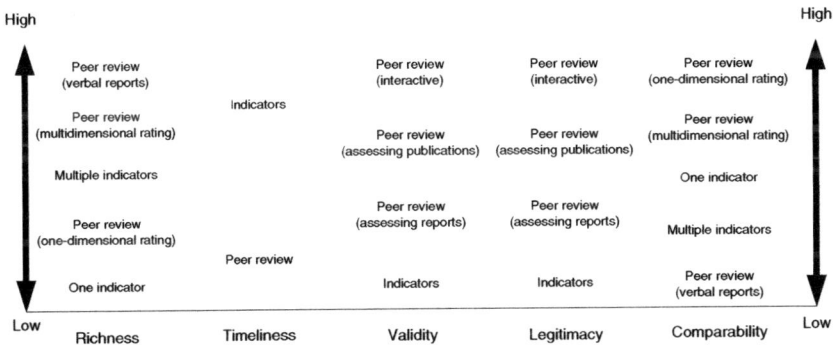

Figure 5.1. Ranking of information properties in the five dimensions

---

[3] These decisions are often only analytically distinguishable. For example, the hiring of a senior academic is likely to simultaneously have resource, structural, and performance management aspects. As we have discussed in the previous section, the extent to which university management has the formal authority to make these decisions varies between countries.

RICHNESS

The richness of information provided by different communication channels is a central concern of the organizational sociology literature. While it is possible to distinguish between the amount of information and the richness of the channels through which it is provided, this is not necessary in our particular analysis. We define the richness of information as the number of different aspects about which information is provided, thereby synthesizing the concepts 'amount of information' and 'richness of communication channels' that are sometimes discussed separately in organizational sociology (Daft and Macintosh 1981: 210; Daft and Lengel 1984: 195–8; 1986).

The ranking of RES according to the richness of information they supply can draw on comparisons of communication media richness (Daft and Lengel 1984: 195–8). Peer reviews have the potential to provide very rich information. In particular, peer reviews—and only peer reviews—can provide information about a unit's potential for future high-quality research. Given the right input (such as research programmes) and procedure (especially site visits), assessors are able to draw inferences about the capabilities of researchers, the suitability of their equipment, and the fit between these conditions and the research programme. However, the design of peer reviews in RES often reduces the richness of information. We rank peer reviews that provide detailed verbal reports higher than those who provide ratings or rankings in several dimensions. A set of indicators that provides information about several aspects of research performance is less rich because the numerical values are lacking the link to research performance, which has to be established by *ex-post* interpretations. Among RES using peer-review-based ratings or sets of indicators, the richness of information decreases as the number of dimensions (indicators) declines.

TIMELINESS

The timeliness of information has received far less attention than the amount of information or media richness.[4] Apparently, the literature implicitly assumes that management always can acquire the necessary information about its technologies, and has to cope with whatever is the timeliness of information provided in the organization's environment. The case of research organizations that obtain information about their

---

[4] But see Choudhury and Sampler (1997) on the 'time specificity' of information.

technology by collaborating with their environment has not yet been taken into account.

In one sense, the timeliness of information can be considered to be an aspect of its validity—the older the information, the less likely it is to adequately depict current research quality. This aspect of timeliness depends on the evaluation cycles of RES, which vary considerably. The longer the period between two assessments, the more outdated the information used in current management decisions becomes.

However, timeliness has a dimension that merits separate treatment: namely, the limitation of timeliness that is produced by the input in evaluations. The material that is evaluated always represents past research. Peer reviews that rely on the examination of submitted publications assess the quality of research whose results were 'written up', reviewed, possibly revised, and thus finally published after months or years. Indicators such as numbers of publications or of Ph.D. completions have the same problem.[5]

VALIDITY

The validity of research evaluation techniques is an issue that is not well understood, and is highly contested and difficult to separate from issues of legitimacy. The major fault line in the discussion about evaluation techniques runs between an argument for peer review as the only way to evaluate the content of research directly and arguments pointing to peer review's opacity and susceptibility to bias, favouring quantitative indicators as 'objective' methods that do not depend on personal judgements of few assessors. In order to be able to include validity as a dimension of informational yield at all, we disregard problems of bias (which reduces validity but is a random rather than systematic property) and opacity (which hinders the measurement of validity rather than reducing it). The validity of management information has not been discussed in the literature except for occasional references to information accuracy (for example, Saunders and Jones 1990: 38). However, it can be linked to information accuracy and one

---

[5] If citation-based indicators were used, the situation would be even worse because for citations to occur some time must pass for a publication to be read, used in other research, and be cited in subsequent publications. The only exception is the indicator 'external funding', which relies on *ex-ante* evaluations and thus refers to research that is being undertaken when the data are collected. This relatively high timeliness is achieved by this indicator's reliance on prospective peer reviews. Contrary to all other evaluation methods and indicators, the indicator 'external funding' reflects an evaluation of plans for the research rather than accomplished research.

of the central themes of management information processing: namely, the equivocality of information. Equivocality can be defined as 'the multiplicity of meaning conveyed by information about organizational activities' (Daft and Macintosh 1981: 211, see also Weick 1979). Both unknown validity and low validity of information about research performance increase the information's equivocality.

If we treat the evaluation embedded in RES as an interrogation of the research about its quality, the ranking of media richness (Daft and Lengel 1984: 195–8) can once more be used to assess the RES. RES based on peer review generally rank higher than those based on indicators, because the former use at least written verbal reports, while the latter just use numbers. Among the peer-review-based systems, those including personal communication between evaluators and the evaluated (interviews or site visits) rank highest because they use a medium of higher richness. The rather simple quantitative indicators currently used in RES partly reflect volume rather than quality (in particular Ph.D. completions and numbers of publications) and in any case reflect factors other than quality (for publications, see Butler 2004; for external funding, see Laudel 2006). Their validity is generally considered to be low (Jauch and Glueck 1975: 70–3; Phillimore 1989; Wood 1989).

LEGITIMACY

Legitimacy is related to the norms and culture of a social context, which is why the legitimacy of the information produced by an RES varies between the contexts of science policy, university management, and academics. For example, quantitative indicators often carry a higher legitimacy in science policy circles because they are considered to be transparent and unbiased (see above, on validity).

For our investigation of the informational yield of RES for university management we focus on the legitimization of management decisions in the scientific community. Here, the legitimacy of information about research quality is strongly tied to the dominant view of the validity of that information, which is that peer reviews produce the most valid information on research quality, and that measures that increase the validity of peer reviews also increase their legitimacy. Indicator-based information on research quality is generally considered a less legitimate basis for decisions than peer review, although some disciplines (especially the biomedical sciences) have begun to rely on citation-based indicators (which are not used in RES).

COMPARABILITY

Since many decisions in universities affect more than one field of research, information on research quality and conditions for research also needs to be comparable. Comparability is highest when peer reviews are designed to produce ratings or rankings. Quantitative indicators are less well suited for comparisons because of their dependence on unit size and field (for example, field-specific publication practices and dependence on external funding). Verbal reports on just one discipline, which are the outcome of some peer-review procedures, have the lowest comparability.

3. Informational Yield of RES and Internal Use of that Information by Universities

Five Case Studies

For our empirical analysis we have selected cases which exhibit different characteristics in terms of RES and autonomy of universities (Table 5.1). Both quantitative indicators and peer-review-based RES are represented in our sample. Australia has the oldest indicator-based system for funding university research, which was introduced in the second half of the 1990s. In Germany, many federal states have begun to introduce indicator-based systems over the last decade. We investigated universities in one of these states (which, in order to protect the privacy of interviewees, we cannot name). The countries featuring peer-review-based RES include those with the two oldest RES—the Netherlands and UK—and the German state of Lower Saxony, which has institutionalized a peer review-based RES.

The second dimension that is important to our analysis is the autonomy of universities, which influences the latter's opportunities to act on the information they receive from RES. German universities still have a relatively low autonomy, while the autonomy of Australian, British, and Dutch

Table 5.1. Cases included in the analysis

	Autonomy of universities	
	Low	High
Evaluation technique		
peer review	Germany, Lower Saxony	UK, Netherlands
indicators	Germany, State X	Australia

universities is high.[6] Table 5.1 demonstrates that all four cells of the cross-tabulation of evaluation technique and autonomy of universities are included. The UK and the Netherlands provide an interesting additional contrast because the Netherlands abandoned the link between evaluation and funding after the first round of evaluations, and have transferred responsibility for the evaluations to the universities.

Peer Review-Based RES and Universities with Low Autonomy: Lower Saxony[7]

In 1997, an 'Academic Advisory Council' (AAC) was established in the German federal state of Lower Saxony, and was tasked with the organization of the evaluation of all of Lower Saxony's university research. The evaluations are conducted as discipline-oriented peer reviews. Units of assessment are departments or institutes within universities and 'research units', which are self-defined by researchers according to local and disciplinary conditions. A 'research unit' can range from a team of scientists (such as in the natural sciences) to an individual chair (such as in the humanities). The usual evaluation procedure is as follows. A short framework paper is provided by the AAC to the evaluated disciplines within the universities to help them prepare a report on the last five years of research activity and future planning. Universities are then visited by the group of evaluators; approximately six professors from the evaluated discipline (but from other German federal states or from foreign universities). These evaluators talk to the university president, the respective dean, each professor of the discipline, some members of scientific staff, and some doctoral students, and discuss their findings. Based on these discussions, a draft report about the discipline and its relative performance at all Lower Saxony's universities is written by the evaluators and edited by the AAC's officer in charge. The evaluated units and individuals are then asked for their comments, which reach the AAC via the president of the university. On this basis, the final

---

[6] There are significant variations within both groups. The autonomy of many German universities has increased since the time of our investigation (2005–6), albeit in an uneven pace that depends on the higher education legislation in the sixteen federal states. The government's control of universities is significantly stronger in the Netherlands than in the UK or Australia. However, the basic distinction between universities that have full control of their recruitment, internal structure, internal allocation of funds, and human resources management (universities in Australia, UK, Netherlands) and those that do not have this control (German universities) still holds.

[7] This section is based on two empirical studies of the evaluation process and responses by universities (Schiene and Schimank 2007; Chapter 7). See Chapter 7 for a discussion of the impact of this evaluation process on authority relations.

report ('assessors' report), which contains evaluations and recommendations, is written and submitted by the evaluators to the AAC. The AAC discusses the report and its recommendations. The report is published with the exception of the evaluations of individuals, which are given to the individuals and their university presidents, with a complete copy sent to the Ministry. There are several follow-ups on the evaluation reports, including an intermediate report after three years.

The evaluation reports include detailed verbal judgements of the research performance and potential for future performance of a discipline in each of the universities of Lower Saxony. The recommendations are also quite specific, and deeply intrude into the structural and resource allocation decisions of universities. The major kinds of recommendation can be listed as follows:

(1) establishment of new professorships, rededication of vacant professorships within the discipline, elimination of vacant professorships or their transfer to a different discipline;
(2) participation of external peers in the recruitment commissions for vacant or new professorships;
(3) additional scientific staff for professorships;
(4) reduction of permanent scientific staff in favour of temporary employment contracts with younger scientists;
(5) study programmes for postgraduates;
(6) additional financial means from the government;
(7) a more performance-oriented allocation of block grants within the university or faculty;
(8) increased acquisition of project grants or research contracts;
(9) infrastructural improvements of buildings, libraries, laboratories; and
(10) intensification of internal and external coordination and cooperation.

The Ministry asked universities to implement these recommendations. The implementation is under way in most cases, although in some it had to be started against the will of the affected faculties or institutes, whose self-perception was completely different from the evaluation. Recommendations were also turned into strategic goals that became part of 'performance agreements' between universities and the Ministry. Even though the evaluated units disagreed with the assessments in some cases, the whole evaluation procedure was perceived as legitimate because it was conducted under the authority of the scientific communities. As a result, the position of the university leadership was strengthened by the double support from the

Ministry demanding the implementation of recommendations and the scientific community legitimizing them.

The paradoxical consequence of the whole evaluation process was that the autonomy, action capabilities, and authority of the university management were strengthened by a procedure that deeply intruded into university matters. The major reason for this paradox is the double limitation of the autonomy of the German university. German universities are not only dependent on the state, which still sets tight frameworks for most of the essential management decisions such as employment contracts and resource allocation, but are also unable to intervene in decisions of their university professors, who are appointed by the state as public servants and have a guaranteed personal budget. If German university professors decide that they do not want change, it is very difficult for a university leadership to achieve it—the more so when the professors of a faculty collectively decide that they do not want change.

In this situation, a demand for change by the Ministry that is legitimized by the scientific community significantly enhances the authority and capabilities of the university management. Assuming that the university leadership wanted to improve research and to build externally recognizable research profiles, and that it would have needed to conduct peer reviews to support these actions, the enforced recommendations of the evaluation procedure could solve most of the problems created by the university management's limited action capabilities.

While the legitimate, detailed, and intrusive recommendations of the peer review in Lower Saxony might seem to be a good solution under the circumstances, several problems need to be mentioned. First, the evaluations and the demand by the Ministry to implement the recommendations were accompanied by severe cuts in the block grants, which not only made the implementation of many recommended changes impossible but also undermined trust in the whole evaluation process. Secondly, the funding cuts occurred in a situation of still increasing teaching loads, which made the implementation of changes for the promotion of research even more difficult. Thirdly, the evaluation procedure effectively disaggregated the universities. Disciplines were evaluated at different times and by different groups of assessors, none of which took the whole university into account beyond the contributions to the process made by the university leadership. Therefore, it is at least open to question whether a more holistic look at the university would have led to a different consideration of context and local knowledge, and thus to different recommendations. As it was, each disciplinary panel by and large defended its discipline in each university. Only one type of recommendation—to transfer a vacant professorship to a different discipline—does not strengthen the evaluated discipline, but another one. Being 'altruistic' from the point of view of the evaluators as disciplinary peers, such a recommendation is of course only given in 'hopeless cases'; that is, very rarely.

Peer Review-Based Devolved RES and Highly Autonomous Universities: The Netherlands[8]

In the Netherlands, evaluations based on peer review are conducted according to a standard evaluation protocol. The previous standard evaluation protocol (valid to 2002) organized disciplinary evaluations at the national level; that is, all disciplinary units (programmes) in a discipline were evaluated at the same time.[9] With the new standard evaluation protocol (from 2003), evaluations are more devolved. Under the current protocol, the universities themselves organize the evaluation of their research. A self-evaluation is prescribed after three years and an external peer review after six years. The procedures for evaluating the units remained the same. The units of assessment—'programmes' (research groups) or university institutes—submit self-evaluation reports and lists of publications to the peer-review committees. These committees form their assessment on the basis of the reports, of an examination of the submitted publications, of interviews with the programme leaders, and (in some cases, particularly in the laboratory sciences) of site visits. The programmes are evaluated in four dimensions (evaluation aspects): scientific quality, scientific productivity, scientific relevance, and long-term viability. These aspects are translated into specific evaluation criteria for each discipline by the peer-review committees. Each programme is rated on a five-point scale (excellent/good/satisfactory/unsatisfactory/bad) in each aspect. These ratings are briefly justified (with half a page of text).

The major change that was introduced with the new standard evaluation protocol from 2003 concerned the comparability of information from the RES. This comparability was already limited with the previous evaluation

[8] This section is based on publications on the Dutch RES and university management by Westerheijden (1997); Jongbloed and Van der Meulen (2006); CPB and CHEPS (2001), VSNU *et al.* (2003); and Meulen (2007). We are grateful to Pleun van Arensbergen, Rathenau Institute Den Haag, for her support of the Dutch case study.

[9] The definition of 'programmes' goes back to the first evaluation exercise that was initiated in 1979 and conducted in the early 1980s. In this first round of peer reviews, the university had to define 'research programmes' whose funding was conditional on an *ex-ante* peer review. Since then, the 'programme' has been the basic unit of the Dutch RES (Meulen 2007).

procedure. Since the evaluations were conducted for each discipline at a different time, the universities never had information of equal timeliness on all of their research. With the new standard evaluation protocol, evaluations are initiated and organized by universities for their research in a certain field. With the exception of a few cases where universities agreed to conduct a joint evaluation for all research in one discipline, most of the evaluations since 2003 produced information on just one unit of assessment, which could only be compared to other evaluation results from different points in time.

Nevertheless, university managers welcome the information provided by the RES as legitimizing the differential treatment of research groups and as a necessary input for this new approach. The actual use of the information has varied widely between universities. One university translated the results into a quantitative measure for the allocation of 10 per cent of its research budget. The formula developed by that university completely disregarded two of the four dimensions (scientific relevance and long-term viability). The other two become synthesized in one weighted measure (75 per cent quality, 25 per cent productivity). The university then ranked the units from all universities in each discipline, and determined the relative position of the University of Tilburg's unit in each discipline by dividing its absolute rank by the numbers of units. The resulting figure was used to compare units from different disciplines and to redistribute resources between them.

This straightforward and simplifying utilization of evaluation results is an exception. In other universities, evaluation results have indirect financial consequences because they are taken into account in budget negotiations between research groups and the faculty. Furthermore, evaluation results are considered when the directions of research and research strategies are developed. Thus, evaluation results are one of several inputs to negotiations about research conditions, such as research budgets or reduced teaching loads.

The flexibility of this approach is illustrated by the case of yet another university. Instead of simply rewarding good and punishing bad research groups, university management took the importance of a field to its discipline into account when decisions about the future of research groups were made. Therefore, high scores in the evaluations were no guarantee of further prosperity. When financial cuts had to be made, even highly evaluated groups were closed in the ensuing reorganization. On the other hand, a group that received a low score but was considered to be important for the discipline by the faculty received extra funding as part of attempts to strengthen the group.

The general perception of researchers is that the results of evaluations do not have severe consequences. Good scores provide a certain protection from administrative intervention (reorganization) and a relative advantage in budget negotiations within the university. Bad scores may lessen this protection but have no automatic consequences either. These findings indicate that the university management at Dutch universities is quite active in terms of restructuring the universities, and that the evaluation scores partly help redefine the targets for these measures. It must be noted, however, that these findings are still preliminary and lack both detail and reliability. The responses by Dutch universities appear to be quite complex and merit further detailed investigation.

Centralized Peer Review and Autonomous Universities: United Kingdom[10]

The British RES was introduced in 1985 as a 'Research Selectivity Exercise' as a response to growing concerns that the quality of the British research base could not be maintained in its entirety, especially after severe cuts in university block grants in the early 1980s (Chapters 2 and 8). For each unit of assessment, universities submit up to four 'research outputs' (mostly publications) of every research active academic and contextual information including data on external funding, prizes and awards, and graduate students. The core of the evaluation procedure is an assessment of the submitted 'research outputs' by the members of the assessment panel. Examination includes reading a significant proportion of the submitted publications. Based upon the examination of publications and the analysis of contextual information, the panel arrives at a judgement about the quality of the research of a unit of assessment. In 2001, assessments took the form of a rating between 5* (international excellence in more than half of the publications, national excellence in the other publications) and 1 (no national excellence in any of the publications). In 2008, the format of the assessment results changed. A slightly changed rating—from 4* ('world-

---

[10] This case study is based on the literature on the RAE including commissioned reviews and reports (Roberts 2003; evidence 2005, 2006), news reports (for example, Curtis 2002; Johnston and Farrar 2003), sociological analyses (for example, Morris 2002; Lucas 2006), reports from academics from various disciplines (for example, Dainty *et al.* 1999), and internal documents of British universities (for example, Queen Mary University 2003; Oxford University 2005; Northumbria University 2007). See also Chapters 2, 8 and 9.

Overall quality profile					
Quality level	4*	3*	2*	1*	u/c
% of research activity	15	25	30	20	10

Research outputs				
4*	3*	2*	1*	u/c
10	25	40	15	10

eg 70% (Minimum 50%)

Research environment				
4*	3*	2*	1*	u/c
20	30	15	20	15

eg 20% (Minimum 5%)

Esteem indicators				
4*	3*	2*	1*	u/c
30	25	10	20	15

eg 10% (Minimum 5%)

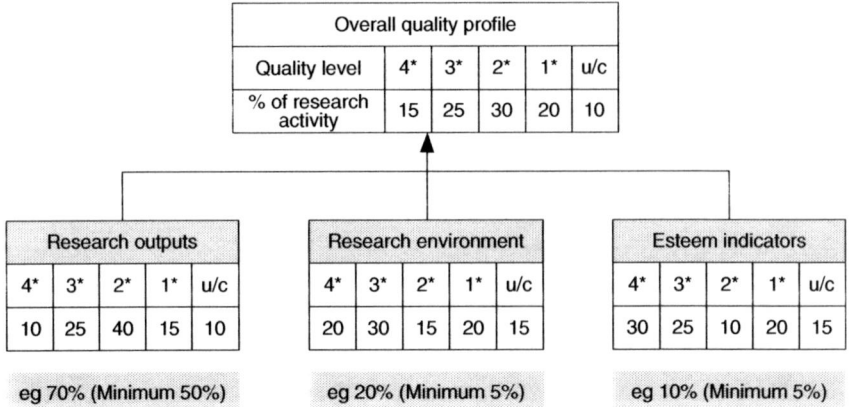

Figure 5.2. The construction of a unit's quality profile in the 2008 RAE (source: HEFCE, 2008: 99)

leading in terms of originality, significance and rigour') to 'unclassified'—was kept. However, instead of one amalgamated rating the panels issued 'quality profiles' that describe the distribution of a unit's research outputs, esteem indicators, and research environment across the quality categories (Figure 5.2). Thus, rather than receiving a '4*' or '4' for a unit of analysis, the university received a statement saying that 30 per cent of a unit's research is at the 4* level, 50 per cent at the 4 level, and so forth. This change in procedure avoided the 'cliff edge' effect produced by the enormous consequences of the small quality differences at the boundaries of rating levels. At the same time, universities received more complex information about the performance of their units of assessment.

Since its inception, the RAE's outcomes have informed the distribution of block grants for university research. Although on average the block grants for research constituted 7.5 per cent of all universities' income in the financial year 2001–02, they varied between 0 and more than 23 per cent.[11] The variation between universities is enormous. Some universities obtain half of their funding from the research block grants, while many others do not receive anything.

---

[11] It is difficult to obtain comparable data on the income of UK universities. Our estimates are based on data from the websites of the UK's Higher Education Statistics Agency (http://www.hesa.ac.uk/index.php/content/view/807/251/, accessed Oct. 2009) and from HEFCE (2004).

Given the actual or potential importance of the research block funding for universities it is not surprising that universities have responded strongly to the RAE.[12] Observers agree that the RAE has moved research into the focus of university management and turned research from an 'unmanaged' into a 'managed' activity (Lucas 2006: 73). This management of research performance has two targets, which can be analytically separated.

A first target of university management is to decide who is submitted in the RAE. Being in full control of the information submitted to evaluation panels, universities try to improve their RAE grades by shaping the material submitted to the evaluation. Careful internal evaluations and 'dry runs' of the RAE are conducted in order to decide on units of assessments to submit and to select the academics who should be submitted as research active. The trade-off faced by universities is that submitting a lower number of academics as 'research active' may improve the grade but, once a grade has been achieved, the amount of funding depends on the number of academics submitted. A second, much-discussed strategy of university management is head-hunting—the search for outstanding academics whose high-quality output is likely to improve a university's grade when added to the submission. Universities also strategically change their structure by closing unsuccessful departments, reshaping and relabelling their research in order to submit it to a different unit of analysis, and combining stronger and weaker units in order to utilize a halo effect of the former.

These three strategies do not only target the evaluation process but are part of the more fundamental attempts of university management to strategically shape the content and quality of research. The strategies applied for that purpose include, foremost, the allocation of recurrent funding. There is a general tendency for universities to follow the results of the RAE—to allocate funding 'as earned'. However, universities also create strategic funds that are used to strengthen weaker units and enable an improvement of their grades. An important aspect of the distribution of resources is the redistribution of time for research. Successful researchers whose output is important for achieving high grades are supported by a reduction of their teaching and administration tasks, which are assigned to

---

[12] There are indications for systematic differences between management approaches of research-intensive and other universities. For example, research-intensive universities submit most of their academics (more than 90 per cent) as research active, thus having no need for the strategic selection of academics to submit, and little opportunity to redistribute teaching and administrative loads. Unfortunately, there is not enough systematic research on university responses to the RAE for these variations to be reliably established, and their impact on research and teaching in universities to be assessed.

their colleagues who are not deemed good enough to achieve the aimed-for grade, and are therefore not categorized as 'research active'.

Finally, British universities apply procedures for managing the research performance of their academics. Research performance is monitored by using quantitative indicators, among which publications play an important role because they are the core of RAE submissions. Research performance is the subject of annual performance talks between academics and their supervisors, and plays an important role in decisions about appointing and promoting academics. In these decisions, research performance appears to have gained a higher weight than teaching performance (Parker 2008).

Indicator-Based RES and Universities with High Autonomy: Australia[13]

Australia has used an indicator-based system of allocating research block grants since the mid-1990s. As it currently stands, the system distributes about 7.9 per cent of the total income of universities according to external competitive funding (indicator weighted at 54.8 per cent), numbers of Masters and Ph.D. completions (29.1 per cent), numbers of publications (8.4 per cent) and current Masters and Ph.D. students load (7.7 per cent) (own calculations based on DEST 2007). The distribution is a competitive zero-sum game in which universities must participate, and the share of the research block grants in their income varies between 0 and more than 15 per cent. All universities have mirrored the external funding formulae in the internal distribution of research funding to faculties and in some cases from faculties to schools. However, while using the indicators from the external formula, some universities have given different weightings to the indicators. This was deemed necessary because within universities the indicators inform a distribution of resources between disciplines. Simply copying the external weightings of indicators would disadvantage the social sciences and humanities because of their systematically lower external funding. This is why this indicator's weight is reduced and the weight of either research student completions or of numbers of publications is increased.

These internal resource allocation systems are of little direct consequence for research projects because the money allocated is largely used to pay part

---

[13] This case study is based on an empirical investigation of the Australian system of indicator-based research block funding of universities (Gläser and Laudel 2007), which is part of a comparative project including Australia and Germany (Gläser *et al.* 2008; see also Ch. 10).

of the academics' salaries and to maintain the basic infrastructure. In most cases there is no recurrent funding of research, regardless of the proportion of a university's income provided by research block grants. Instead, internal grant schemes have been set up by most universities. The grants are rather small and usually limited to one year. They have the function of recurrent funding insofar as they are intended to be used as preparation to acquire external funding, even though they are the sole source of funding for many researchers.

The major effects of the formula-based system concern the attempts of the universities to maximize their income by structural changes, targeted investments, and (to a lesser extent) individual performance management. Structural changes and targeted investments are mainly focused on the single most important indicator used in the funding formula: the amount of external funding. The universities attempt to create research units that are likely to acquire external funding. A characteristic way of doing this is to provide seed funding for 'centres' that is used to buy specialized research equipment or to reduce teaching loads of the leading academics involved. This funding is provided temporarily with the aim to create a centre that can exist on the basis of its external funding after some time (usually three to five years). Since this is rarely possible, it is quite common that the centres disappear after the university withdraws its funding. Apart from these approaches, which are purely quantitative in the sense that they are focused on external grant applications, there are only very few attempts to use detailed information on research quality in internal management decisions. Only two of the seven universities in our sample used peer review-based evaluations of their sub-units (schools or centres). Others applied the quantitative indicators in a rather superficial manner, or had no internal evaluation policies in place at all.

While research evaluation of sub-units played a minor role in most universities, procedures of individual performance management were in place in all of them. At the time of our investigation, the annual performance appraisals of academic staff were inconsequential because they affected neither incremental pay rises nor the distribution of work loads. More thorough performance appraisals were emerging during 2005 and 2006 because the federal government had made additional funding available for universities that implemented them.

The application of the quantitative indicators in performance evaluation schemes across all disciplines caused some problems because they are not equally applicable to all fields. In one university, the classification of academics who did not perform in two out of the three indicators external grants, publications, and Ph.D. supervision as 'research inactive' caused a protest by the department of mathematics. Here, some internationally renowned mathematicians were suddenly classified as research inactive despite their many important publications, because they had no external funding and no Ph.D. students to supervise. We also detected some perverse effects of the indicator-based individual performance assessments, such as academics applying for research grants they did not need for their research or supervising Ph.D. students outside their core area of expertise.

The assessment of individual research performance has played a significant role in decisions on tenure (which took place in one university) and in decisions on promotions in all universities. The major approach to individual performance evaluation for promotions included the performance indicators used in the funding formula but was not restricted to them. Academics who applied for promotion had to submit information on external grants, supervision of Ph.D. students, and numbers of publications. These data were often looked at in context, and supplemented by other data that supported the case for promotion. Universities often use external peer reviews for decisions on promotion—at least where higher levels (associate professor and professor) were concerned.

Indicator-Based RES and Universities with Low Autonomy: Germany—State X[14]

The case of the German universities in federal state X—we call them university A and B—is both special and instructive because the universities in question had limited autonomy and therefore could not use information on research quality, regardless of its content and form. The limitations of autonomy were the same as already described in the case of Lower Saxony: namely, the subordination of universities to the state and the autonomy of university professors.

The universities received a part of their block grant according to a formula that includes indicators of research performance: namely, external funding (1.7 per cent of the block grant) and completed Ph.D.s (0.4 per cent). They thus had an information base for assessing the contributions by their faculties, institutes, and academics to their income. However, this information was not used by the university management for steering or managing

---

[14] This case study is based on the investigation of German universities in the collaborative German–Australian project. For a detailed report on one of the German cases, see Lange (2007).

purposes because there were no legitimate grounds for doing so.[15] Therefore, management activities which take into account the quality of research did so informally. 'The university leadership knows its good performers and treats them accordingly', as the head of administration of university B expressed it.

The *internal funding* of faculties by the university had not changed since the introduction of the funding formula. Funds were still allocated according to a system that was kept secret. The funding of faculties needed to honour the agreements between the state and each individual professor, which guarantee basic supplies consisting of posts for research and teaching associates and assistants, equipment, consumables, and travel.

One faculty in university A and the university leadership in university B had begun to implement performance-based funding schemes that redistributed a small proportion of the recurrent funding.[16] The only instrument at university level was introduced by university B, which rewarded external grant acquisition according to sources of funding. Professors who were successful with their proposals for large competitive grants from the Deutsche Forschungsgemeinschaft (Germany's most important funding agency for university research) received a reward that amounts to 5 per cent of the external grant in the previous year. Successful proposals for graduate schools and funds from selected private foundations could be rewarded with up to 2.5 per cent, and grants from the Federal Ministry for Education and Research received a reward of 1 per cent on top of the grant budget.

In the faculty of social sciences at university A, some money was taken away from professors and turned into small awards for teaching or research initiatives. The rewards for research performance varied between 750 € for presenting a paper at an internationally recognized conference and 8,000 € for leading a successful bid for a large collaborative research grant. Most interesting in the German context is the attempt to steer the publication behaviour in the faculty by rewarding publication in an international top journal with 6,000 €, in a high-ranking journal with 4,000 €, in a good journal with 2,000 €, and in an applied journal with 1,000 €.

The major new instrument for performance-based funding in universities was not implemented in both universities until 2007. A change in higher

---

[15] There are some exceptions. A natural science institute organized peer-review evaluations on its own to boost its international visibility and reputation, while neither the university nor the faculty had established any evaluation procedures.

[16] In both universities the faculties for medical sciences led the way by using internal resource allocation regimes that rewarded both external funding and publication behaviour.

education legislation introduced in 1998 made it possible for the university management to make the basic supplies of a newly appointed university professor subject to a renewal every five years. The renewal shall depend on the professor's performance in teaching and research. However, only a few of the professors appointed after 1998 in universities A and B were aware of such a performance evaluation procedure or the performance criteria for this evaluation.

The universities also attempted to build their research profiles by creating research centres. The purpose of these *structural changes* was to create 'critical mass' or simply to increase the visibility of a certain research area. They were not aimed at increasing research performance, be it measured by the indicators or by any other means. The new centres were by and large administrative layers added to the traditional structure of the professoriate, and had little or no resources of their own. According to our interviews, they did not much influence the content of research. In the natural sciences they were deemed to be useful by professors for the purpose of pooling staff, laboratory equipment, and other resources. In the humanities they had not yet had any effect on research.

Apart from the few attempts at performance-based funding of professors mentioned above, no system of *individual performance management* had been developed within the university. The more recent introduction of performance-based salaries applies only to very recent appointments, and thus falls outside the scope of our investigation.

4. Comparison of the Cases

Informational Yields of RES Compared

'Informational yield' is only one of the properties that need to be taken into account in the design of a RES. The following discussion should therefore not be read as an assessment of RES. In particular, RES utilizing peer reviews are much more costly than indicator-based systems, and this property appears to affect political decisions on the design of RES.

The information outputs of the five RES are obviously different, with another significant variation between the 2001 and 2008 rounds of the British RAE. Our comparison of informational yields is based on the listed five properties of the information and their ranking. Figure 5.3 summarizes the comparison by providing 'informational footprints' for each of the five RES. We use the empirical information from the case studies to 'rank' the

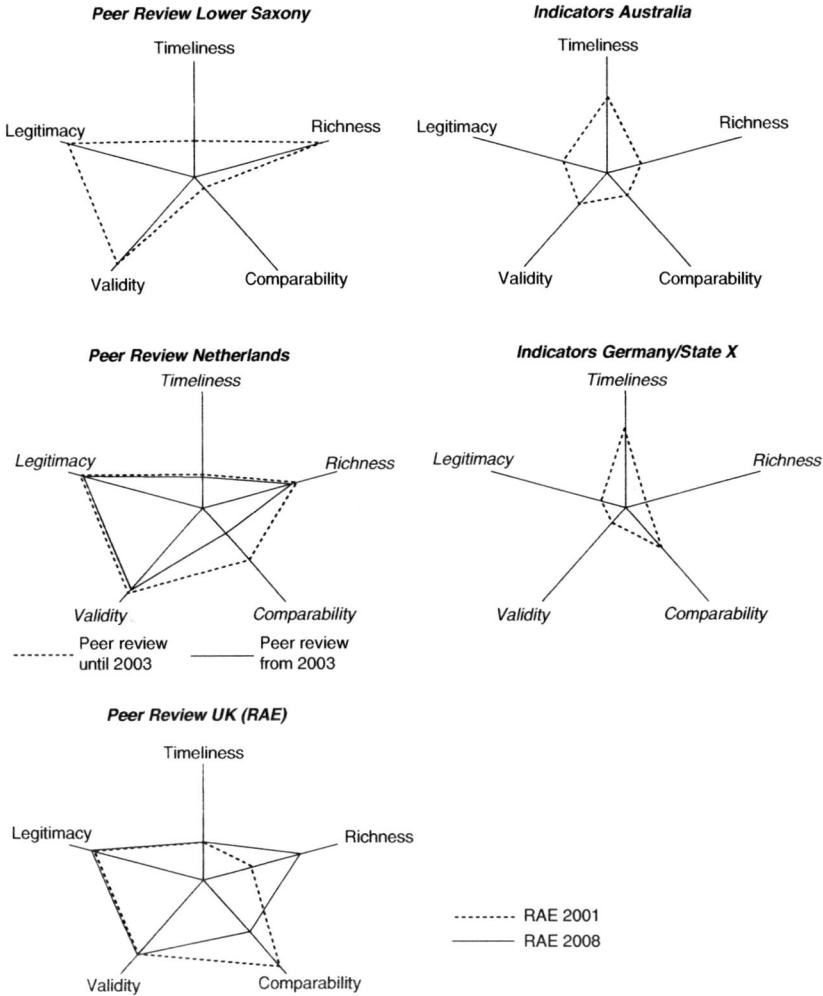

Figure 5.3. A comparison of five RES' informational yields for intra-university decisions

five systems in each of the five dimensions, and summarize the rankings in characteristic 'footprints' for each RES.

RICHNESS

The kind of rich information that is characteristic of peer reviews—detailed, multidimensional, and assessing past performance as well as potential—is only provided by the peer review conducted in Lower Saxony and its 'open-ended' evaluation reports. Since the richness of information increases with the number of dimensions in which units are rated, the four-dimensional rating in the Netherlands is richer than any one-dimensional rating. Quantitative indicators offer little information at all, but the three indicators used in Australia provide richer information than the one-dimensional rating provided by the RAE until 2001. The quality profiles of the 2008 RAE are much richer than the outcomes of previous rounds. They consist of the proportions of research activity that fall into each quality category in three dimensions (quality of outputs, indicators of esteem, and research environment). As a result, it has become impossible to unambiguously rank the universities even in one unit of assessment (see below, comparability). The two indicators used in the German state X provide the least rich information.

TIMELINESS

The timeliness of information provided by peer review-based RES depends on the evaluation cycles. The peer review in Lower Saxony was conducted only once. Meanwhile, the state government decided not to repeat it— much to the regret of all universities' central administrators. The RAE started with a three-year cycle and then moved to distances of four, five, and now seven years between evaluations. The Dutch peer reviews are to be conducted once in six years. In all these cases the information lag already contained in the submitted material becomes aggravated, and universities are soon left with outdated information. The time lags are less severe in the case of indicator-based systems, because information collection is relatively cheap and already occurs annually in many universities.

VALIDITY

In terms of validity, peer-review-based systems are superior to the indicator-based systems currently in use (but see our remarks above on opacity and bias). The more information is gathered in a peer review, the more intense the analysis, and the more the assessors interact with the evaluated academics in order to validate information, the higher the validity.

The rather simple quantitative indicators used in Australia and in the German state X (which also is representative of the indicator-based systems in the other German states) do not achieve any satisfying level of validity. The information provided by these RES, while tolerated as the basis of resource allocation to universities, is therefore ill-suited for internal decision-making about conditions for research (which does not preclude them from being widely used for precisely that purpose).

LEGITIMACY

The legitimacy of the interactive peer reviews in Lower Saxony can be considered to be the highest, while the legitimacy of the information provided by the British RAE is also very high even though it does not include interactive elements. Indicator-based information on research quality as it is currently produced by the RES with their simple indicators is not considered a legitimate basis for decisions. However, indicator-based information carries its own legitimacy when it stems from RES that are used for resource distribution. In the Australian universities the fact that the university receives funding according to the indicators legitimized decisions of the university management about their use for internal resource allocation, while no such argument could be observed in the German universities yet.

COMPARABILITY

The comparability of information was highest for the 2001 RAE, which produced a one-dimensional rating. The outcome of the Dutch RES is a rating in four dimensions, which makes comparisons difficult. The introduction of quality profiles with the 2008 RAE demonstrates the trade-off between the richness and the comparability of information. The introduction of a second dimension (the proportion of research that attained each quality level) makes comparisons dependent on a selection of information or on aggregation procedures (Travis 2009). The comparability of the indicator-based information is low because the numbers are field-specific, and the reference values that are necessary for comparisons do not exist.

The footprints in Figure 5.3 show how peer review-based and indicator-based systems produce quite different kinds of information. The outcomes of peer reviews are considered valid and legitimate. Their timeliness is problematic, mainly due to the large intervals at which they are conducted. Peer reviews have the potential to produce rich information about research, which leads to a trade-off between richness and comparability.

Indicator-based RES are not considered valid and carry little legitimacy. The numbers produced by these RES are neither comparable, nor do they contain rich information about research. Their only advantage is that some of the quantitative information can be produced quickly and cheaply, which enables shorter evaluation cycles and increases the timeliness of information about research.

The Use of Information by Universities

Decision-making in universities is of course a political game in which the information provided by RES is just a resource. Nevertheless, it is possible to say something about the usability of this resource in that game. *Internal resource distribution* is best supported by RES that render their units of assessment comparable both within and between disciplines in one dimension. This is clearly expressed in the attempt of a Dutch university to turn the outcome of the Dutch peer review into a basis for internal resource allocation. The only way to achieve this was apparently to further collapse the multidimensional outcomes of the peer reviews into a single figure.

The importance of comparability for resource distribution that is indicated by this practice is due to an important difference between RES-based funding of universities and funding of sub-units within universities. The former is based on the implicit assumption that universities represent similar mixes of disciplines, while internal funding includes a *redistribution of resources between disciplines*. This aspect of universities' responses to RES has yet to be fully appreciated. The traditional use of peer reviews is limited to internal evaluations and redistributions of resources *within fields* exactly because research is evaluated by *peers*. When units of assessment are rendered comparable across disciplines, resources can be redistributed between disciplines on the basis of the comparable attribute (ratings in one dimension or a synthetic quantitative indicator). Since this attribute is some version of research quality, disciplines become differentially funded within universities on the basis of their quality as measured by the current RES. This makes universities major sites of a new process whose aggregate effects have yet to be investigated. From our cases we can draw the conclusion that quantitative indicators are ill-suited for redistributing resources between disciplines because of their dependency on size and field-specific research and publication practices. However, even if valid information from peer reviews is used, the redistribution of resources on the basis of only one criterion may have unanticipated aggregate effects.

Three of the RES we investigated (those of Lower Saxony, the UK, and Australia) are highly consequential for universities, and thus created strong incentives for universities to *change the structures of their research*. This task is best supported by traditional, dedicated peer reviews that take into account local conditions and potential and result in extensive verbal recommendations. Even simple approaches such as the combination of 'weak' with 'strong' groups or the disguise of weak groups by relabelling them cannot be easily based on the comparable quality ratings or quantitative indicators provided by most RES, because this information does not enable any projection of what will result from such mergers. This is why Australian universities often used *ad hoc* peer reviews in order to assess the necessity of structural changes.

The most detailed information base for restructuring is provided by peer reviews as conducted in Lower Saxony. The detailed recommendations do not leave the university much choice beyond implementing or rejecting them. This kind of peer review thus restricts the opportunities for the university management to make its own decisions. There is a trade-off here between the richness of information and the room of manoeuvre for university management. The more formalized and parsimonious the outcomes of peer reviews are, the wider is the room of manoeuvre they provide for university management. If peer review outcomes are reduced to a one-dimensional rating, its function is reduced to legitimization, and it provides the widest range of options for university management. On the other hand, none of these options is any longer supported by information and legitimization from the peer review.

A third task for which information produced by RES provides an input is *individual performance management*. Since individual performance management is turning into a ubiquitous activity in universities, considerations of efficiency and practicability become very important. Australian universities tend to use the simple indicators of the funding formulae as simple indicators in yearly performance appraisals. The basic approach is benchmarking—one has to be as good as the average colleague from other universities. Another variant that is spreading is labelling. In order to count as 'research active', academics have to score on one or two of the three major indicators. Both practices rely on the use of quantitative indicators at the individual level, where they are largely invalid. Promotions are handled with greater care and include peer reviews for promotions to higher levels (professors). In other countries individual performance management appears to be less widespread. However, it will be interesting to see how individual performance will be measured in Germany, where it potentially could have severe consequences for both the resource base and the salary of a professor.

An important aspect of individual performance management is linked to academic identities (Henkel 2000, 2005). The RES significantly differ in their definition of academics' contributions to the income of the university. The peer-review-based system in the UK makes universities categorize academics as either good researchers who are 'breadwinners' for their university, or bad researchers who are not. This categorization and the ensuing differential treatment of academics have been shown to create enormous stress and tensions in universities. The formula-based systems with their simple indicators are less disruptive insofar as they enable academics to contribute to their university's income with relatively small 'research' contributions. A refereed conference paper of 3,000 words earns the Australian university more than 2,000 AU\$, and the co-supervision of a Ph.D. student earns research money as well. This situation makes it possible for most academics to 'save their identities', but also deprives universities of an effective control instrument, which is why the dichotomy of 'research active' and 'research inactive' academics is becoming increasingly popular among Australian university managers.

5. Conclusions

The university's position in the system of authority relations concerning research is constrained by the specific nature of the research process. The authority of university decision-makers is limited because they cannot define research goals and because the relationships between conditions, aims, and outcomes of research processes are opaque to anybody except researchers in the same field. The authority of university decision-makers is also contingent, among other factors, on the information about the directions, quality, and efficiency of 'their' research they possess. This is why RES are not only important as stimuli for universities to improve their research, but also as sources of information that can be instrumental in the management of research quality within universities.

Our chapter has considered how information provided by RES can be, and is, used to manage the research process. We compared three peer-review-based and two indicator-based systems according to their informational yield and to the use to which the information is put by university management, and uncovered clear differences between the informational yields of varied RES for universities and between the ensuing usability of information for different purposes. The clear superiority of peer reviews was demonstrated by the range of changes it enabled in Lower Saxony, as well as by the distortions resulting from the internal use of quantitative indicators in Australia. Further indirect confirmation stems from the widespread *ad hoc* use of peer reviews in structural decisions, and decisions about promotions to professorial levels in Australia. The only advantage of quantitative indicators (at least of the simple ones that are currently used in RES) is that they are cheap and timely.

In all our cases the information provided by RES is used by universities for managing their research. Thus, RES affect authority relations in science not only by stratifying the organizational field of higher education and by providing incentives for universities but also by strengthening (to varying extents) the authority and action capabilities of university management *vis-à-vis* its academics. However, there is an interesting trade-off between richness and action capabilities of university management. The richer the information, the more it might suggest a particular solution to a problem, which then becomes the only legitimate solution in the eyes of the academic community. The more advice management receives from peer reviews, the more difficult it becomes to act against that advice. If this information is ambiguous, the room of manoeuvre for university management increases again because it can choose which information to use.

In the case of peer-review-based RES, the internal use of the information by university management contributes to another change of authority relations. These RES do not only affect universities, but also the relationships between researchers and the scientific elite of their field. Since it is the elite who decide what is good research, RES employing peer review increase the dependence of researchers on the norms and assessments of their scientific elite. By using the same assessments internally, universities increase this dependency, thus relatively weakening the researcher's authority and relatively strengthening the elite's authority. This effect is strengthened by the fact that, except for the very rich information provided by the non-comparative RES in Lower Saxony, RES provide little support to university management when the latter needs to handle exceptions—cases of research that deviates from the mainstream and is of unknown quality.

Finally, comparing the five RES suggests some conclusions for the design of RES. Taking into account the internal use of external evaluations considerably increases the complexity of the impact of RES. It becomes obvious that the quest for the holy grail of the best RES is futile. The trade-offs between various aspects of the informational yield of an RES, on the one hand, and the usability of RES for other tasks, on the other hand, make any design decision a compromise, which in itself is a product of the authority relations in science.

References

Bekhradnia, Bahram (2008) *Evaluating and Funding Research through the Proposed 'Research Excellence Framework'*: http://www.hepi.ac.uk/466-1324/Evaluating-and-funding-research-through-the-proposed-Research-Excellence-Framework.html (accessed Oct. 2009).

Butler, Linda (2004) 'What Happens When Funding is Linked to Publication Counts?', in Henk F. Moed, Wolfgang Glänzel, and Ulrich Schmoch (eds.), *Handbook of Quantitative Science and Technology Research: The Use of Publication and Patent Statistics in Studies of S&T Systems* (Dordrecht: Kluwer), 389–405.

Choudhury, Vivek, and Sampler, Jeffrey L. (1997) 'Information Specificity and Environmental Scanning: An Economic Perspective', *Management Information Systems Quarterly*, 21: 25–53.

Cohen, Michael D., March, James G., and Olsen, Johan P. (1972) 'A Garbage Can Model of Organizational Choice', *Administrative Science Quarterly*, 17: 1–25.

Curtis, Polly (2002) 'Leicester's Communications Department to be Axed', *Guardian* (4 July).

CPB (Netherlands Bureau for Economic Policy Analysis) and CHEPS (Centre for Higher Education Policy Studies, University of Twente) (2001) *Higher Education Reform: Getting the Incentives Right*, CPB Bijzondere publikaties, 29: http://www.cpb.nl/nl/pub/cpbreeksen/bijzonder/29/ (accessed Oct. 2009).

Daft, Richard L., and Lengel, Robert H. (1984) 'Information Richness: A New Approach to Managerial Behavior and Organizational Design', in L. L. Cummings and B. M. Staw (eds.), *Research in Organizational Behavior*, vi (Homewood, Ill.: JAI Press), 191–233.

—— and —— (1986) 'Organizational Information Requirements, Media Richness, and Structural Design', *Management Science*, 32: 554–71.

—— and Macintosh, Norman (1981) 'A Tentative Exploration into the Amount and Equivocality of Information Processed in Organisational Work-Units', *Administrative Science Quarterly*, 26: 207–24.

Dainty, Roger, Williams, Syd, and Brown, Paul (1999) 'Current Trends and Changes to Research Strategy in UK Universities and Medical Schools', paper presented at the 50th Pittsburgh Conference on Analytical Chemistry and Applied Spectroscopy, Pittsburgh, PA.

DEST (Department of Education, Science, and Training) (2007) *Higher Education Report 2005* (Canberra: Department of Education, Science, and Training): http://www.dest.gov.au/NR/rdonlyres/5213D64C-C3BA-4D8A-B139-7906281D27AD/16972/HigherEdReport2005FINAL.pdf (accessed Oct. 2009).

Geuna, Aldo, and Martin, Ben R. (2003) 'University Research Evaluation and Funding: An International Comparison', *Minerva*, 41: 277–304.

Gläser, Jochen (2006) *Wissenschaftliche Produktionsgemeinschaften: Die soziale Ordnung der Forschung* (Frankfurt a. M.: Campus).

—— (2007) 'The Social Orders of Research Evaluation Systems', in Richard Whitley and Jochen Gläser (eds.), *The Changing Governance of the Sciences: The Advent of Research Evaluation Systems* (Dordrecht: Springer), 245–66.

—— (2008) *Evaluationsbasierte Managementsysteme für universitäre Forschungsleistungen* (Report to the German Ministry for Education and Science): http://www.bmbf.de/pub/evaluationsbasiertes_managementsystem_universitaere_forschungsleistung.pdf (accessed Oct. 2009).

—— and Laudel, Grit (2007) 'Evaluation without Evaluators: The Impact of Funding Formulae on Australian University Research', in Richard Whitley and Jochen Gläser (eds.), *The Changing Governance of the Sciences: The Advent of Research Evaluation Systems* (Dordrecht: Springer), 127–51.

—— Lange, Stefan, Laudel, Grit, and Schimank, Uwe (2008) 'Evaluationsbasierte Forschungsfinanzierung und ihre Folgen', in Renate Mayntz, Friedhelm Neidhardt, Peter Weingart, and Ulrich Wengenroth (eds.), *Wissensproduktion und Wissenstransfer* (Bielefeld: transcript), 145–70.

HEFCE (Higher Education Funding Council for England) (2004) *Higher Education in the United Kingdom* (Bristol: HEFCE): http://www.hefce.ac.uk/Pubs/hefce/2004/HEinUK/HEinUK.pdf (accessed Oct. 2009).

—— (2008) *RAE 2008: The Outcome:* http://www.rae.ac.uk/pubs/2008/01 (accessed Oct. 2009).

Henkel, Mary (2000) *Academic Identities and Policy Change in Higher Education* (London: Jessica Kingsley).

—— (2005) 'Academic Identity and Autonomy in a Changing Policy Environment', *Higher Education*, 49, 155–76.

Isenberg, D. J. (1984) 'How Senior Managers Think', *Harvard Business Review*, 62: 81–90.

Jauch, L. R., and W. F. Glueck (1975) 'Evaluation of University Professors' Research Performance', *Management Science*, 22: 66–75.

Johnston, Chris, and Farrar, Steve (2003) '100 New Chairs Created in Bid to Lift RAE Scores', *Times Higher Education* (12 Dec.).

Jongbloed, Ben, and Meulen, Barend van der (2006) *Investeren in Dynamiek: Endrapport Comissie Dynamisering*, 2nd edn. (Twente: Universiteit Twente, CHEPS).

Lange, Stefan (2007) 'The Basic State of Research in Germany: Conditions of Knowledge Production Pre-Evaluation', in Richard Whitley and Jochen Gläser (eds.), *The Changing Governance of the Sciences: The Advent of Research Evaluation Systems* (Dordrecht: Springer), 153–70.

—— and Schimank, Uwe (2007) 'Zwischen Konvergenz und Pfadabhängigkeit: New Public Management in den Hochschulsystemen fünf ausgewählter OECD-Länder', in Katharina Holzinger, Helge Joergens, and Christoph Knill (eds.), *Transfer, Diffusion und Konvergenz von Politiken: Sonderheft der Politischen Vierteljahresschrift* (Wiesbaden: VS Verlag für Sozialwissenschaften), 522–48.

Laudel, Grit (2006) 'The "quality myth": Promoting and Hindering Conditions for Acquiring Research Funds', *Higher Education*, 52: 375–403.

Lucas, Lisa (2006) *The Research Game in Academic Life* (Maidenhead: SRHE/Open University Press).

Meulen, Barend van der (2007) 'Interfering Governance and Emerging Centres of Control: University Research Evaluation in the Netherlands', in Richard Whitley and Jochen Gläser (eds.), *The Changing Governance of the Sciences: The Advent of Research Evaluation Systems* (Dordrecht: Springer), 191–203.

Morris, Norma (2002) 'The Developing Role of Departments', *Research Policy*, 31: 817–33.

Northumbria University (2007) *2008 RAE Code of Practice: Northumbria University* (Newcastle upon Tyne: Northumbria University).

Oxford University (2005) 'The University's Resource Allocation Method (RAM) (Updated for 2005–6)', *Oxford University Gazette* (20 Oct.).

Paradeise, C., Reale, E., Bleiklie, I., and Ferlie, E. (eds.) (2009) *University Governance: Western European Comparative Perspectives* (London: Springer).

Parker, Jonathan (2008) 'Comparing Research and Teaching in University Promotion Criteria', *Higher Education Quarterly*, 62: 237–51.

Phillimore, A. J. (1989) 'University Research Performance Indicators in Practice: The University Grants Committee's Evaluation of British Universities, 1985–86', *Research Policy*, 18: 255–71.

Queen Mary University of London (2003) *Research Strategy: 2003 to 2008:* http://www.qmul.ac.uk/research/policies/docs/s-research0308.pdf (accessed Oct. 2009).

Roberts, Sir Gareth (2003) *Review of Research Assessment* (London: Higher Education Funding Council for England): http://www.ra-review.ac.uk/reports/roberts.asp (accessed Oct. 2009).

Saunders, Carol, and Jones, Jack W. (1990) 'Temporal Sequences in Information Acquisition for Decision Making: A Focus on Source and Medium', *Academy of Management Review*, 15: 29–46.

Schiene, Christoph, and Schimank, Uwe (2007) 'Research Evaluation as Organisational Development: The Work of the Academic Advisory Council in Lower Saxony (FRG)', in Richard Whitley and Jochen Gläser (eds.), *The Changing Governance of the Sciences: The Advent of Research Evaluation Systems* (Dordrecht: Springer), 171–90.

Schimank, Uwe (2005) '"New Public Management" and the Academic Profession: Reflections on the German Situation', *Minerva*, 43: 361–76.

Travis, John (2009) 'Research Assessment: U.K. University Research Ranked; Funding Impacts to Follow', *Science* (2 Jan.): 323–4.

Van Raan, A. F. J. (1996) 'Advanced Bibliometric Methods as Quantitative Core of Peer-Review Based Evaluation and Foresight Exercises', *Scientometrics*, 36: 397–420.

VSNU (Vereniging van Nederlandse Universiteiten), NWO (Nederlandse Organisatie voor Wetenschappelijk Onderzoek), and KNAW (Koninklijkje Nederlandse Akademie van Wetenschappen) (2003) *Standard Evaluation Protocol for Public Research Organisations* (Utrecht: VSNU, NWO, and KNAW): http://www.qanu.nl/comasy/uploadedfiles/sep2003-2009.pdf (accessed Oct. 2009).

Weick, Karl E. (1979) *The Social Psychology of Organizing* (Reading, Mass.: Addison-Wesley).

Weingart, Peter, and Maasen, Sabine (2007) 'Elite through Rankings: The Emergence of the Enterprising University', in Richard Whitley and Jochen Gläser (eds.), *The Changing Governance of the Sciences: The Advent of Research Evaluation Systems* (Dordrecht: Springer), 75–99.

Westerheijden, Don F. (1997) 'A Solid Base for Decisions: Use of VSNU Research Evaluations in Dutch Universities', *Higher Education*, 33: 397–413.

Whitley, Richard (2008) 'Constructing Universities as Strategic Actors: Limitations and Variations', in L. Engwall and D. Weaire (eds.), *The University in the Market* (London: Portland Press), 23–37.

—— and Jochen Gläser (eds.) (2007) *The Changing Governance of the Sciences: The Advent of Research Evaluation Systems* (Dordrecht: Springer).

Wood, F. Q. (1989) 'Assessing Research Performance of University Academic Staff: Measures and Problems', *Higher Education Research and Development*, 8: 237–50.

6

Changing Authority Relations within French Academic Research Units since the 1960s

From Patronage to Partnership

Severine Louvel

1. Introduction

Much recent research has focused on how intensified competition for resources and increasing demands for relevance and accountability have affected patterns of authority relations between academics and various stakeholders (the state, companies, research councils, and so on). Such effects may be visible at the bottom level of individual researchers or research teams (see Chapter 9), which are the elementary units of scientific production (Knorr-Cetina 1999), as well as at a more aggregated level—that which the institutions teams belong to, and whose nomenclature and characteristics vary across countries: university departments (Morris 2002), institutes and research centres (Stahler and Tash 1994; Geiger 1990; Etzkowitz and Kemelgor 1998), and research units or laboratories (Joly and Mangematin 1996). These studies of 'Organized Research Units' (to draw on the generic term proposed by Stahler and Tash 1994) usually focus on how ORU cope with external pressures and defend their professional autonomy against external claims on the products of their research. In contrast, there are only a few investigations on how external drivers for change affect authority relations within ORU, and more specifically between research teams and the administrative head of the unit.

This chapter focuses on the reconfiguration of these intra-organizational authority relations as a result of structural changes affecting the public science system. It argues that the understanding of these intra-organizational dynamics is crucial as the management of research is still a decentralized and distributed process. Because ORU do not simply undergo external pressures but also develop their own strategies, such intra-organizational authority relations affect the way ORU resist these pressures or, on the other hand, how fast they adapt to them.

This study addresses these issues in a specific subtype of ORU: the French 'mixed research units' (*unités mixtes de recherche*). It rests on three longitudinal case studies over three to four decades in the life sciences. The life sciences are particularly interesting because they are highly symbolic of the structural reorganization of the public sciences: the growing requirements for relevance, accountability, and management of academic research (see Chapter 8); the increased emphasis on project-based funding from public and private stakeholders (Poti 2001); and the rapid growth of evaluation schemes and performance measurements (Whitley 2007). Finally, France provides a good example of a stratified 'state-shared' public science system (see Chapter 1) in which the central state and to some extent scientific elites constituting academic oligarchies traditionally had the most influence on research strategies and performance standards, but where the authority of employment organizations and funding agencies has increased during the last decades.

The chapter argues that there has been a shift from the 'patronage' type of authority relations within ORUs, predominant in the 1960s in France, to the 'partnership' type as a result of three major transformations of the public science system concerning (a) the recruitment and promotion procedures of academics, (b) the funding of research, and (c) evaluation procedures. It explains how these transformations have modified the roles of directors of research units and of team leaders, and consequently intra-organizational authority relations. It also shows how, in turn, these changing authority relations affect the scientific strategies of the mixed research units.

The chapter is structured as follows. The next section gives some background information about the French research system, and continues with a brief description and justification of the approach adopted. The third section describes the two types of intra-organizational authority relations and summarizes the transition process from 'patronage' to 'partnership' within the three research units studied. The fourth section outlines the structural conditions supporting the establishment of each ideal type of authority relations and the causes of change—particularly the structural changes in career systems, funding, and evaluation, which gained overall importance in the 1980s and 1990s and gave rise to the partnership type of authority relations. Finally, the fifth and last section analyses the impact of these changes on the research unit's strategies.

2. Three Longitudinal Case Studies

An Overview of the French Research System: French Mixed Research Units

After the Second World War, the state reorganized the French public science system around large national public research organizations (PROs) which are either generalist (such as the interdisciplinary CNRS with 32,000 staff in 2008) or mission-oriented. These PROs manage almost all ORUs which are the cornerstone of French academic research. In comparison with the research units to be found in other countries, French research units are distinctive in three major ways: their relations with academic departments, their staff composition, and their internal organization.

RELATIONS WITH ACADEMIC DEPARTMENTS

PROs were created outside universities (in order to remedy the weakness of university research), so that most research units were originally only managed by PROs and were independent from universities. Since the 1960s however, as argued by P. Larédo and P. Mustar (2001), this traditional view of the French research system as a dual one (with strong research units administered by PRO on the one side and weak university research units on the other) has been challenged. The most significant change deals with the rise of 'mixed research units', managed in common by a university and one or several PROs. The CNRS started such partnerships in the 1960s, and other PROs followed the same direction. Mixed research units are thus the new organizational standard, blurring the boundaries between universities and PRO, as Table 6.1 illustrates.

STAFF COMPOSITION

French academics are civil servants of the French state. The main difference between France and other countries is that the share of permanent staff is much higher in France (80 per cent among which 15 per cent are technical staff, as shown in Table 6.2). Another interesting feature is that the permanent research staff of mixed research units is composed of university academics (appointed by the university, having teaching duties and spending

Table 6.1. Number of CNRS research units (managed by the CNRS only) and number of mixed research units since 1992

	1992	1993	1994	1995	1996	1997	1998	1999	2000	2001	2002
CNRS units	237	222	204	192	198	190	183	161	136	109	108
Mixed units	100	117	134	273	385	522	521	624	743	936	1060

Table 6.2. Average composition of research units linked to CNRS

	No.	%
University academics	14	29
CNRS researchers	9	18
Researchers from other PROs	2	4
Other permanent research staff with postgraduate degrees	4	8
Other technical personnel	10	20
Doctoral researchers and post-doctoral researchers	10	21
Total	49	100

Source: Larédo and Mustar 2001.

their research time in the research unit) and PRO researchers (appointed by the PRO, with no teaching duties)

INTERNAL ORGANIZATION

The average size of CNRS mixed research units is around fifty members—a figure which however covers a broad diversity in terms of size, composition, and activity profile (Joly and Mangematin 1996). All mixed research units affiliated to the same PRO fall under the same statutory framework defined by the PRO and the university. They are first created for a limited period of time (usually for four years) after going through a selection procedure carried out by the PRO and the university. They receive recurrent funding as well as administrative, technical, and scientific staff. Their affiliation can be renewed for four-year periods so that a significant proportion of research units are quite long-lasting organizations. To take an example from the CNRS, 31.1 per cent (in chemical sciences) and 13.2 per cent (in life sciences) of the research units were created more than fifteen years ago.[1]

The relations between PRO/universities, the director of the research unit, and the research unit staff, also follow formal rules. The *director of a research unit* is an academic nominated by both the PRO and the university (after proposals made by the research unit council). He or she is the official representative of the research unit. As the research unit does not appoint the permanent research staff (university academics and PRO researchers), the director is unable to exert formal authority over them. Most research units are organized into research teams managed by *team leaders*. Teams are, as mentioned above, the elementary units of scientific production (Knorr-Cetina 1999; Poti 2001). Each team is also 'mixed' in the sense that it is

[1] CNRS database (2004), data analysis Séverine Louvel.

composed of PRO researchers and university academics. Team leaders launch research programmes, manage research contracts, and hire Ph.D. students and post-doctoral researchers. However, they have no more formal authority over the permanent research staff than do directors of research units. Finally, a *research unit council* is composed of representatives from the administrative, technical, and research staff (permanent and contractual researchers). It plays a consultative role in proposing guidelines and advising the director on research orientations.

The Study

Longitudinal case studies are particularly appropriate to capture intra-organizational dynamics extending over several decades (Siggelkow 2007). In this study, field-work was conducted between 2003 and 2005 in the second largest French academic centre (11,000 academics spread over more than 200 research units). Preliminary interviews with natural scientists revealed striking differences in authority relations and allowed identification of two polar cases (here called White and Green). Authority relations between the directors and the teams in these polar cases differed greatly up to the 1990s, but seemed to converge in recent years. Because of this, White and Green appeared as particularly favourable cases for retracing the transformation of authority relations. A third case, here named Red, was added during the study. Red was going through a deep internal crisis, which led to its break-up shortly after the end of the study. This research unit thus presented an excellent opportunity for evaluating the consequences of changing authority relations on scientific and organizational strategies. Table 6.3 summarizes the principal characteristics of the three cases. The research material is mostly composed of 103 interviews with members of the research units (and also with scientists who left them), and of archives (minutes of meetings, assessment reports, correspondence with PRO, universities, and so on). I also gathered additional material for White and Green through hallway discussions, and attendance at meetings and other events—such as Ph.D. defences—as I stayed full-time in these research units during several months.

Interviews were first compared to records in order to limit retrospective shifts (Leonard-Barton 1990) such as the over-evaluation or under-evaluation of the importance of an event. Oral and written sources were also cross-checked during interviews and data analysis. Data interpretation was oriented towards the way directions of research units exert authority over their teams since the research units' creation—authority being defined here as the influence on their strategic autonomy and capabilities, notably pertaining to

Table 6.3. Details of research teams studied

	White		Green		Red	
	1966	2003	1974	2003	1996	2003
Size and structure teams	4	5	1	5	1	2
Permanent research staff	8	26	3	24	5	13
Doctoral and post-doctoral researchers	2	35	–	13	3	6
Technical and admin staff	–	27	–	12	1	4
Institutional affiliations university	Yes		Yes		Yes	
PRO	CNRS		CNRS, INRA (National Institute for agronomic research), and CEA (French Atomic Energy Commission)		Inserm (National Institute for Health and Medical Research)	
Scientific field	Macromolecular chemistry and glycobiology		Plant physiology		Neurosciences	

research priorities, resource allocation, and attribution of reward and recognition (see Chapter 1). Authority relations were related to the direction's formal authority well as to the informal mandate (Hughes 1971) granted by team leaders. Comparison within and between the cases reveals two distinct ideal types of authority relations, described here as 'patronage' and 'partnership'. It also appears that all three research units fall currently under the partnership type of authority relations, and they all experienced the patronage type of authority at one point before the 1990s.

3. Types of Authority Relations: From Patronage to Partnership

Patronage Type of Authority Relations

In the 1960s and 1970s, directors of French research units played the role of 'patrons' towards their teams (Clark 1973). Clark defines patronage as an academic system in which the power is centralized in the hands of several professors, who dominate both the scientific production market and the market of academic positions. These patrons supervise a small number of *protégés* who are destined to succeed them, and control their access to work means, to scientific recognition, and lastly to career progression. Clark designed his model to describe the insular system which characterizes the Paris Faculty of Arts up to the middle of the twentieth century. The expansion of French academia which occurred throughout the second half of the twentieth century, and which was remarkable during the 1960s and the 1970s,[2] may have relaxed patronage relationships, yet it did not put an end to them at that time. Defined as a patron, the director of the research unit draws his authority upon domestic principles (Boltanski and Thévenot 2006) where the research unit appears as a family: his relationships to the team leaders are highly hierarchical and personal. As a patriarch, the director of the research unit enforces the rights and duties of each team, ensures their prosperity, and imposes on them mutual aid and generosity.

HIGHLY HIERARCHICAL RELATIONS BASED ON SCIENTIFIC LEGITIMACY AND ON MENTORING

In this type, the director keeps, sometimes years after team leaders formally become his equal, scientific authority over his former *protégés* and has *scientific legitimacy* in deciding which research tracks have priority and in orienting the professional commitments of the teams. The difficulty of contesting his or her legitimacy is reinforced by the fact that every criticism means disavowing a tutelary authority. Moreover, most team leaders are former doctoral students of the director, locally recruited and promoted thanks to his support. Academic inbreeding has been particularly salient at Green as well as at White, although to a smaller extent, as seen in Table 6.4. The director positions himself here as *team mentor*, and at the same time scientific counsellor and career protector. He protects each team's development using his wide-ranging powers in the academic world: 'Dave's policy was to favour people's career. Green was really like a "rank A publications factory", which meant that senior people have really extraordinary CVs' (former team leader at Green from 1973 to 1995).

HIGHLY PERSONALIZED RELATIONSHIPS: LOYALTY DEBTS AND PROTECTION

Relationships between the team leaders and the director are characterized by strong loyalty debts, the team leaders staying indebted for their research apprenticeship and for direct help in their academic career. The director

---

[2] At that time a dramatic increase in the number of students was followed by a spectacular increase in the number of academic positions. Between 1961 and 1971 university academic staff went from 7,901 to 30,546, or an average annual growth of 14.5 per cent. After 1971 the annual rate of growth oscillates between 0 and 4 per cent. CERC (1992).

Table 6.4. Recruitments and careers at White and Green (1970–2003)

	Junior researchers		Senior researchers	
	Former Ph.D. students	Other Ph.D. holders	Former junior researchers	Other junior researchers
White				
1970–90	18	8	4	1
1991–2003	4	20	6	3
Green				
1975–90	5	1	2	0
1991–2003	4	7	3	1

feeds these highly personalized relationships and these loyalty debts by protecting teams against external pressures. This protection rests notably on the banning of competition between teams by installing each team in a scientific niche:

We always thought here that people should not compete against each other. In my opinion that's the role of a good boss, to know how to distribute themes so that people are not in competition with each other and at the same time let them room for manoeuvre. Because in research there is room for everyone. We can work on the same theme without walking on each others' toes. (Former team leader at Green from 1973 to 1995.)

Furthermore, comparison of team's relative performance has always been as a taboo subject. Introducing internal discussions on this topic would be putting a fox in the hen house. Finally, some directors promote *financial cooperation* between teams while merging all third-party money or taking a significant percentage of it. Green's teams, for example, merge the entirety of their contractual resources. This rule pinpoints that the director attaches a great importance to the teams' mutual financial aid. It was established from the very beginning, at a time when project-based funding was still limited and considered a bonus improving the environment for research. The director was also the main contract manager: by bringing the largest contribution to this 'common pot', he was the first to demonstrate the virtues of generosity. Reciprocity rests here on the principles of social exchange (Blau 1967) in which gifts and counter-gifts are financial or non-financial (services, investment of collective interest in the administration, training . . .) and given to any member of the community.

Green, White, and Red were not subjected at the same time and in the same way to the patronage type of authority relations. Consequently, partnership emerged at different times in these three research units. Patronage was strong at Green from until the 1970s to the 1990s: the research unit was then successively managed by two founders who belong to the small elite of their scientific community and who had a free hand to exert their authority as patrons. Moreover, these two directors were deeply committed to patronage, which they considered the most efficient way to run a research collective. For instance, the second director still encouraged financial cooperation and mutual aid between the teams in the 1990s, after each team gained autonomy in competing for project-based funding. The move to the partnership type of authority relations occurred at Green in the 1990s, and was closely related to a generational change among directors and team leaders. The research unit was no longer managed by one of its founders (who are, however, still working in it), but by scientists who had just been promoted to a senior position. The director's institutional position no longer gave him the far-reaching powers needed to exert patronage over his team leaders. Furthermore, the director as well as team leaders were now highly critical of the domestic principles upon which patronage relies. Among the bones of contention with the founders was, for example, the director's legitimacy in orienting team leaders' professional commitments in return for his career mentoring.

Red experienced a similar move in the 2000s. Until then, it was managed by its founder who was considered as the protective patriarch of a small family. Patronage was suddenly replaced by partnership with the appointment of a new director in 2001. Unlike Green's new director, he was an 'outsider' who had not previously worked in the research unit. His intention to get rid of patronage, which he considered to be counterproductive, caused overt conflicts with the former director as well as with other members of the research unit.

Finally, White appears as a hybrid type from the 1970s to the 1990s before completely fitting into the partnership type of authority relations. Indeed, the research unit was then successively managed by two founders who are 'patrons' in the sense that their high status in the scientific community granted them high scientific legitimacy and their institutional position enabled them to exert control over team leaders' careers. However, team leaders benefited at that time from considerable autonomy in managing their research contracts so that, for example, the teams' financial cooperation remains limited. Patronage completely disappeared in the 1990s with a generational change among team leaders and with the arrival of a director who was—as at Red—an 'outsider'. As was the case in the two other research units, the institutional conditions were no longer favourable to the patronage type of authority relations, and patronage fell into disfavour with the director and the team leaders.

Partnership Type of Authority Relations

Since the 1990s, relationships between directors and team leaders have become closer to the ideal type of collegial relationships (Waters 1989) insofar as directors are looked on as *primus inter pares*. However, directors still have some authority over the team leaders, which can now be characterized as a partnership type of authority relations. Directors draw their authority on their ability to promote the teams' interests and do not usually interfere in their scientific strategy.

WEAK HIERARCHY BASED ON STRATEGIC LEGITIMACY

In this type of authority, research teams form an *association of equals*: each team leader directs an autonomous research jurisdiction with regard to scientific strategy and management of research contracts. Contrary to the patronage model, teams no longer acknowledge the scientific authority of the director. However, insofar as the director plays a bridging role with their environments, the team leaders may grant him or her *strategic legitimacy*. Whenever large scientific operations are at stake, they recognize in him the capacity to determine the most profitable projects and to filter teams' demands according to their strategic potential. The director has thus legitimacy to support projects that fit best into the policy objectives of PRO and universities and that protect the research unit against drastic financial downturns. However, the director loses the mentor role for team leaders. This function may be taken up by team leaders at the team level.

OPPORTUNISTIC RELATIONSHIPS

As the director no longer mentors team leaders, the latter do not feel indebted towards him as *protégés* were towards their patron. On the contrary, their relationships can be described as opportunistic, in the sense that direction and team leaders are not unconditionally bound by mutual aid. Thus, duties of protection (avoidance of competition between teams, protection against outside criticism, mutualization of all third-party funding, and so on) which prevailed in the patronage type of authority relations have faded. Given that research has become hypercompetitive, directors consider that every team must make do with the available resources and be accountable for its performance. However, there is a common agreement that the director can, in some circumstances, buffer teams' activities from excessive pressures, and that there are still, in that sense, limited forms of protection and cooperation at the level of the research unit. White's example notably shows that *mutual protection* still matters in research units. In 2003 a CNRS visiting committee evaluated White and recommended closing down a team. The position of the director was controversial, as he agreed with the committee that the team had failed in its mission to train Ph.D. students and to publish in leading journals. White's researchers could not agree whether the director should have defended the team (or at least stayed neutral) or whether he was right to denounce the teams' poor performance.

In Red's case, the brutal transition from patronage to partnership, or from a 'family' to an association of competing teams, generated a general outcry at the beginning of the 2000s. In particular, the teams could not tolerate the new director telling the visiting committee that some researchers are 'slowing down'. ('Instead of publicly defending his researchers ... he completely destroyed them'; researcher at Red). They blamed him with having established a trading relationship with the committee in which he 'sold' a team in exchange for the committee's support for his own project.

Furthermore, some forms of *financial cooperation* still have their place, even though the relationships may be mostly characterized as opportunistic. From the 1970s White's director imposed a 5 per cent tax on teams' contracts. He convinced them that there is a collective interest in keeping small budgetary surpluses without one team being a free-rider benefiting from another's contracts. Financial cooperation is thus set up on an economic exchange basis (Blau 1967), the terms of which are precisely defined: the debtor teams reimburse their financial debt with the help of their contracts. Green's teams still pool all contracts, but mutualization no longer relies on generosity: every researcher has the duty to seek contractual resources—the 'common pot' appearing as a collective insurance against uncertainty.

In the three cases studied, the move to partnership coincided with a generational change, the research units' founders handing the reins to a younger generation of scientists. This new generation no longer acts as 'patrons', because the institutional background became less and less favourable to patronage in the 1990s. Indeed, changes pertaining to career systems, funding, and evaluation have increasingly conflicted with the patronage type of authority relations during the last two decades.

4. Institutional Causes of Changing Authority Relations

More Institutionally Managed and Competitive Career Systems

Considering first changes to scientists' careers,[3] universities and PROs increasingly use them to implement their science policies. As French universities emerged more and more as strategic actors for research in the 1990s, they increasingly determined position openings in terms of research priorities, and not only on the basis of teaching requirements, as was for example common during the first massification wave of higher education in the 1960s. Consequently, job announcements generally specify not only an area of teaching, but also a research specialization and the research unit where the research has to be carried out.

As for research staff, they are appointed by PROs. Positions are allocated to the scientific departments of which each PRO is composed. There are then further negotiations within departments to decide upon the scientific sub-field to recruit researchers (for example, the department of humanities and social sciences will open two junior positions for sociologists). This procedure furthers the PRO overall policy which aims at developing scientific fields and not at supporting designated research units. This policy also means that vacant positions (after retirement, and so on) go back to the department and then potentially to any of its research units. This scientific policy has been strengthened over recent last decades. For instance, CNRS job announcements since 1990 do not indicate in which research unit the newly recruited researcher will work, so that candidates can select any relevant research unit affiliated to the scientific department. Since the 1990s there has also been a growing number of so-called 'signposted' positions which designate the precise subject in which the research has to be performed (for example, among the two sociologists, one specialized in industrial relations and the other in gender relations at work). PROs present these 'signposted' positions as a way of implementing targeted national research strategies.

Second, informal criteria for recruitment and promotion have been subjected to noticeable changes since the 1990s. The first informal change regarding recruitments and promotions concerns the formalization and toughening of criteria. It is particularly striking in the life sciences as publications in leading journals and post-doctoral positions have become the

---

[3] At least until the study. New legal rules organizing the recruitment of university academics were enacted in 2007.

norm for obtaining a junior position (Sabatier 2008; Robin and Cahuzac 2003).[4] Moreover, criticism of academic inbreeding (its presumed effect on clientelism, and on the stagnation of certain scientific fields) has led PROs and universities to counter it through informal rules. At the CNRS, mobility has become an informal requirement for a position as a junior researcher: the percentage of junior researchers who were not recruited in the research unit where they completed their Ph.D. went from 65 to 75 per cent between 2001 and 2004.[5] With regard to university academics, inbreeding has unevenly evolved across universities: some of them—mainly teaching-oriented—mostly recruit their former Ph.D. students, whereas those which are deeply engaged in scientific competition have almost banned inbreeding.[6] Finally, PROs have implemented policies favouring the mobility of junior researchers after their recruitment. For example, the CNRS Life Sciences department introduced 'Thematic Actions and Incentives by Project' (ATIP or Action Thématique et Incitative sur Programme) in 1990 which are 'starting grants' for young principal investigators who set up in another research unit (the CNRS grants financing of 140,000 € and a post-doctorate over two years).

The Rise of Project-Based Funding

The funding of French public research, as well as the form of state involve-ment, has also changed considerably over the last three decades. Until the 1980s, science policy was driven in many sectors by the dominant role of the Colbertist state (Mustar and Larédo 2002) where the state was not only the funder, but also the initiator and the quasi first user of large technologi-cal programmes. These programmes were mainly launched in aeronautics, electronics/computer, space and civil aeronautics, telecommunications, and electronuclear. In other scientific fields, research was predominantly funded by long-term core funding from the state and by a limited share of project-based funding. At that time state's annual subsidies covered the general infrastructure as well as the majority of research materials.

---

[4] The percentage of life-sciences Ph.D. holders in post-doctoral positions rose from 20 per cent in the mid-1980s to 32 per cent in the mid-1990s (Robin and Cahuzac 2003) and finally to 47 per cent in the 2000s (Giret 2006). Moreover, the percentage of newly recruited *maîtres de conférences* having worked in post-doctoral positions (abroad or in France) went from 43 per cent in 2002 to 60 per cent in 2007 (French Ministry of Higher Education).

[5] CNRS, follow-up of the recruitment policy of CNRS researchers.

[6] After the law on autonomy, universities have to indicate in their contract with the state a target proportion of external recruitments and promotions. This should means that inbreeding will decrease to a greater extent.

Table 6.5. Sources of funding in the biosciences (%)

	All	France	Germany	Italy	Spain	Sweden	UK
Long-term core funding	25	44	28	18	13	25	27
National funds (project basis)	38	17	47	36	55	35	19
Foundations	16	15	9	25	5	22	35
EU programmes	6	7	6	2	5	5	7
Regional funds	5	4	3	8	10	5	1
Contracts with industry and consultancy	8	12	7	9	8	6	11
Other	2	NS	NS	1	3	2	NS

Source: Larédo 2001.

Moreover, the salaries of permanent staff (who represented up to 80 per cent of the research unit staff, see above) were paid on separate budgets. Since the 1980s there have been significant changes: the large technological programmes have either disappeared or become marginal, meaning the near disappearance of the Colbertist state (ibid.). Moreover there has been an increase in project-based funding by the state but also by companies, EU, regional governments, and intermediary agencies.

In the life sciences, the share of project-based funding now exceeds that of long-term core funding by the state, as shown in Table 6.5, although these data do not include the salaries of permanent staff. The rate of project funding is still lower in France than in other European countries, although Thèves and colleagues (2007) argue that decisions about the allocation of human resources by French PROs are similar to project-based funding ones, as research units compete on the basis of their scientific programmes and results.

Monitoring: A Limited 'Deinstitutionalization' of French Research Units

Until 2006, mixed research units are evaluated by national peer committees from the PRO to which they are affiliated (the very same committees which recruit and promote PRO researchers).[7] These committees are often located at 'the crossroads of the disciplines and of the organization' (Vilkas 2001) as they both distribute scientific recognition and rewards and implement the PRO science policy. They evaluate the research units after discipline-based

---

[7] In 2006 the creation of the evaluation agency AERES modified the institutional frame which was effective at the time of this study. AERES is in charge of the evaluation of higher education and research institutions, of research units, and of higher education curricula. It is not entitled to evaluate individuals (PRO researchers or university academics).

peer reviews, reporting on the overall performance of the research unit and also taking into account how the research units' projects fit into the PRO scientific policy. Since the 1980s the heads of PROs have repeatedly attempted to 'deinstitutionalize', to some extent, French research units. These have indeed tended to remain long-lived—longer than their German or English counterparts (Hollingsworth 2006)—and be evaluated in terms of their compliance with a few institutional norms, such as meeting global requirements of journal publications, Ph.D. training, research contracts, and sometimes relations with the economic world (Fixari *et al.* 1993).

A limited process of deinstitutionalization was initiated in the 1990s when PROs started to evaluate each team independently and according to a formalized procedure which made explicit the teams' strengths and weaknesses. For example, the outcome of CNRS teams' evaluations was a letter (from A to E) according to a set of criteria. In the 2000s, this process of deinstitutionalization expanded through an upward trend in the use of bibliometric indicators pertaining to research units as well as individual researchers. The National Research Evaluation Committee noticed a breakthrough in the use of bibliometric analysis in the natural sciences (CNER 2003), especially in the life sciences. Even if France is still behind the Anglo-Saxon and Nordic countries, this change has created strong controversies among French life-sciences researchers (French Neuroscience Society 2006).

Consequences for the Directors' Authority Relations over the Team Leaders

FROM A DIRECT TO AN INDIRECT AND LIMITED INFLUENCE ON CAREERS

Patrons used to 'rule the roost', exerting a direct influence on recruitment and promotion procedures. They were part of the scientific elite which had extended powers at the local and national levels, and was thus in a position to negotiate position openings and to directly influence the selection procedures. This was the case at White until the beginning of the 1990s—the director notably holding important responsibilities at the CNRS Scientific Department but also at the university level: 'At that time I was also dean of the teaching department; I was member of the academic council at the university; I was also in charge of a master programme for more than twenty years' (director of White from 1984 to 1996). The same holds true for Green and Red until the beginning of the 2000s. The institutional background facilitated the directors' extended powers over careers. Being a member of the scientific elite is relatively easy when the discipline can still be considered as a 'small world' as it was precisely the case for the French life sciences in the 1960s and the 1970s.[8]

Moreover, the negotiation of position openings benefited from close inter-personal relationships between the direction of the research unit and the head of the PRO. At that time, PROs were still relatively small and their bureaucratic structures were not much developed, so that most directors could have a chance to negotiate directly with PRO heads: 'Papon, Feneuil,[9] you could still meet them. Then it was no longer possible because the bureaucracy became so heavy' (director of White from 1984 to 1996). Finally, the high level of institutionalization of PROs and universities enabled directors of research units to defend poorly performing scientists. This seems to have played a role in the research units' early years: 'Would you like to know how Liz got the job? She was rival with a guy who was just brilliant. But she got the position because the head wanted her to join the research unit, everybody knows that' (senior researcher at White, relating how another senior researcher obtained her first position at the research unit at the end of the 1970s). Other interviews, as well as external accounts of the period from the 1960s to the 1980s,[10] indicate that the last statement should not be disregarded as simply reflecting a settling of scores between colleagues.

As PROs became bureaucratized, and as the number of research units increased dramatically,[11] directors began to have only an indirect influence on position openings, which depended on their 'lobbying activity' within PRO and universities. Consequently, since the 1990s the directors of Green, Red, and White emphasize that being a member of boards is crucial in order to pursue their research programmes to the top of the university/PRO priorities, and in order to obtain position openings. This lobbying activity is increasingly distributed among a few senior scientists: 'We should really

---

[8] For instance, the CNRS life-sciences dept was only composed of 47 research units in 1966 (31 CNRS research units and 16 mixed research units).

[9] Pierre Papon was the head of the CNRS from 1982 to 1986. Serge Feneuil succeeded him in 1987.

[10] For instance, Pierre Tambourin, a prominent life-scientist who notably held important positions at CNRS as well as at Inserm, also makes severe judgements on the poor quality of some researchers recruited in the 1970s: 'At that time patrons made it a point of supporting their *protégés* no matter how good they were. And so we had a lot of people who were promoted without having published a single paper' (Pierre Tambourin, on the history of Inserm: http://picardp1.ivry.cnrs.fr/Tambourin2.html, tr. Séverine Louvel).

[11] Between 1991 and 2003, 378 research units have been associated with the CNRS life-sciences department (CNRS database, data analysis Séverine Louvel). In 2008, the department was composed of 296 research units and 11,510 permanent staff (PRO and universities) (CNRS database, Labintel).

have a full professor from Green sitting on the university boards. This is a quite simple argument; the university would show more consideration. We've had a hard time since Dave [the former director] no longer takes part in these boards' (Green's former team leader, 1979–2000, and director, 1990–2000). The formalization of recruitment and promotion criteria and the close monitoring of inbreeding starting at the end of the 1990s[12] also changed the directors' role in the application procedures. They may only defend 'local' applicants if they prove to have outstanding publication records. However, they also have to practise networking to find and attract the best candidates for junior as well as for senior positions. At White some scientists thus ironically describe their director as a 'sales and marketing person' who tries to convince any good researcher he gets to know to join the research unit.

SHARING THEIR MEDIATING ROLE WITH OTHER PROJECT MANAGERS

Patrons exerted strong control over resource allocation, and consequently monitored research priorities closely, which was facilitated when the research unit obtained most of its funding through state core support and when the patron managed the limited amount of third-party funding. This was the case for Green until the 1990s, when the director's role was clearly that of a unique mediator between the research unit and various stakeholders: 'He acted as a go-between with the outside. He was our unique mediation with the outside' (Senior researcher at Green, speaking about the first director from 1979 to the 1990s). Green's situation is typical of many life-sciences sub-fields which were long kept at distance from research contractualization. Red's creation in the mid-1990 is concomitant with the rise of project-based funding in this discipline. By contrast, the autonomy from which White's team leaders benefited from the beginning as project managers, and the early distribution of the mediating role between them and White's direction, reflect the early start of project-based funding in chemistry (White's main discipline along with life sciences).

With project-based funding becoming the sinew of war (Louvel 2007), the brokerage function became distributed among senior as well as junior researchers. Third-party funding represented between half and two-thirds of White and Green teams' budgets in the 2000s, the remaining part being

---

[12] Between 1997 and 2003, inbreeding pertains only to 25 per cent of the recruitments and promotions at the local university, which is much below the national average. See also Table 6.5 for the radical change in the origin of the junior researchers at White and Green.

PRO and university core funding.[13] Consequently almost all junior and senior researchers manage research contracts. However, the director's role still shows some distinctiveness. As the research unit's official representative, he conducts the institutional negotiations (with PRO, universities, regional governments, and so on) which are from time to time needed to develop the research unit (building an extension at White in the late 1980s).

THE DECLINE OF THEIR ADVOCACY ROLE

Patrons had no difficulty protecting teams against external evaluations when research units were highly institutionalized. First of all, most evaluations did not put teams in danger as they were considered as formalities, as suggested by the following ironical description: 'The visiting committee arrived by train in the morning, did a walk-through of the research unit, had a nice lunch and chatted with the senior director and a few senior scientists. Then they went back to the rail station and everyone was happy' (senior researcher at White). Moreover, even if one team was criticized for not having reached an acceptable level of scientific performance, the patrons could easily mitigate this negative appreciation by arguing for the research unit's conformity to broad institutional norms: 'I think John's team who deals with the MNR spectroscopy has always done engineering jobs more than basic research. But as long as the others teams published it did not matter so much' (senior researcher at Green).

The end of the 1990s was a turning point in the three research units with regard to the ability of the directors to protect their teams. Two elements of the limited deinstitutionalization process—the evaluation of each team separately, and the introduction of bibliometric criteria—made the director's protection less significant: indeed, he cannot defend a team which does not fulfil the performance criteria fixed by the PRO. At White, the new assessment procedure led notably to the splitting up of one team in the 2000s, which had not occurred since the creation of the research unit in 1966.

The changing authority relations finally have consequences on the kind of scientific strategy that research units are more likely to promote. First, the move from patronage to partnership has an impact on the conditions for research in the research unit; that is on research priorities, recruitment policy, and so on. In turn, these distinct conditions for research may have

---

[13] Louvel (2007). Budgets do not include salaries of permanent staff.

dissimilar—although ambiguous—effects on the scientific outputs of the research unit, its performance, and its ability to innovate.

5. Consequences for Scientific Strategy

Conditions for Research

Patronage first implies that the director limits the team's growth. By doing so he preserves his scientific legitimacy as well as his mentor function. This restrictive policy relies on several instruments depending on the research unit. In some research units the director imposes strict limitations on the number of Ph.D. students in each team. He also establishes that scientists should only be recruited among former Ph.D. students, and he selects these candidates to academic positions (informally vetoing other applications). These rules were in use at Green until the 1990s. Other research units followed more flexible ones insofar as academic inbreeding is not always the rule for recruitments and promotions.

The director may, for example, decide that recruitments and promotions are done by turns in each team, which is also a way to control their expansion and thus to keep the team leaders under his authority: 'The wish being to maintain an equilibrium between White's teams, we will seek to obtain: in section 28 (team B), the recruitment of X; in section 19 (team C), the recruitment of Y, these two teams currently only have one CNRS researcher' (verbal proceedings of White's research unit council from 27 April 1978). Green also turned to this policy in the 1990s: 'The director does what he can so that teams are homogeneous and so that there are no favourites among them, so the recruitments are done in turns in the different teams...The downside of this policy is that I was basically told—if you know how to read between the lines—you already had your researcher recruitment for this year and you won't have a second' (senior scientist at Green). Finally, patrons can also restrict the number of 'external' senior scientists joining the research unit, as they would potentially start a new team and thus challenge the internal balance of power.

Second, patrons keep an overarching view over all research programmes. This condition for research is quite crucial for the director to preserve his scientific legitimacy. Here again, directors may implement more or less strict informal rules. Some directors may keep the research programmes within a strictly delineated mono-disciplinary frame. Following this perspective, Green's directors preserved strong scientific coherence until the mid-1990s: 'We launched the research unit to study the interactions between plant cell compartments. So we've had a very strong scientific coherence around that until the team working on cytoskeleton arrived [in 1997]. Now the coherence is not so visible, but it is still there somehow' (Green's former team leader, 1979–2000, and director, 1990–2000). Some researchers are still nostalgic about that time: 'We can't remain a single research unit with that many people. The problem when there are this many groups and different themes, it's that there is not one person at the head who is capable of understanding all the themes...Dave and Julia could give interesting and constructive judgements on all the thematics but I think that it's no longer the case...' (assistant professor at Green).

In other cases, the director promotes pluridisciplinarity within the research unit but firmly subjects the scientific evolutions to his own strategies, and thus leaves little room for evolutions he has not initiated. This is the way White's directors start new themes in the 1980s and in the 1990s: 'Paula played a genuine scientific leadership role, even if it was dictatorial. One can criticize her actions, but at least she proposed themes. In particular it's she who launched material chemistry and microbiology in the mid-1980s' (professor at White). The arrival of the team working on cytoskeleton at Green in 1997 also responded to the director's scientific vision.

The move to the partnership type of authority relations implies that the director no longer exerts control over the teams' growth. Team leaders freely decide to hire Ph.D. and post-doctoral students, to have several Ph.D. holders applying for a permanent position, and so on. As his authority does not rely on scientific legitimacy, there is also no need for him to gain an overarching view of the research programmes. As his strategic legitimacy relies on his ability to support the teams' projects and to help them acquiring funding, the research unit's expansion indicates his achievement. The director thus encourages the teams' growth even if it threatens the scientific coherence of the research unit or if it contravenes the *ex-ante* formulation of scientific goals. There are recurrent debates at White about the possibility of designing and following a clear scientific policy, with some senior scientists expressing concern about the lack of such a policy and the director arguing that it would be counterproductive: 'It's so hard to recruit people, you need pragmatism and flexibility. You must not be blocked by a scientific committee by saying that we must recruit in such-and-such team. Decisions must be taken quickly. When a good candidate appears, you have to decide right away' (White's director, meeting of the research unit council in 2003).

Scientific Performance and Innovation

Patronage can be considered as a double-edged sword in terms of scientific outputs. Indeed, the director may use his far-reaching powers to set up and protect poorly performing teams. This is the reason why patronage has been highly criticized for being anti-meritocratic. The director's control over scientific strategies brings another negative side effect: it may either point in the wrong direction or favour conservatism and mainstream research to the detriment of innovation. For example, some senior scientists argue that their research unit has somehow 'missed turning points' because of the director's monopoly on research orientations: for example, new themes were introduced too late (molecular biology at Green in the mid-1990s) or should not have been introduced at all (microbiology at White in the mid-1980s).

On the other hand, patronage is also quite efficient in protecting teams' professional autonomy against hierarchical pressures. In this sense it may preserve a long-term view for scientific research: this happened in the late 1980s when the head of a PRO suddenly decided to deeply restructure the area of plant biology in a sense that would have meant dramatic funding cuts for Green.[14] Clearly, patronage here limited the negative consequences of an authoritarian policy which would be relaxed in the 1990s. The mutualization of third-party funding also spared researchers fund-raising practices and thus may have enabled them to engage in innovative research: 'Everyone knows there will always be money to work... With programme X, the contracts were our main source of funding; it was more than the money given by PRO. That allowed everyone to work' (director of Green).

Similarly to the patronage form of authority, partnership has neither obvious nor unique consequences for the research unit's scientific production and its ability to innovate. In fact, one can argue that it furthers meritocracy as it gives room for manoeuvre to every researcher who shows initiative. On the other hand, the opportunistic relations between the director and the teams, as well as the high level of pragmatism with which the director determines the research unit's scientific orientations, have less clear consequences for the research unit's performance. For example, White's director took the opportunity to launch a team on 'glycobiology' in 1996 while convincing three researchers to join White. This strategy

---

[14] See Vilkas (2001) for the analysis of the scientific policy of the CNRS life-sciences department from 1988 to 1992.

may foster scientific performance and innovation in the long term. The 'glycobiology' team's scientific achievements are unanimously recognized as outstanding and its collaborations with other teams have also advanced knowledge cross-fertilization within White.

On the other hand, as the arrival of the team stirred up internal competition, some researchers at White consider it as a 'huge waste of human resources'. Their harsh judgement may partly be explained by the fact that the director's widespread opportunism conflicts with his mandate to mitigate competitive pressures. However, their diagnosis also accounts for the numerous conflicts arising after the team's arrival, resulting in the departure of ten researchers in eight years: an exceptional number of exits for French research units.[15] Faced with a similar situation, Red was too small to withstand a high number of departures: it broke up in 2003 after nearly half of the researchers left to demonstrate their disapproval of the director's policy. Mobility between research units has surely positive consequences for research system performance. However, a high number of hastened departures puts a sudden end to projects and collaborations, and they mean a very bad return on scientific investment for the research unit.[16] One may then argue that these exits do not provide fertile ground for the capitalization and transmission of knowledge in the research unit.

6. Conclusions

This chapter casts light on some intra-organizational consequences of the structural transformations which have affected the French public science system since the 1960s—particularly how authority relations between the directors of research units and their team leaders have shifted from the 'patronage' type dominant in the 1960s to the 'partnership' type prevailing today. Changes in career systems, funding sources, and assessment procedures have greatly affected director's authority over their research teams: more open and competitive academic labour markets have restricted his direct influence on recruitments and promotions; a gradual increase of third-money funding brings an end to his monopoly over contract management; and the formalization of assessment procedures limits his

---

[15] These exits are all the more striking as there was almost no exit during the previous thirty-five years.

[16] In both cases, the research units had acquired costly equipment and appointed technicians to work with the researchers who left the unit.

advocacy role on behalf of his teams. Additionally, while the patronage type of authority implied a limited growth of the research unit and a controlled diversification of research areas, which both enabled patrons to maintain their strong authority and their scientific aura, the partnership form is consistent with the research unit's expansion and with the multiplication of research areas, which both demonstrate the director's ability to defend the teams' interests in a competitive environment.

The authority of directors over their teams seems to have faded along with this transition. One could here argue that the locus of power in French academia has moved from the professors. The trend here would be comparable to the one observed in Germany where 'professors are challenged from outside by international competitors, and from inside as their mini-cosmology is infiltrated by junior professors' (Harley *et al.* 2004; Chapter 7). However, our analysis shows that one should not hastily conclude that hierarchical authority has disappeared in French research units. Directors still play a mediating role between their teams and the environments, and provide teams with strategic resources (work space, equipment, technical personnel, PRO budgets, reputation, networks, and so on). Furthermore, they partly contribute to the teams' ability to bridge and buffer with their environments (Meznar and Nigh 1995).

The transition observed here can be considered as a micro-sociological expression of a broader trend also occurring in other countries such as Switzerland (see Chapter 3), which can be characterized as the evolution of state-coordinated public science systems towards more state-shared ones, and even perhaps moving to a limited form of state-delegated ones (Chapter 1). In this type of public science system, research agencies (still in limited numbers even in France) and PRO administrative centres restrain the authority of organizational scientific elites (such as the research units' directors). Therefore one should expect that institutional changes subsequent to this study (such as the creation of the national research agency (ANR) in 2005 or the national agency for the evaluation of research (AERES) in 2006, as well as the law on university autonomy (LRU) adopted in 2007), which might encourage the shift towards a more state-delegated system, will reinforce the partnership type of authority relations.

Finally, our study indicates the limitations of such a global picture by showing that the impact of macro-sociological trends varies to a great extent from one research organization to another. Indeed, the pace of change relies on disciplinary and institutional factors, but also on more idiosyncratic features, such as the generational balance within research units, the director's position among the scientific elite and towards PRO administrators, or his commitment towards certain values and principles.

References

Blau, P. (1967) *Exchange and Power in Social Life* (New York: Wiley & Sons).

Boltanski, L., and Thévenot, L. (2006) *On Justification: The Economies of Worth* (Princeton: Princeton University Press).

CERC (1992) *Les Enseignants-chercheurs de l'enseignement supérieur: Revenus profession-nels et conditions d'activité* (Document, 105; Paris: Centre d'études sur les revenus et les coûts).

Clark, T. N. (1973) *Prophets and Patrons: The French University and the Emergence of the Social Sciences* (Cambridge, Mass.: Harvard University Press).

CNER (2003) *Evaluation de la recherche publique dans les établissements publics français* (Paris: La Documentation française).

Eisenhardt, K. M. (1991) 'Better Stories and Better Constructs: The Case for Rigor and Comparative Logic', *Academic of Management Review*, 16: 620–7.

Etzkowitz, H., and Kemelgor, C. (1998) 'The Role of Research Centres in the Collectivisation of Academic Science', *Minerva: A Review of Science, Learning and Policy*, 36: 271–88.

Fixari, D., Moisdon, J.-C., and Pallez, F. (1993) *Gérer en évaluant: Le Rôle du Comité National de la Recherche Scientifique* (Paris: Centre de Recherches en Gestion, École des Mines).

French Neuroscience Society (2006) 'Bibliométrie: L'Avis des Présidents de section', *La Lettre des Neurosciences*, 14: 7–18.

Geiger, R. L. (1990) 'Organized Research Units: Their Role in the Development of University Research', *Journal of Higher Education*, 61: 1–19.

Giret, J. F. (2006) *Rapport sur l'insertion professionnelle des jeunes docteurs* (Marseille: Cereq).

Glaser, B., and Strauss, A. (1967) *The Discovery of Grounded Theory: Strategies of Qualitative Research* (London: Wiedenfeld & Nicholson).

Harley, S., Muller-Camen, M., and Collin, A. (2004) 'From Academic Communities to Managed Organisations: The Implications for Academic Careers in UK and German Universities', *Journal of Vocational Behavior*, 64: 329–45.

Hollingsworth, R. (2006) 'A Path Dependent Perspective on Institutional and Organizational Factors Shaping Major Scientific Discoveries', in J. Hage and M. Meeus (eds.), *Innovation, Science, and Institutional Change: A Research Handbook* (Oxford: Oxford University Press), 423–42.

Hughes, E. C. (1971) *The Sociological Eye: Selected Papers* (New Brunswick, NJ: Transaction Books).

Joly, P. B., and Mangematin, V. (1996) 'Profile of Public Laboratories, Industrial Partnerships and Organisation of R&D: The Dynamics of Industrial Relationships in a Large Research Organisation', *Research Policy*, 25: 901–22.

Knorr-Cetina, K. (1999) *Epistemic Cultures: How the Sciences Make Knowledge* (Cambridge, Mass.: Harvard University Press).

Larédo, P. (2001) 'Benchmarking of R&D Policies in Europe: Research Collectives as an Entry Point for Renewed Comparative Analyses', *Science and Public Policy*, 28: 285–94.

—— and Mustar, P. (eds.) (2001) *Research and Innovation Policies in the New Global Economy: An International Comparative Analysis* (Northampton, Mass.: Edward Elgar).

Leonard-Barton, D. (1990) 'A Dual Methodology for Case Studies: Synergistic Use of a Longitudinal Single Site with Replicated Multiple Sites', *Organization Science*, 1: 248–66.

Louvel, S. (2007) 'Le Nerf de la guerre: Relations financières entre les équipes et organisation de la coopération dans un laboratoire', *Revue d'anthropologie des connaissances*, 297–322.

Meyer, J. W., and Rowan, B. (1977) 'Institutionalized Organizations: Formal Structure as Myth and Ceremony', *American Journal of Sociology*, 83: 340–63.

Meznar, M. B., and Nigh, D. (1995) 'Buffer or Bridge? Environmental and Organizational Determinants of Public Affairs Activities in American Firms', *Academy of Management Journal*, 38: 975–96.

Morris, N. (2002) 'The Developing Role of Departments', *Research Policy*, 31: 817–33.

Musselin, C. (2005) 'European Academic Labor Markets in Transition', *Higher Education*, 49: 135–54.

Mustar, P., and Larédo, P. (2002) 'Innovation and Research Policy in France (1980–2000) or the Disappearance of the Colbertist State', *Research Policy*, 31: 55–72.

Poti, B. (2001) 'Appropriation, Tacit Knowledge and Hybrid Social Regimes in Biotechnology in Europe', *International Journal of Technology Management*, 22: 741–61.

Rip, A. (1997) 'A Cognitive Approach to Relevance of Science', *Social Science Information*, 36: 615–40.

Robin, S., and Cahuzac, E. (2003) 'Knocking on Academia's Doors: An Inquiry into the Early Careers of Doctors in Life Sciences', *Labour*, 17: 1–21.

Sabatier, M. (2008) 'Do Female Researchers Face a Glass Ceiling in France? A Hazard Model of Promotions', *Applied Economics* (5 Dec.): 1466–4283. Available at: http://informaworld.com/smpp/search#db=all?searchtitle=713684000&searchmode=advanced&term1=sabatier&field1=all&ssubmit=true.

Siggelkow, N. (2007) 'Persuasion with Case Studies', *Academy of Management Journal*, 50: 20–4.

Stahler, G. J., and Tash, W. R. (1994) 'Centers and Institutes in the Research University: Issues, Problems, and Prospects', *Journal of Higher Education*, 65: 540–54.

Thèves, J., Lepori, B., and Larédo, P. (2007) 'Changing Patterns of Public Research Funding in France', *Science and Public Policy*, 34: 389–99.

Vilkas, C. (2001) *L'Art de gouverner la science dans le système public français: Le Cas du CNRS. Représentation, évaluation, direction de quatre disciplines* (Paris: IEP de Paris).

Waters, M. (1989) 'Collegiality, Bureaucratization, and Professionalization: A Weberian Analysis', *American Journal of Sociology*, 94: 945–72.

Whitley, R. (2007) 'Changing Governance of the Public Sciences: The Consequences of Establishing Research Evaluation Systems for Knowledge Production in Different Countries and Scientific Fields', in R. Whitley and J. Gläser (eds.), *The Changing Governance of the Sciences: Sociology of the Sciences Yearbook* (Dordrecht: Springer), 3–27.

7

Mission Now Possible

Profile Building and Leadership
in German Universities

Frank Meier and Uwe Schimank

1. Introduction

Deliberate and successful attempts to build distinctive collective research strategies, or 'profiles', that involve reallocating resources to particular areas constitute a new phenomenon in the German university system. Rectors and presidents have come to consider such profile-building as an important task, which their increased authority makes it possible for them to push forward. In this chapter we discuss how this significant shift in organizational leadership and identity came about, focusing on the key elements of the 'new public management' (NPM) that were decisive for this development, and on the factors explaining variations in profile-building activities and their success at German universities.

Our empirical cases come from the state of Lower Saxony, which was one of the early adopters of NPM for its universities. Not only are governance changes towards NPM more pronounced in Lower Saxony than in other parts of Germany, but performance measurement by evaluation was institutionalized in Lower Saxony's university system in a much more systematic manner than in other German states. Lower Saxony established a special agency for the measurement of research performance of this state's universities, the Wissenschaftliche Kommission Niedersachsen (WKN), which started its work in 1999 (WKN 1999). In contrast to many other evaluation procedures the WKN's evaluations do not end with the assessment of research performance but add recommendations for im-provements.[1] Since profile-building figured prominently among these re-commendations, Lower Saxony's universities offer very suitable empirical cases for our analysis.

We analyse only four cases of profile-building here. They have in com-mon that the creation of a collective research profile for a particular disci-pline or scientific field at the respective university was a prominent goal of the university leadership and that these leaders were satisfied with their achievements. Obviously, the small number of cases cannot cover the logical combinations of possible outcomes and explanatory factors. Our results are therefore predominantly exploratory. However, we can identify distinct outcomes of profile-building which are perceived as successes by university leadership, and show how NPM reforms changed authority rela-tions such that the feasibility of top–down initiatives of profile-building in German universities in general has increased. In particular, we can point out some combinations of factors that produced specific outcomes.

The cases we studied are from two disciplines belonging to the social sciences: one natural science, and one transdisciplinary research field which incorporates both natural and social sciences. These cases are located at three universities. For our data we rely, first, on documents from WKN evaluations: the self-reports of the respective institutes or faculties as well as the reports of their evaluators. The WKN's evaluation procedure is the source of original reports and follow-up reports which document the state of affairs four years later. Secondly, we conducted ten expert interviews with involved persons at the universities.[2]

We present our findings in five steps. First, we show that the traditional authority relations of German universities did not enable top–down profile-building. Secondly, we describe our four cases of successful top–down profile-building as new phenomena that need to be explained. Thirdly, the major step of explanation refers to the new position of university leadership and its new relationship to national scientific elites as a conse-quence of NPM reforms. This dyad of organizational leaders and scientific

---

[1] In Richard Whitley's (2007) categorization it is a 'weak' evaluation regime (Schiene and Schimank 2007).

[2] We thank the representatives of the selected universities and cases, especially our interview partners, for their readiness to make available to us the mostly confidential documents and to talk with us about sometimes still controversial matters. For reasons of anonymity we cannot describe our cases and interview partners in more detail here. In addition, we thank officials of the WKN for additional information and their help in the selection of cases. Finally, we should mention that the second author of this chapter belonged to a group of WKN evaluators for the disciplines of sociology and political science in 2003.

elites is the principal driving force behind profile-building activities. Fourthly, we turn to the other two actors that also must be taken into account: the ministry of science and cultural affairs, and the targets of profile-building activities. Fifthly, an important contextual factor is the teaching mission of universities.

2. Traditional Authority Relations: No Possibility of Top–Down Profile-Building

The traditional authority relations of German universities were characterized by a dual power structure (Schimank 2005; Schimank and Lange 2009). On the one hand, strong state authorities maintained an extensive and detailed bureaucratic regulation of university affairs, especially with regard to financing, personnel structures and teaching programmes. On the other hand, the academic professionals constituted a powerful self-governing organizational scientific elite. University professors as a collectivity, as well as each one of them individually, were highly autonomous with respect to academic issues, while the bodies of academic self-governance, in particular faculty councils and academic senates, dominated intra-organizational decision-making. This decision-making was strongly consensus-oriented, which meant that decisions which violated the interests of particular professors were rare and there was a marked tendency to uphold the status quo.

This dual power structure made universities 'hollow organizations' (Whitley 2008: 26–31) in the sense that they were subjected to decision-making by the state or the academics but were not decision-making actors themselves. In other words, traditional German universities were organizations without 'actorhood' (Meier 2009). State regulation and academic self-governance left hardly any room to manoeuvre for university managers. The most prominent function of organizational leadership, setting collectively binding goals for all members of the organization, was structurally prevented by this governance regime. From the point of view of the academic profession, this was not a fault. Organizations without 'actorhood' are both not able and not legitimized to tell professors what to do. Instead, a professor regarded his university as a commons he shared with all other professors at this university, and they all competed in the exploitation of their university's resources. With regard to collective profile-building in teaching, as in research this often meant that strategic decisions were not taken. University leaders were powerless, the bodies of academic self-governance practiced mutual non-interference among professors, and the ministries responsible for the university system only sporadically and incrementally intervened in the academic affairs of the universities.[3]

The research agenda of an institute or faculty to which a number of professors belonged traditionally was nothing more than a list of disconnected choices made by individual professors with respect to their theoretical and methodological interests and beliefs. This did not preclude professors actively coordinating their research activities in ways that could lead to large externally funded research agglomerations, but it was nevertheless strictly based on voluntary bottom–up coordination. Additionally, something like a local tradition with respect to particular objects of research, theoretical or methodological preferences could evolve over time, leading to scientific 'schools' dominating certain organizations.

In this situation, university managers were restricted to supporting or inducing bottom–up activities by soft measures, such as the initiation of contacts and talks among potential partners of research coordination or cooperation, or the use of small amounts of financial or other incentives. To emphasize the needs of collective profile-building too strongly in the formulation of the qualification profile of a professorship would have been perceived as a violation of individual 'academic freedom' as well as an intrusion into disciplinary territory. Thus, university managers' room to manoeuvre with regard to collective profile-building was very limited.

3. The New Thing: Top–Down Profile-Building

Against this traditional understanding, commitment to proactive profile-building is now not only a legitimate task of university management, but is demanded from it. A coherent collective research profile of a discipline or transdisciplinary field at a university is an indicator of research quality and, at the same time, of good leadership. This shift in the normative and evaluative dimension of research management in universities corresponds with numerous attempts of top–down initiated and directed profile-

---

[3] To be sure, especially in the late 1960s and early 1970s, state authorities cultivated some planning rhetoric with regard to higher education and the establishment of research priorities (Meier 2009: 201–3). But the decisions actually made were rather incremental than strategic, and in effect did not lead to clearly defined profiles of universities or disciplines. Still, historically the discourse on educational and research planning marked a farewell to the traditional ideal of any single university being an organizational representation of science as a whole.

building including the four cases analysed here. These cases exemplify a variety of results:

- Case A illustrates the most obvious variant. This discipline's research made— as the WKN evaluators testify almost euphorically—'an outstanding development'; and they refer especially to the priority-setting which they see as 'probably the most successful one in Germany' in the respective discipline, and to large-scale research cooperations such as 'special research areas' of the Deutsche Forschungsgemeinschaft (DFG), Germany's most important funding agency, and a cluster of the recent 'excellence initiative'[4] which extends to other universities and institutes. A closer look shows that the top–down profile-building by the university leadership built on a number of long-standing bottom–up initiatives by individual professors or groups and well-established traditional research priorities.

- In case B, after earlier bottom–up initiatives had failed, the university leadership is now confident that it is 'on a good way' towards a collective research-profile for one of its disciplines, which was assessed to have a mixed performance and no discernible profile by the WKN evaluators. They recommended establishing a commission of mainly external experts to conceive a feasible and convincing profile. This commission's recommendations were mostly accepted and are now being implemented step by step, which will take a long time because some measures require the retirement of existing personnel. Still, vacant chairs have been respecified, and one professor has been recruited who has started to work as a bridge to a neighbouring discipline with which fruitful future cooperation is expected.

- Case C concerns an institute that was assessed as very weak in its research performance by the WKN evaluators. In addition they found fault with the lack of certain core sub-fields of the discipline in terms of professorships. Several professorships of retiring professors were respecified to cover these core fields, and new professors with a very good research standing were recruited. Taken together, these professorships do not automatically amount to an integrated collective research profile of the discipline. However, this may be a matter of time;

---

[4] This is a new competitive funding programme introduced by the federal government and the states in 2004, with a budget of 1.9 billion € for the first two rounds, which supports graduate schools, clusters of excellence, and outstanding concepts for the future development of a university as a whole.

moreover, each of the new professors had already started making contacts with colleagues from other disciplines, so that another scenario might be an integration of the professorships in different transdisciplinary research profiles.

- Finally, case D concerns a transdisciplinary field. Here, WKN evaluators recommended the integration of various bottom–up initiatives into a comprehensive university-wide research profile institutionalized as a central research institute. This institute was created, but everybody admits that until now not much progress has been made towards research coordination or even cooperation beyond disciplinary boundaries, because the research topics and types of research are too diverse. The institute succeeded in producing an application for a large project consisting of a number of sub-projects from the different disciplines which would not have been achieved without the institute; but the application failed, and the emerging transdisciplinary communication and coordination broke down again. Optimists hope that such profile-building will happen after some time; others are more sceptical but still claim that the institute is a useful instrument to make the university-wide research activities in this transdisciplinary field more visible. At least the expectations of the ministry were satisfied, which is advantageous to the university at large and potentially to each of the disconnected activities which are covered formally by the institute.

On the one hand, it is clear that the absolute level of what the university leadership achieved with respect to profile-building is quite high in case A, high to medium in case B, and medium to low in cases C and D. On the other hand, the starting points were quite different, so that the relative achievement in case C, for instance, may be even higher than in case A where the leadership could rely on a lot of groundwork already done. Initially, university managers might be content with establishing a new formal structure that simply signals profile-building, as in case D, or even with window dressing where an institute or faculty presents a coherent profile on paper, but behind this façade research remains as disconnected as before. In case C it was apparent to the university leadership that the paper in which the respective institute outlined its research profile was nice rhetoric without substance. But while being satisfied with the successful implementation of other measures to improve this institute's research profile, the president tactfully overlooked this decoupling of 'talk' and 'action' (Brunsson 1989). In addition, he hoped that perhaps over time the 'backstage' activities would adapt to the 'frontstage' presentation because consistent false impressions become wearisome in the long run. So, all in all there was a realistic appraisal of the outcomes from the point of view of the university leadership in all four cases.

These brief descriptions of the cases are sufficient basis for us to ask our twofold question. How was this kind of intentional top–down initiated and directed organizational change in general possible, and how can the specific variants of change that we find in our cases be explained? To answer these questions we look at the authority relations of Lower Saxony's universities, and the changes of these relations with the recent NPM reforms. Our focal actor is the university leadership, and we study this actor in its relation to three other actors: the scientific elites from which the WKN evaluators are recruited, the ministry responsible for the universities in Lower Saxony, and the institutes or faculties at the universities which are the targets of profile-building activities. It is the interplay of these four actors which explains that top–down profile-building in general now occurs more often at Lower Saxony's universities than before, and also explains the particular outcomes of these efforts.

4. Authority Relations Part 1: University Leadership and Scientific Elites

A key part of the recent NPM governance reforms in Germany, and in Lower Saxony in particular, was a *strengthening of university leadership*.[5] The formal responsibilities of the deans and presidents were extended, and they gained formal powers in decision-making that were relocated both from the state and from the bodies of academic self-regulation. Furthermore, a longer stay in office increasingly becomes the rule, and presidents need not be professors of the respective university but can be recruited from outside. These governance changes not only allow for a more managerial approach to university leadership but also make the leaders more independent from the academics at their university. This new independence is also formally underlined by the fact that university rectors or presidents are now often appointed by university boards, which include stakeholders from outside the university, whose choice needs only

---

[5] As an extensive general overview of governance changes in the German university system since the 1990s see Kehm and Lanzendorf (2006); Lanzendorf and Pasternack (2009) note developments of the last years.

approval from the senates.[6] Also, in line with the NPM blueprint, state authorities have partly retreated from detailed regulation and favoured more global forms of steering from a distance. Under the heading of 'university autonomy' the specification of financing procedures and teaching programmes was reduced, and university boards, mission contracts between universities and ministries, a performance-based allocation of basic funds, and the accreditation of study programmes by special agencies, were introduced as substitutes.

The changing authority relations in the German university system do not only reflect new distributions of power but also a new institutional model of the university as an organization. Universities are now perceived as organizational actors in the sense of coherent and compact entities that are in control of, and accountable for, their own actions (Krücken and Meier 2006; Whitley 2008; Meier 2009). In this new understanding, teaching and research are not only individual but also organizational tasks. This is the context in which profile-building has become a central task of university leaders. Not all of them have already adopted the new understanding of their role within the new model of the university, but for our cases we chose only those leaders who were willing to push forward profile-building.

If a university leadership identifies with its new role, it is rational for several reasons that profile-building in general is high on its agenda. First of all, profile-building is one of the 'myths' of rationality (Meyer and Rowan 1977) of the current higher-education reform programme. It is broadly believed in the science policy-making scene that profile-building is a good thing, as it is supposed to improve research quality—so even if a university leader did not share this view he had better behave as if he did, because his corresponding activities are legitimized and non-activities in this direction raise critical questions. But most university leaders seem to be honestly persuaded of the desirability of profile-building.

This broad acceptance is certainly reinforced, secondly, by the fact that collective profiles are, almost by definition, much more visible than dozens of small-scale individual research lines. University leaders often need to present something to impress others such as the ministry, the university board, funding agencies, international partners, potential sponsors from business, or journalists, favourably with regard to their university. So leaders are happy with profiles because they help them in their 'impression management' (Goffman 1959).

---

[6] If this approval is not given, the board has the last word. However, in Lower Saxony university boards are only advisory bodies.

Thirdly, profiles are more susceptible to managerial authority than are other relevant impression 'myths', such as a large number of publications in international peer-reviewed journals. Managers have more direct means at hand to bring about a collective research profile of an institute or a faculty—especially in the short run—than to produce an excellent publication output. At least, *formal structures* that signal profile-building are comparatively easy to establish by leadership decisions.

This general commitment of most university leaders to profile-building goes along with more or less specified ideas about what the profile of a particular institute or faculty should look like. In case A the university president had a rather clear concept for the research profile of the university, whereas in case B the university leaders were aware that the lack of a profile for the particular discipline was a problem which was only underlined by the WKN evaluation—but they had no substantial idea of their own about in which direction profile-building might go. Cases C and D were in-between: the university leadership had some ideas, but these were not very precise, and it was also open to other suggestions.

A second determinant of a university leadership's success in profile-building is how much attention it is willing and able to devote to it. As loosely coupled organizations with diffuse goals and ill-defined core technologies, universities are predisposed to 'garbage-can decision-making' (Cohen *et al.* 1972; Cohen and March 1974) in which quite often unexpected decision alternatives pop up as 'simply a result of timing' (March and Olsen 1976: 31) and decision agendas for the leaders are generally unstructured and overloaded. Confronted with too many issues for a limited span of attention, university leaders are well advised to select a small number of issues to which they are able to pay constant attention and to adhere to them persistently (Cohen and March 1974: 207–8; Musselin 2007: 77). In all four cases which we studied, this was what the leadership did.

In particular, the university presidents showed a high personal engagement. This manifested itself in their regular presence at the meetings of the special commissions and in their active participation in the discussions of the issues at hand. The engagement had a symbolic effect as well, showing to everybody that the president had a personal interest in the issues which motivated those who also wanted to achieve changes, and discouraging those who otherwise would have shown stronger resistance. Although we have studied no such case in depth, we are quite sure that management-initiated profile-building, as any other ambitious project of change, is doomed to failure if there is no-one in university leadership who feels personally responsible for pushing this issue forward until an irreversible result is reached. In this sense, the personal factor really matters.

From this perspective, the period when our cases happened was problematic, because German universities in general were overwhelmed with diverse simultaneous reforms. After decades of delayed reforms, now everything started simultaneously: the far-reaching governance reforms of NPM, the Bologna process as a comprehensive reform of the study system, and reforms of research which could be seen as the implementation of 'mode 2 of knowledge production' (Gibbons *et al.* 1994) alongside the traditional mode. An important reason why policy-makers, in spite of their limited capacities of decision-making and implementation, started all these ambitious reforms at the same time was simply that they did not know how long the 'policy window' to change the status quo at universities in these respects would remain open.[7] This was an understandable rationale. Nevertheless, one can easily imagine that the other side of the coin was turbulence caused by reform measures which were implemented simultaneously without attention to interaction effects. But, despite the fact that university managers were confronted with many reforms at the same time, quite a number of successful cases of management-initiated profile-building can be identified—which documents the high priority of this issue for university leaders.

But even if the university leadership is committed to profile-building and it can invest enough time and energy to adhere to a profile-building process, it is in need of another essential resource: information. Information is needed in two respects: first, sufficient and valid information about the actual state of research and about viable paths of profile-building; and secondly, information to legitimize decisions. Even if university leaders are aware of the strengths and weaknesses of research at their university and have a clear concept of the kind of research profile they regard as appropriate and desirable, they need some support from the bodies of academic self-governance, and even more support by the individual professors, to implement their plans.

In spite of their new formal powers, university leaders are by no means autocratic rulers. And even where it is legally and technically possible to

---

[7] One reason perceived by policy-makers as well as university leaders why the window of opportunity to bring about any kind of change will close again soon is the large-scale generation change of professors which started some years ago and will be over in a few years. The retirement of many professors means that professorships become generally disposable, not just in a coincidentally scattered manner, and can be reallocated, or their denominations can be changed.

ignore the views of an institute, a faculty, or the senate, this is perceived not to be in accordance with the traditional academic culture of decision-making, and produces anger, resistance, or simply demotivation. The latter is critical because in the end the university leaders depend on the compliance of professors and the scientific staff in general. University leadership may decide on its own what a collective research profile should look like, but it cannot simply command professors to conduct coordinated and cooperative research which fits this profile. Dedicated and creative work which realizes the potential of a profile cannot be enforced.

With regard to both aspects, research evaluations can provide valuable informational resources (see Chapter 5). This held particularly true for the twenty-five evaluations of disciplines or transdisciplinary scientific fields at all universities in Lower Saxony undertaken by the WKN between 1999 and 2005 (WKN 2006; Schiene and Schimank 2007).[8] The interplay of university leadership and WKN was an exemplary demonstration of the new connection of local organizational leaders with national *scientific elites*, and of the driving force originating from this connection with respect to organizational change:[9] University leaders were provided with a well-founded estimate of the relative standing of their university's institutes or faculties compared to corresponding institutes or faculties elsewhere; they received information about research strengths and weaknesses, and about the causes of both. On the basis of this knowledge they can conceive and legitimate specific changes such as an extension or a reduction of a discipline's research capacity at their university, compensation for weaknesses, or a systematic search for professors with special qualifications.

---

[8] In 2006 the WKN decided to continue its work with more *ad hoc* local evaluations, on the one hand, and on the other, more general assessments and recommendations of larger scientific fields, each of which comprises a number of disciplines.

[9] There was a standard procedure for the WKN evaluations (WKN 1999; Orr 2003: 28–35). Based on self-reports of all locations and visits, the group of evaluators wrote a report which contained both evaluations of research quality and recommendations. The published evaluation report gave a detailed overall and comparative picture of a discipline or transdisciplinary field at all universities in Lower Saxony. It assessed not only the activities of each research unit, but also discussed the discipline's situation at each university visited and the overall situation of the discipline in Lower Saxony in general. Follow-up reports after three or four years documented what improvements had been attempted and achieved as a consequence of the initial report. This evaluation procedure of the WKN scores high in four of out of five quality dimensions: richness, timeliness, validity of information provided, and legitimacy of the judgements based on this information (see Chapter 5). However, with respect to the comparability of evaluations of different disciplines this procedure is of limited use. Thus, it seems to be a very powerful procedure as long as comparisons remain within a discipline or transdisciplinary research field.

In our case A, the university leadership stated without hesitation that the recommendations of the evaluators were absolutely in line with their own intentions and that the evaluation made it much easier to implement their reforms. Here the legitimation aspect was central. In two other cases—C and D—the evaluation provided the university leaders with both direction and legitimation. They were provided with new ideas on how to reorganize an institute or a faculty with respect to its research profile, and with good reasons to justify reforms in this direction. Though in these cases the recommendations produced at least some new and valuable insights for the university leaders, the interviewees reported that the general direction of the evaluation had not come as a surprise to them. Still, they very much welcomed the recommendations not only as a support of their concrete reform plans but also as a longed-for occasion to become active. Even in case B, where the university leadership had no concrete plans and the recommendations of the WKN remained too vague to instruct concrete changes, the president could use the external pressure by the WKN to initiate a collective search process for a research profile by the establishment of a special commission.

In case C the evaluators' recommendations explicitly differed from initial ideas of the university president, but an open conflict was avoided because these ideas were still rather vague, and the president retired from office before the change process actually started. An interesting situation would occur if recommendations by the WKN and stated intentions of the university leadership overtly clashed. Something like that happened at the university of case B in the evaluation of a different discipline. There the leadership adopted the position of the evaluated professors that the evaluation gave an inadequate picture of their work and ambitions and silently ignored the evaluation. Another possible scenario when the plans of the university leaders differ from what is recommended by the evaluators could be that the respective institute or faculty perceives the latter's intentions as the greater evil, and thus a 'common enemy' is constructed which might increase the support for what the leaders want. Even this is a means by which the WKN could legitimate profile-building activities of the university leadership.

The ultimate provider of these informational resources is the group decision-making of members of the scientific elite in the disciplinary evaluation commissions of the WKN. Thus, the scientific elite has become very influential in these processes of profile-building.[10] Here it is very important

---

[10] See also Whitley (Chapter 1), who repeatedly emphasizes an increased influence of the national—or even international—scientific elites in the new authority relations of national university systems.

that the scientific elites differ from the 'rank-and-file' members of their disciplines with respect to their opinion on the desirability of collective profile-building (Schiene and Schimank 2007: 182–3). For various reasons, the elites share, beyond disciplinary boundaries, the 'myth' of profile-building, and they do it with a basically functionalist reasoning according to which local mutual cognitive stimulation—which presupposes positive coordination of research lines—and even more local cooperation are in more and more research fields not only supportive but necessary elements of research quality. This transdisciplinary convergence of elites' views is not shared by the majority of professors in many disciplines, especially in the social sciences and humanities, who still follow individual research agenda.[11] Thus, the WKN evaluations are part of a transdisciplinary research policy avant-garde which challenges the traditional understanding of research quality with a new 'myth'.

Some aspects of the description we have given of the relationship between academic elites and university leaders might, on first glance, give the impression that university leadership has become nothing but the executive body of the scientific elites, who actually decide what to do about certain disciplines or transdisciplinary fields at particular universities. However, such a view disregards the autonomous interests and information of the leadership. A more realistic characterization of the new authority relations sees the university leadership at the centre of a set of actors among whom scientific elites certainly are very important. Two additional actors will be discussed at greater length in the next part: the Ministry and the targets of profile-building.

Additionally, the university leadership is embedded in authority relations with other actors, such as micro-political coalitions with various groups within the university, the university board, funding agencies, accreditation agencies,[12] and regional stake-holders, especially from the business world, or journalists. Our cases show that such other players may exert their influence on university managers, thereby modifying the influence of the WKN evaluators. Moreover, the leadership can actively use such cross-pressures to neutralize some, so that it need not act strictly according to any one of them. From such constellations, managers gain limited autonomy.

---

[11] This preference can be seen, for instance, in the predominance of individual projects funded by the DFG in these fields.
[12] Even though their domain is teaching, their recommendations can have important side effects on research, as we will show later, and these effects can be strategically used.

The leadership's scope of discretion consists especially in the following aspects of the evaluation procedure and change processes:

- University leaders may suggest, from their perspective, specific problems or chances of a discipline or research field at their university to the evaluators. As in case B, this may influence the results of the evaluation process, especially with respect to the recommendations given.[13]

- There is also the need to interpret the recommendations given, including some scope to choose which recommendations are prioritized. This can be shown in all four cases we studied. The scope of interpretation becomes even greater in the implementation of concrete measures. For example, when in case B the evaluators recommended the establishment of a special commission of mainly external experts on the structural renewal of an institute, the university president influenced the selection of these experts.

- The modification of, or even deviation from, recommendations by the university leadership is possible if good reasons can be articulated. For instance, a lack of financial resources can be declared; and it is usually difficult to demonstrate that the money would be available if the leadership really wants to achieve the particular recommendation. Another good reason is if the recommendations from other WKN evaluations contradict a particular recommendation, as we mentioned in case C.[14]

Up to this point we have mainly focused on the dyad of university leadership and scientific elites. We see that profile-building as a 'myth' is present on both sides, and that university leadership in general puts a strong emphasis on this issue and often has its own ideas about how the research profile of an institute or faculty should appear. This more or less omnipresent combination of factors—to which must be added the new powers of leadership and the installation of performance measurements—explains that in all cases substantial progress was made with top–down initiated and directed profile-building.

Looking now at case A, the decisive factors in explaining the very successful profile-building there seem to be located in this dyad: an energetic

---

[13] By now, the WKN has modified its procedures so that instead of regular evaluations for which the date is fixed by the WKN the university leadership may take the initiative to ask for an evaluation of a particular discipline or research field at its university. This right of initiative certainly is an important element of the leadership's autonomy.

[14] As will be discussed, the pressure imposed by the ministry to implement changes, clearly restricts the possibility of deviating from the recommendations.

leadership giving a high priority to the issue of profile-building and having concrete views of the profile, and an evaluation which strongly reinforced this policy of the leaders. In contrast, in case B the leadership was not that engaged and had no clear ideas, and the evaluators, too, did not give sufficient recommendations in this respect. In case C there is even a contradiction between the university leadership's initial ideas of profile-building and the evaluators' insistence on the higher priority of teaching concerns. So it is understandable that profile-building made less progress in these two cases compared to case A.

However, case D also shows an engaged leadership and recommendations from the evaluators which underline the leaders' ideas—and still profile-building has as yet produced not much more than a façade. One answer why this is so has already been given: the recommendations here conflicted partly with recommendations of other evaluations. But this is not the whole explanation. This case rather draws attention to two other actors in the authority relations of Lower Saxony's universities: the Ministry and the targets of profile-building.

5. Authority Relations Part 2: The Ministry and the Targets of Profile-Building

The dyad of university leadership and scientific elites we have analysed until now is clearly a key element of the new constellation of actors relevant for top–down profile-building in universities. When we turn now to Lower Saxony's *Ministry* of science and cultural affairs that is responsible for the universities of this state this may come as a surprise, since this Ministry had declared that it was giving the universities self-governance. So is it still an important player?

With the foundation of the WKN the Ministry had committed itself not only to accepting but to promoting the recommendations of the WKN evaluators.[15] Moreover, the ministry shared the 'myth' with regard to profile-building. Accordingly, first of all it supported university leadership with an informal external pressure to implement changes. This pressure could be used by university leaders as a legitimation of their activities. They could even put the blame for what they did on the Ministry, against whom 'any resistance is futile'. Beyond this diffuse support of university

---

[15] This is the same logic of 'outsourcing' of competencies as with the installation of university boards.

leadership there was also specific support of particular measures by the Ministry. In cases A and C the university leadership could refer to a concrete informal threat from the Ministry: if certain changes were not implemented the respective disciplines' very existence at the particular universities was at stake. That in cases B and C the specifications of new professorships, which were cornerstones of profile-building, had to be approved by the Ministry were detailed demonstrations of this pressure; and the Ministry continued to play an important role in the background.

Besides pressure, the Ministry sometimes—as in case C—was also helpful in supporting changes with additional resources given to the university. Provided the respective institute or faculty did what it was asked to do it received, under unusually favourable conditions, an additional professorship, or—more typically—some money for infrastructural purposes such as its library or laboratories. Under some circumstances, though, an overall resource scarcity which the Ministry is not willing to compensate may legitimize drastic measures, as in cases A and C. If large-scale cost reductions are inevitable, they may be used to reshape an institute's or faculty's structure of professorships after a long period of directionless growth. In case C this opportunity was very strategically taken by a new university president. On that occasion necessity surely was the mother of organizational innovation.

An unintended influence of the ministry on profile-building resulted from the so-called *Hochschuloptimierungskonzept* (higher education optimization concept) which started in 2003 and consisted of significant enforced overall cost-reductions for university budgets—as for other public expenditures—as a consequence of general deficits of the public budget of Lower Saxony. This negative-sum game reduced a university leadership's room to manoeuvre if it wanted to initiate an institute's profile-building, for example, by recruiting a good performer on a newly established professorship, hoping that s/he might carry along the whole institute after a while. In case C, although new personnel could be recruited as a consequence of the retirement of professors, this opportunity for change was very much impaired by the sudden announcement that the institute had to give away another professorship as its contribution to the *Hochschuloptimierungskonzept*. This sharply increased the level of conflict and prevented all efforts to create an atmosphere of constructive discussion about the future profile of the institute.

Overall, the Ministry acts like a 'big brother' that the WKN and the university leaders can mobilize in support, despite its delegation of formal competencies and the corresponding declaration that it would refrain from intervention. Since the targets of profile-building usually have no direct contacts with the Ministry, the WKN and the university leaders can bluff a lot. They can threaten vaguely but effectively with this 'higher power'; and some demonstrations of its punishing or supportive local intervention can be pointed out as demonstrations of the Ministry's willingness to intervene if needed. The overall scarcity of money underlines only the importance of keeping good relations with the Ministry, and rare events of financial generosity show how obedience may pay off. Thus, the Ministry predominantly worked as a reinforcement of the driving force in the interplay of university leadership and scientific elites.

The *targets of profile-building*—the evaluated institutes or faculties—differ strongly in both their interests and in their affinity to profile-building. Their capacity for collective action, which is essential when assessing their role in the new authority relations, also varies considerably. The first relevant factor here is the professional and epistemic culture of a discipline,[16] especially how disciplinary cultures differ with respect to the acceptance or even habitualization of all kinds of performance measurements.[17] Members of those disciplines where research is strongly dependent upon project funding and where a widespread peer review of manuscripts submitted to journals exists, are accustomed to frequent assessments of the quality of their work. The higher a discipline's institutionalization of performance measurements,[18] the less irritation or resistance is shown by its members if they are subjected to measures which aim at an improvement of their research performance—including profile-building. Of course, it still makes a difference whether such quality control is exercised by the scientific community or by the leadership of one's university, especially because applications for project funds or submissions of manuscripts are done on a voluntary basis, which is not the case when one is evaluated.

However, as the recent decision of German historians not to participate in the Science Council's national 'research rating', in comparison to the

---

[16] There is no coherent body of knowledge from sociology of science with respect to this factor. See, as diverse contributions which focus on different elements of a discipline's culture, Whitley (1984) and Fuchs (1992) on the nexus of cognitive and organizational features, Becher (1989) on 'academic tribes', or Knorr Cetina (1999) on 'epistemic cultures'. See also Ch. 11. With respect to sociology Buroway (2005) sketched a picture of the different audiences of its knowledge production which—with adaptations—can be applied to other disciplines as well.

[17] See Schröder (2004) for a German case study focusing on the acceptance of quality assurance.

[18] This, of course, somehow correlates with the degree of consent with regard to performance indicators.

unquestioning compliance of chemists with this evaluation, shows, there are significant disciplinary differences; and a university leadership certainly meets less resistance to the consequences it draws from research evaluations from disciplines with an explicit culture of quality measurement. This cultural factor was favourable for the measures of university leadership in case A, and partly also in case D, whereas case C showed the typical negative reaction of most social scientists.

If clear criteria of performance and quality are broadly accepted, the university leadership can use performance-based funding to sharpen the profile. The advantage of this strategy does not only rest on the comparatively low transaction costs. By using formal criteria the leadership avoids being drawn into academic controversies (Musselin 2007: 78), it makes invisible the decisions underlying the dynamics of profile-building, and thus enhances its legitimacy. In Case A the university leadership clearly regards the introduction of a performance-based funding scheme as a central element of its profile-building efforts.

Part of a discipline's epistemic culture is its inherent cognitive pressure towards 'critical masses' of research capacities, in the quantitative as well as in the qualitative sense.[19] Many research processes in the natural and engineering sciences require a cooperative effort of several and sometimes even hundreds of scientists who submit to a division of labour and intensive coordination. Performance criteria in these disciplines often emphasize the necessity of 'critical masses'. This approach is uncommon in the social sciences and humanities, where the bulk of research work is still done by individual researchers. Thus, activities of the university leaders towards profile-building may find a very favourable response in some disciplines— case A is an example—whereas other disciplines may be quite reserved or even irritated, as case C demonstrates. There is a considerable likelihood of mere window-dressing in the latter disciplines, for which case C is also an illustration.[20]

Whereas the discipline's culture may be more or less favourable to collective research profiles, unambiguous indicators of the target's positive attitude to profile-building are individual or, even better, collective bottom–up initiatives in this direction. Sometimes these grassroots initiatives have developed over time to quite substantial results, so that university

---

[19] This cognitive pressure should not be reified. In fact, it expresses nothing but an historically given fit of the discipline's social organization and cognitive shape which, however, can be very stable for a long time (Fuchs 1992).

[20] This may be the most adequate reaction to an inadequate 'one size fits all' approach by the WKN evaluations and university leaders.

leadership has nothing more to do but to support further steps on a well-established path. This was the situation in case A when the new university president came into office. A number of large-scale and visible cooperative research activities already existed so he just needed to further advance and sharpen the university's pre-existing research profile.

If such bottom–up activities go precisely in the direction the university leaders want to go, they have just to support them in diverse ways. In other cases such initiatives have to be adjusted to the goals of the university leaders—provided they have formulated concrete goals. More often it will happen the other way round: a university leadership adapts its initial ideas for profile-building to what emerges bottom–up because the probability that an improvement of performance is reached is higher this way, and for the leaders profile-building is not an end in itself but a means to the end of higher research quality or, more simply, a good public standing of their university. Besides, if the university already has established big and visible externally funded research cooperations, the university leaders have hardly any legitimate alternative to continuing support of these activities.

The opposite of bottom–up initiatives is resistance to profile-building by the target, be it motivated by general fears that 'academic freedom' is in danger or by more specific individual interests. Such resistance can be overcome by the university leadership if the target is weak in terms of its collective capacity to act. An element of a discipline's epistemic culture which must be taken into account here is its inherent cognitive homogeneity or heterogeneity.[21] An extreme expression of this cultural factor is strong intra-disciplinary paradigmatic conflict among different theoretical or methodological 'schools' represented by local professors. Case C showed ambivalent effects of this on the change dynamics. On the one hand, this institute's collective capacity to act was so weak that it could not show much resistance to the measures executed by the university leadership. On the other hand, however, a sustainable implantation of an agreed-upon collective profile cannot succeed as long as this dissensus continues to exist. As a consequence, change activities must attempt to bring about a greater cognitive homogeneity of the local representatives of the discipline.[22] In a sense, this is profile-building—but with the risk that a fad which sooner or later is a declining 'school' is locally cultivated.

---

[21] See also Geser (1975) on 'paradigmatic consensus' or 'dissensus'. Again, this feature of a discipline is to a large part a social construction.

[22] This holds true at least as long as one insists on only one coherent profile.

If profile-building must be achieved with the same personnel, especially on the level of professors, there is no other chance but to motivate them by incentives or persuasion to join in the cooperative research activities or to remove obstacles for such cooperation. This is comparatively simple if the main obstacle to profile-building is missing resources—as long as a university leadership has some disposable money which it can use to initiate, accompany, or speed up changes, as exemplified by cases B and C. For example, it is often appropriate that certain structural elements such as a new professorship are installed at a particular point in time, but a post that can be used to establish this professorship becomes available only some time later. In such a situation it is very helpful if the university leadership can bridge this time span by a pre-financing of this professorship. Another frequent use of such money is the support of the elaboration of a joint project application, which is often a lot of work.

Up to now we have discussed resistance to profile-building as something which for good reasons of scientific progress should be overcome. However, resistance may also be motivated by the fact that no real necessity for a collective profile exists—in other words, the format this 'myth' prescribes is overstretched. This is the explanation for the outcome of profile-building in case D. Bottom–up initiatives in different disciplines had produced scattered small- and medium-scale collective profiles in this transdisciplinary research field. The top–down attempt to build an integrative institute resulted not in all but in most respects in nothing but a façade. All researchers involved could live with the common label under which their work was subsumed; but their research topics and types of research were still so disconnected that they had no reasons for more communication, let alone coordination or cooperation. Since open resistance was imprudent because this research field was high on the public and political agenda, researchers behaved as if they were part of a new transdisciplinary community in which everybody profits from everybody else.

Summing up what the analytical inclusion of these two actors—the Ministry and the targets of profile-building—adds to the understanding of top–down profile-building and its outcomes, we can say that both moderate the driving force of the activities of university leaderships backed by the scientific elites. The Ministry mainly strengthens the driving force, and the targets can strengthen or weaken it depending on disciplinary culture and local opportunities or restrictions. So the activities of both actors are additional explanatory factors in cases A, B, and C; and the target even provides a decisive factor in case D, where the driving force of leadership and evaluators is almost stopped, at least for the moment.

6. Research Profile-Building and Teaching

Up to now we have discussed universities as research organizations. But they have teaching as a second mission alongside research; and this duality of missions affects profile-building.[23] The closer the coupling of teaching and research, the stronger considerations of teaching intervene in the profile-building process.[24]

This connection is very pronounced in case C. Here an institute showed a very weak research performance; and in addition, in terms of specifications of professorships it lacked several basics of the discipline which are core elements of any teaching programme. To improve the research quality of this institute it is not necessary to install professorships for these basics with which every trained researcher is well acquainted; indeed, to carve a clear-cut research profile of the institute it would be better to do without the basics in favour of specialties which complement each other. Nevertheless, the WKN evaluation group recommended strongly—and not only in this case—that the first thing to do was to establish professorships for these basics; and university leaderships often followed this advice.

Case D is a similar example. Here a transdisciplinary research field had profited in the past from several professorships of various disciplines. When the professors retired and successors were recruited this thematic connection to the transdisciplinary field was deliberately neglected in favour of recruiting new professors who had a very good teaching and research standing in the basics of the respective disciplines; and the main reasons given for this—in this case against the recommendation of the evaluators of the transdisciplinary field, but supported by the evaluators of the disciplines—were teaching concerns.

It is surprising that evaluators whose explicit task was to look after research quality partly neglected this mission and instead first of all took care of the requirements of teaching.[25] This neglect, moreover, happened without much discussion, which indicates that the precedence of teaching concerns seems to be almost a self-evident priority even among these members of the disciplinary elites who won their personal reputation

---

[23] For the medical schools of universities or for medical universities, patient care is a third task which may be even the dominant one under certain circumstances (Braun 1991: 37–57).

[24] See Meier and Schimank (2009) for tendencies of a decoupling of teaching and research in the wake of NPM reforms in different countries.

[25] In this respect, Lower Saxony is a good case because there the evaluation of teaching and the evaluation of research are executed by two different organizations with different groups of evaluators for the same discipline who publish independent reports. Responsibility for the evaluation of teaching lies with the Zentrale Evaluations- und Akkreditierungsagentur (ZEVA).

with research, not teaching activities. Despite the fact that the evaluators often criticized the time pressure of the teaching load on research work they did not use the opportunity to take a stand for research quality but, on the contrary, worried about the education of future members of the respective disciplines. The priority of teaching over research appears to be anchored in the professional identity of disciplinary elites who have more contact with the discipline's stakeholders than ordinary members of the disciplinary community. As a consequence, university leaders whose initial focus is on profile-building did not receive the orientation and legitimation they wanted. However, in the two cases mentioned they changed their own original plans and followed the recommendations of the WKN evaluators.

Taken together, teaching concerns are of a high priority to the scientific elites as well as to university leadership—as long as there are no universities with a profile giving clear priority to research in Germany, which may come over time with the recent 'excellence initiative'—and, finally, also to the Ministry, which wants to provide more and more young people with a diploma from a higher-education institution and needs the Bologna reforms to be a success story. If research profile-building conflicts with such teaching concerns it is not surprising that the latter comes first.

7. Conclusion

We have studied two elements of NPM reforms in the German university system and their interplay in this chapter: the strengthening of hierarchical self-governance of universities, and the introduction of evaluations of research performance. These changes of the governance regime brought about a corresponding change in authority relations in universities. The traditional dual power structure dominated by the Ministry, on one side, and the academic community at large as well as each individual professor, on the other, was transformed into a new dual power structure of university leadership and national scientific elites, with the latter becoming a crucial supporter of the former's activities of profile-building.

The establishment of the WKN in Lower Saxony institutionalized this support function of national scientific elites in an unprecedented way, and not only with respect to Germany. The WKN evaluations offered university leadership not just assessments but recommendations—more precisely, recommendations based on assessments. This served as a key resource of orientation and legitimation of top–down initiated and directed profile-building. In this way the general proliferation of profile-building activities all over Lower Saxony can be explained. Still, specific variants of this interplay of leadership and scientific elites as well as additional factors—the Ministry, the targets of profile-building, and teaching concerns—have to be taken into account in the explanation of specific outcomes of profile-building.

Looking beyond Lower Saxony, by now everywhere in the German university system profile-building is attempted, and everywhere university leaders play an important role in these endeavours. Of recent prominence is the 'excellence initiative'. Convincing profile-building is the heart of a successful application in all three funding lines of the 'excellence initiative'. Compared to what was recommended by the WKN, the profiles demanded in the 'excellence initiative' were much more ambitious. They were usually transdisciplinary and included, especially in the third funding line, not only research but also teaching and other activities of a university. However, from a different perspective profile-building in the 'excellence initiative' was easier than in a number of cases in Lower Saxony, as strong incentives were at work and the high performers involved in the 'excellence initiative' could usually build on many previous activities. Then again, university leadership in the 'excellence initiative' lacked the systematic backing by evaluators whose recommendations gave both orientation and legitimation of the measures taken by the leaders. In this respect, the WKN and the universities in Lower Saxony have a very special relationship, and this gives rise to the question of what are potential functional equivalents for the WKN in other states and other exercises of profile-building.

It is surely worthwhile to focus future studies on different kinds and constellations of profile-building and to compare the results, so that by and by a more thorough understanding of success factors may be achieved. A further question we could not deal with in this chapter concerns the real effects of successful profile-building on research quality. To begin with, it is clear that profile-building is only one of several factors that determine the quality of research. The general causal link which WKN evaluators perceive, and which is the core of the conventional 'myth', combines an inner-scientific and an extra-scientific dynamic: The problems which nowadays are being studied in more and more disciplines are increasingly on a larger scale than those capable of being researched by the traditional individual scientists; and the societal problems for which answers are sought from science, be it climate change or international terrorism, have also often become too big in their manifold implications for disconnected individual research activities (Schiene and Schimank 2007: 180–1).

An additional argument for local profile-building could be drawn from Stephan Fuchs's (1992) thesis that the cognitive anomie of many humanities and social sciences, compared to the natural sciences, is a consequence of the low task interdependence of their research activities. A positive coordination of research, and even more so research cooperation, could lead to a more codified theoretical and methodological body of knowledge in these disciplines. These—and possible other—assumed functional effects of profile-building on research quality must be confronted with suspected dysfunctional effects. The other side of the just-mentioned disciplinary codification is a narrowing down of cognitive diversity to mainstream research which produces incremental but no radical innovations.[26] And the ability produced by profile-building to handle large-scale research problems may lead to a marginalization of all those approaches and topics for which disconnected individual research is not only a possible but the most adequate format. Thus, the changes we studied in this chapter are certainly at present and perhaps lastingly ambivalent with respect to the future advancement of scientific knowledge. This is another reason to keep them under close observation.

References

Becher, Tony (1989) *Academic Tribes and Territories: Intellectual Enquiry and the Cultures of Disciplines* (Milton Keynes: SRHE/Open University Press).

Ben-David, Joseph (1971) *The Scientist's Role in Society: A Comparative Study* (Englewood Cliffs, NJ: Prentice-Hall).

Braun, Dietmar (1991) *Die Einflußmöglichkeiten der Forschungsförderung auf Strukturprobleme der Gesundheitsforschung in der Bundesrepublik* (Bremerhaven: Wirtschaftsverlag NW).

Brunsson, Nils (1989) *The Organization of Hypocrisy: Talk, Decisions, and Actions in Organizations* (Chichester: Wiley).

Buroway, Michael (2005) 'For Public Sociology', *American Sociological Review*, 70: 4–28.

Cohen, Michael D., and March, James G. (1974) *Leadership and Ambiguity* (Boston: Harvard University Press).

————and Olsen, Johan P. (1972) 'A Gargabe Can Model of Organizational Choice', *Administrative Science Quarterly*, 17: 1–25.

di Maggio, Paul J., and Powell, Walter W. (1983) 'The Iron Cage Revisited', *American Sociological Review*, 48: 147–60.

---

[26] See Ch. 1 for constellations which may bring about such results.

Fuchs, Stephan (1992) *The Professional Quest for Truth: A Social Theory of Science and Knowledge* (Albany, NY: State University of New York Press).

Geser, Hans (1975) 'Paradigmatischer Konsens in Forschungsorganisationen', in Nico Stehr and René König (eds.), *Wissenschaftssoziologie* (Opladen: Westdeutscher Verlag), 305–27.

Gibbons, Michael, Limoges, Camille, Nowotny, Helga, Schwartzman, Simon, Scott, Peter, and Trow, Martin (1994) *The New Production of Knowledge* (Beverly Hills, Calif.: Sage).

Goffman, Erving (1959) *The Presentation of Self in Everyday Life* (New York: Anchor Books).

Kehm, Barbara, and Lanzendorf, Ute (2006) 'Germany: 16 Länder Approaches to Reform', in Barbara Kehm and Ute Lanzendorf (eds.), *Reforming University Governance: Changing Conditions for Research in Four European Countries* (Bonn: Lemmens), 135–86.

Knorr-Cetina, Karin (1999) *Epistemic Cultures: How the Sciences Make Knowledge* (Cambridge, Mass.: Harvard University Press).

Krücken, Georg, and Meier, Frank (2006) 'Turning the University into an Organizational Actor', in Gili S. Drori, John W. Meyer, and Hokyu Hwang (eds.), *Globalization and Organization: World Society and Organizational Change* (Oxford: Oxford University Press), 241–57.

Lanzendorf, Ute, and Pasternack, Peer (2009) 'Hochschulpolitik im Ländervergleich', in J. Bogumil and R. G. Heinze (eds.), *Neue Steuerung von Hochschulen: Eine Zwischenbilanz* (Berlin: Edition Sigma), 13–28.

March, James G., and Olsen, Johan P. (eds.) (1976) *Ambiguity and Choice in Organizations* (Bergen: Universitetsforlaget).

Meier, Frank (2009) *Die Universität als Akteur: Zum institutionellen Wandel der Hochschulorganisation* (Wiesbaden: VS).

——and Schimank, Uwe (2009) 'Matthäus schlägt Humboldt? New Public Management und die Einheit von Forschung und Lehre', *Beiträge zur Hochschulforschung*, 31(1): 42–61.

Meyer, John W., and Rowan, Brian (1977) 'Institutionalized Organizations: Formal Structure as Myth and Ceremony', *American Journal of Sociology*, 83: 340–63.

Musselin, Christine (2007) 'Are Universities Specific Organizations?', in G. Krücken, A. Kosmützky, and M. Torka (eds.), *Towards a Multiversity? Universities between Global Trends and National Traditions* (Bielefeld: transcript).

Orr, Dominic (2003) 'Verfahren der Forschungsbewertung im Kontext neuer Steuerungsverfahren im Hochschulwesen: Analyse von vier Verfahren aus Niedersachsen, Großbritannien, den Niederlanden und Irland', *HIS Kurzinformation*, A1/2003: 16–74.

Schiene, Christof, and Schimank, Uwe (2007) 'Research Evaluation as Organizational Development: The Work of the Academic Advisory Council in Lower Saxony (FRG)', in Richard Whitley and Jochen Gläser (eds.), *The Changing Governance of the Sciences: The Advent of Research Evaluation Systems* (Sociology of the Sciences Yearbook, 26; Dordrecht: Springer), 171–90.

Schimank, Uwe (2005) ' "New Public Management" and the Academic Profession: Reflections on the German Situation', *Minerva*, 43: 361–76.

——— and Lange, Stefan (2009) 'Germany: A Latecomer to New Public Management', in C. Paradeise, E. Reale, I. Bleiklie, and E. Ferlie (eds.), *University Governance: Western European Comparative Perspectives* (Dordrecht: Springer), 51–75.

Schröder, Thomas (2004) 'Der Einsatz leistungsorientierter Ressourcensteuerungsverfahren im deutschen Hochschulsystem: Eine empirische Untersuchung ihrer Ausgestaltung und Wirkungsweise', *Beiträge zur Hochschulforschung*, 26(2): 28–58.

Whitley, Richard (1984) *The Intellectual and Social Organization of the Sciences* (Oxford: Clarendon Press).

———(2007) 'Changing Governance of the Public Sciences: The Consequences of Establishing Research Evaluation Systems for Knowledge Production in Different Countries and Scientific Fields', in Richard Whitley and Jochen Gläser (eds.), *The Changing Governance of the Sciences: The Advent of Research Evaluation Systems* (Dordrecht: Springer), 3–27.

———(2008) 'Universities as Strategic Actors: Limitations and Variations', in Lars Engwall and D. Weaire (eds.), *The University in the Market* (London: Portland Press), 23–37.

WKN (Wissenschaftliche Kommission Niedersachsen) (1999) *Research Evaluation at Universities and Research Institutions in Lower Saxony: Procedure Outline* (Hanover: WKN).

———(2006) *Forschungsevaluation an niedersächsischen Hochschulen und Forschungseinrichtungen. Bewertung des Evaluationsverfahrens* (Hanover: WKN).

III

Reorganizing Scientific Fields

Changing Authority Relations and Intellectual Innovations

8

Authority Relations as Condition for, and Outcome of, Shifts in Governance

The Limited Impact of the UK Research Assessment Exercise on the Biosciences

Norma Morris

1. Introduction

The UK Research Assessment Exercise (RAE) is the punk celebrity among the rather grey population of British science policies. It has had the power to shock and alienate, and has faced the charge of undermining academic culture as we know it by suborning academics to an insidious and morally suspect audit culture and Foucauldian surveillance (Morley 2005; Shore and Wright 1999). It has its devotees and cryptic followers in political and policy circles, and in universities and other research institutions. Its fame and influence have spread abroad, with similar polarized reactions. Over time it has become institutionalized. It loses power to shock and becomes 'part of the wallpaper' (Keenoy 2005). A positive contribution to the knowledge production industry is claimed—not least the value of its once-shocking ranking system as a marketing tool for academic institutions (HEFCE 2008; Royal Society 2006; Georghiou 2008).

The RAE is a periodic nationwide exercise that assesses the research performance of any university in the UK that opts to put itself forward, and a prime example of a fully institutionalised strong Research Evaluation System (Whitley 2007). Research infrastructure funds are awarded selectively to universities as a block grant on the basis of the scores achieved by their individual departments—or, more accurately 'units of assessment' (for a fuller account of the workings of the RAE see Chapter 2). Although the RAE has been around for nearly thirty years and many areas of its design, operation, and effects have been intensely explored, little attention has been given to its differential impact in different academic fields. In this chapter I focus on interactions at the level of research performers (drawing on previously collected empirical data, published articles, and archival sources) to provide a practical illustration of how differences in pre-existing authority relations in different academic fields, particularly natural science and arts/social sciences fields, significantly shaped their intellectual organization and policy-adaptability, and hence how the RAE was perceived and its effects managed within universities.

Framework for Analysis

In Chapter 1, Whitley sets out a general framework for analysis of governance changes, such as the RAE. He identifies six major changes in the period since the Second World War in 'the conditions governing the production, coordination, and control of formal scientific knowledge' which have had significant consequences for authority relations within systems. The RAE is a well-known example of one such governance change: namely, the institutionalization of procedures for assessing research organizations' performance and auditing their outputs. The RAE was both a culture-shock to British academia, and a potent instrument of governance on account of the direct link of its assessments with funding decisions. This assured its reputation as a key player in bringing about shifts in authority relationships—particularly those governing the selection of scientific goals and standards for evaluation of results in universities. As Whitley points out, however, the effects of changes in governance on authority relations in public science systems may be substantially modified by the way the system is organized, funded, and coordinated in different national contexts. Similarly, at the micro-level of the university or academic field, local political, social, and scientific contexts can influence the outcomes of specific interventions. Empirical evidence available on the RAE gives scope to map its impacts on authority relations governing knowledge production against field-specific power structures, practices, and external relations, and strategies developed by research leaders and individual scientists to manage these demands.

For analytical purposes university research is understood as part of a *multi-level research system* comprising a macro-level (government), a meso-level (intermediary institutions such as higher-education funding councils or research councils), and a micro-level of research performers. The micro-level may itself be further subdivided, as it comprises both institutions such as universities, and individual researchers/research groups working within these institutions. All are held together by mutual interdependencies. Each level has its characteristic attributes and dynamic, but the development of each is influenced by and influences the development of the others (Rip and van der Meulen 1996). Introduction of the RAE into this system added a new and potentially destabilizing element to these dynamics, and involved shifts in authority relations both between system levels and within them.

In this chapter I first examine the role of the UK natural sciences community in the genesis of the RAE and how this meshed with the distinctive organizational model and patterns of authority relations already prevalent in many UK academic science departments, contrasting this with the situation in other fields. The middle section illustrates how the practices of this community both shaped the impact of the RAE at the micro-level and were themselves shaped by it. The concluding discussion considers the degree of assimilation of the RAE, and what this means for authority relations and how far academics succeed in creating personal and professional space within a managerialist and competitive system.

2. The RAE in Context: Policy Issues around Sustainability of Science

To understand the impact of the selective policies implemented through the RAE, it is important to consider both the material circumstances and the policy concerns and discussions taking place in the years preceding its introduction. The general background of massive expansion, followed by economic downturn and political commitment to 'new public management', has already been extensively analysed, both in the context of the RAE (see, for example, Henkel 2000; Phillimore 1989; Tapper and Salter 2002; Chapter 2) and in relation to higher education and research more generally by, among others, Blume (1982), Kogan and Hanney (2000), and Whitley (Chapter 1). My focus is rather on how these developments reacted differentially with the different substrates of resource-intensive science faculties and faculties of arts and social sciences, with the effect that the former were better prepared for the RAE when it came on the scene in 1986.

The immediate backdrop to the RAE was the convergence in the 1970s and 1980s of financial retrenchment and demands for greater economic 'relevance' of science (factors shared with many other national economies including the USA—see, for example, Hackett 1987; Teich 1990) combined with a broader government commitment to reform the public sector on managerialist principles. Concerns in the (natural) scientific community about the financial sustainability of the research system had, however, surfaced earlier, based on their perception of changing needs in science. A key issue was the emergence of 'big science' in the sense of science requiring large teams and expensive facilities. This international trend was manifest mainly in physics in the 1950s and 1960s but spread to the biological sciences by the 1970s (Hagstrom 1964; Ziman 1983).

The perception was that the trend towards bringing multiple skills to bear on an investigation and to use ever more sophisticated technologies and facilities was driving the cost of research above inflation (Berger and Cooper 1979; Blume 1980; Cohen and Ivins 1967; CSP 1967: 26). There was also recognition that the exponential rate of growth of science since the Second World War was unsustainable (Price 1963). By 1970 the UK Science Research Council (SRC) had begun to implement policies of selectivity and concentration as necessary 'to sustain viable research groups' (quoted in Blume 1982: 23). Even the conservative Medical Research Council likewise set out a policy for some degree of selective funding of priority areas in its university support portfolio (MRC 1970: 5–7). A debate and publications sponsored by an influential charitable trust further articulated the scientific establishment's commitment to selectivity as the way forward for research (Oldham 1982), given the fear that rising research costs would otherwise outstrip any conceivable public-funded budget and the resultant chronic underfunding would spell death to high-quality scientific research.

The science community—as represented by the vocal scientific research elite, working with and through the publicly funded research councils[1]— had no doubt that selectivity and 'critical mass' was the way forward for research if Britain was to remain globally competitive in science. Government circles in most smaller economies took a similar view, though the USA could still sustain a view that it was big enough not to need to set priorities (Cozzens 2003: 510). The model for science championed by the disciplinary elites was resource-intensive, collectivized, and managed. Such a model

---

[1] The working alliance between the scientific elite and selective grant-awarding bodies such as research councils (Braun 1993; Rip 1994) was particularly well developed in the natural sciences and strengthened the authority of both parties within and across system levels. Government reforms to the structure of the funding bodies in the 1980s and 1990s (DES 1987; OST 1993), designed to draw them closer to government, diluted, but did not eradicate the alliance, based on mutual need and interest (Morris 2003).

made the scientific community (both the elite scientists and their numerous contract-funded retinues) highly dependent on competitively won funding from external sponsors. Their political energies were therefore focused on shaping the funding system to sustain this vision of the social and intellectual organization of the field.

It is apparent that the model implies particular patterns of internal and external authority relationships that will not necessarily be reflected in other academic fields, nor in some sub-fields within the sciences. The implications of 'big science' or collectivization for researchers' autonomy had already been remarked on—for example:

Recent changes in the organisation of research within certain specialties have brought about a definite increase in social relationships involving formal authority [...] Although an attempt is made to maintain some freedom of problem selection, each individual's choice is considerably restricted by the requirements of the group as whole. (Mulkay 1972: 22)

Earlier still, Hagstrom (1964) noted the rise of 'modern forms of scientific teamwork', stemming from technologization, and increasing need for specialized skills and interdisciplinary approaches. In this situation 'decisions about the selection of experiments are necessarily made collectively [so the individual scientist has] his independence restricted' (Hagstrom 1964: 251). Hagstrom also notes, however, that at the time he was writing (1964), working in groups 'in which authority is centralised' was still the exception rather than the rule.

Some twenty years on, Ziman (1983) confidently proclaimed (addressing a scientific rather than a sociological audience) that a radical, and general, transformation in ways of scientific working had already taken place: 'The crucial decision in the research process—"this is the question to be studied by this method"—now rests in the hands of organized groups of people, rather than in the hands of independent scientists [...] the traditional individualism of the academic mode of research has been decisively and irreversibly curbed.' The claim here that a significant change in authority relations—the curbing of individual autonomy, often attributed to the RAE—had already been brought about by a way of doing science well before the introduction of the RAE, provides a first indication of how field-specific differences in impact might arise. Demeritt (2000: 316) draws attention to such differences: 'The model of professor and graduate student apprentice, common in the social sciences and humanities, is being displaced in the sciences by large, hierarchically organized research groups [...] This organizational form is necessary to compete for research grants and capitalize on expertise.' Demeritt refers also to Etzkowitz and Kemelgor's (1998) article on the development of the collectivized model in the United States.

Targeting the University Block Grant

At this point it is necessary to recapitulate the basics of the funding system for UK public science. There are two complementary sources for funding for research in universities,[2] the so-called 'dual support system': one arm based solely on the principle of competitive project funding, the other providing block grants for institutions. The Research Councils fund selectively on the basis of defined project proposals: in parallel, the higher education funding councils[3] distribute block grants to universities to cover the costs of teaching and research infrastructure. Prior to the introduction of the RAE the block grant sum for each university was based almost entirely on student numbers, and took no account of the actual level of research activity (let alone its quality) in the university. About 40 per cent of the total block grant was notionally 'for research'. Although the dual support system is highly regarded in the UK as providing for a degree of flexibility in the public funding system, the lack of coordination in research provision it implies can prove contentious at times of funding shortage and conflicting priorities—as arose in the 1970s.

Research in universities was a particular concern to scientists and to the research councils because the universities' straitened block grant funding in the 1970s was inadequate for the needs of research infrastructure once other costs, mainly relating to teaching, had been met. The University Grants Committee (UGC, the body then responsible for block grant funding) acknowledged this in its Annual Survey for 1975/6 (quoted in Beverton and Findley 1982) but took no action until agreeing in 1980 to set up a Joint Working Party with the Advisory Board for the Research Councils (ABRC, which represented all the research councils), with the remit, broadly speaking, to make the case for the selective allocation of the research element of the UGC block grant in order to sustain the dual support system (Tapper and Salter 2002).

---

[2] In addition there is contract funding available from government departments, and outside the Public Science System (PSS) substantial support from industrial R&D and charities.

[3] In the 1980s there was a single body for universities, the University Grants Committee (UGC), briefly succeeded by a University Funding Council (UFC) and subsequently (until the present day) by separate Higher Education Funding Councils (HEFCs) for England, Scotland, Wales, and Northern Ireland.

The Working Party's report (Merrison 1982) recorded the UGC's commitment to being 'more directive in promoting a greater concentration of effort in the university system' and maintaining a system where 'the needs of research were specifically taken into account'. It would be the responsibility of university management to ensure that this selective allocation was appropriately passed on to researchers and concentrated into selected areas. Thus was born the need for a separate mechanism for determining the UGC allocation of money for research—the 'Research Selectivity Exercise', later known as the RAE (Research Assessment Exercise). It seems reasonable to assume that the arguments of the predominantly science-oriented research councils in favour of selectivity were influential in reaching this conclusion, which gave the science community the opportunity of laying claim to a larger share of the one pot of public money that could be argued still to have some slack in it—the notional 40 per cent 'for research' of the UGC block grant. This is confirmed by one of Henkel's elite respondents who claimed that 'selectivity was science driven' (Henkel 2000: 113).

The benefits of such a policy for the arts and social sciences, where research was seldom conducted on the collectivized, semi-industrial model of many sciences, were less obvious. According to Henkel (2000: 114):

The imperative was to install selectivity in the funding of science research, and the decision to include arts and the social sciences as well was taken only after consultation with representatives of these fields. They indicated that to exclude them would send a signal to the public that these forms of research were unimportant, and so played their part in making the RAE a comprehensive system.

There is a piquancy to this story, given that arts and social science departments were subsequently among the most bitter and enduring of the exercise's critics. They have been a major source of material for discipline-specific studies of the effects and implications of the RAE, demonstrating negative social and intellectual consequences (for example, Harley 2002; Keenoy 2005; Lee 2007; Shore and Wright 1999; Strathern 2000). Empirical studies in natural science departments, however, indicate that many of these criticisms are not necessarily shared by natural scientists.

Implications for Reception and Assimilation of the RAE

This brief historical excursion shows how the leaders of the natural science community, in alliance with the research councils, were already by the early 1980s advocating the principles of selectivity and concentration that lay behind the RAE. Rather than selectivity being an imposition from outside, it could be considered as a victory for the science lobby—not only reinforcing the position of the elite institutions within the university community (which did not at that time include the polytechnics), but also helping to assure the epistemic development of the sciences along the laboratory-based, collective, resource-intensive, shared facility basis favoured by the dominant voices in the community.

While selectivity might well more easily appeal to an elite than to the rank and file, the daily exposure of a large tranche of the scientific community to collective ways of working (Ziman 1983, 1994) was likely to mitigate the shock of any curtailment of individual autonomy implicit or explicit in the RAE. Scientists were also helped by the system basing itself on scientific research practices (for example, in choice of indicators such as numbers of publications, external grant income). But probably the most critical circumstance was that by this period selectivity had become entrenched in the reality of scientists' research funding. The scale of the science considered necessary in the physical and biological sciences already ruled out survival on the basis of block grant funding, so university scientific researchers came increasingly to rely on competing for external grants to fund their work. Science faculty were thus immersed in a competitive culture in contrast to any academics whose research required only modest resources, largely covered by block grant. Once the rapid funding growth of the 1960s ended, competition for grants increased steeply. Staff seeking research council resources were subject to research council authority: they had to become inured to scrutiny and rejection, and to shaping their applications and aspirations in ways that might increase fundability. Internal authority relations were also affected, since coordination within and between researchers might be needed, calling for internal management. Thus many of the features of the RAE which most troubled their arts/social science colleagues presented many natural scientists with little that was new, though there were exceptions in the case of mathematicians and other researchers who were used to working alone.

This is not to claim that the RAE posed no challenge to scientists. The extent to which the process was made a public affair, with publication of the university rankings, introduced a novelty that went beyond the question of policies of selectivity and concentration, or familiarity with failure in the grants competition. Nevertheless, the prior exposure of the majority of scientists to the rationale and practice of selectivity and the authority relationships it entailed helped prepare them for dealing with its demands and coloured their view of how far conditions of knowledge production were changed.

Other possibilities for variation in impact have been briefly noted in the literature. Power (1999) for example comments on how some disciplines (he instances accounting and economics) seem to 'fit' better with the RAE. Henkel (2000: 134) includes the comment that 'the RAE had a bigger impact on the structures and cultures of humanities and social science departments' than on science departments,[4] since in the latter, 'already more structured, with more explicit division of labour and authority relations' the RAE simply took its place as one of a number of policies and pressures on their work. In light of the discussion above, we can add to this that the RAE represented implementation of a resource allocation policy (selectivity and concentration) that the leaders of the science community had been advocating and from which all research-active departments could expect collectively to benefit. The expectation of benefit among scientists was based not only on confidence in their own research strengths, but also on the expressed policy preference at government level for support of science and technology over other areas. Additionally, the structures and strategies already in place for resource mobilization in the competitive grant-awarding environment, and the management skills this implied, could be applied (and honed and further developed) to achieving success and maintaining morale on the new testing-ground of the RAE.

3. Co-constructing the RAE

In this section I draw on empirical work mainly in life sciences to discuss how scientists collectively and individually, at departmental and research group level, handled the pressures associated with the RAE (and related pressures). The first part discusses reactions and adaptations at departmental (or equivalent) level; the second part focuses on the strategies developed by individuals to manage their own and their group's research future.

University–Department Authority Relations

An important perceived effect of the RAE is its role, along with other pressures, in pushing universities to adopt a more corporate management style, thus bringing about a shift in authority relations between departments, individuals, and central university management. Two significant

---

[4] Henkel (1999), however, posits possibly more serious consequences for science departments in the longer term.

areas for such change where field-specific variations may be seen were internal restructuring and development of research strategy. While much of this activity related to the draconian cuts in funding of the 1980s, and the requirement put upon all public-funded institutions to move to more managerialist-corporate organizational models, some elements can be related specifically to the RAE.

The literature records some variability in views on the extent of the shifts in authority relations resulting from the RAE. Henkel's (2000: 124) case study of the RAE (which was based on evidence gathered in the period 1995–7 from departments of biochemistry, chemistry, physics, economics, sociology, history, and English) concludes that it resulted, overall, in a shift of authority from departments to the centre, particularly in the areas of research strategy and recruitment policy. The composite report produced by HEFCE in 1997 (based on commissioned studies) comments however that 'strategic development of research policies and plans takes place at different levels in different institutions' and 'former UFC[5] institutions have a greater departmental or sub-institutional focus for strategic discussion' (para. 53). A number of more limited case studies, discussed below, provide more detail about how such shifts in authority relations might be contested, countered, or co-constructed.

Unpublished data from a study of my own on life scientists' response to science policies (see also Morris 2002) provide an illustration of how authorities may be interpreted and negotiated. The example here concerns the formation of a 'centre' to bring together scattered research strengths in a managed entity that would form a stronger 'unit of assessment' to present to future RAEs. The initiative usually came from central university management, apparently signalling a shift in authority relations between central management and departments. The head of such a centre in a 'new' university, however, gives a slightly different account of its formation. He describes how, as research director of a biology department with poor research overall but some strengths in environmental biology, he proposed to colleagues in the environmental sciences department:

that we join forces, and get the good environmental biologists together, and put in a joint bid... So this also coincided with the Vice Chancellor in the university announcing that he wanted people in the university to apply to create the service centres, because he wanted to raise the [research] profile. Well, we talked about that, but university politics being what they are, if you, if you *ask* to do something... So

---

[5] The older or 'pre-1992' universities (1992 being the year when the polytechnics joined the university system).

we decided we wouldn't do anything—we'd wait until *they* asked *us*. And eventually they did [*laughing*]. They said, please will you set up an environmental research centre? We said, OK [*laughing*] [...] That seemed an important strategy.

Formally in this case the initiative was with the university, but clearly this biologist sees such strategic restructuring as a natural part of his own game plan, and claims to have no problem with such centre-led restructuring as long as he can make it clear in appropriate contexts that he retained control throughout (unpublished interview, 1999).

A similar story is told by a professor at an 'old' university. In this case it was a question of restructuring and amalgamating existing departments rather than setting up a centre:

What happened was that biological sciences in the middle to the early 80s was clearly at a low ebb [...] And the one major contribution the vice chancellor made which I think was a crucial one, was to indicate his discontent [...] and his indication that, you know, if you don't do something about it, then don't come to me for resources, so there was a kind of a, a pistol. Now that actually was very helpful... That stimulated debate and there [was a small group of] professors who actually were at the core of saying, well, how're we going to get out of this? How're we going to persuade the rest of the professoriate to go along with this, not to oppose us and then how're we going to persuade the staff to follow us, [...] There was a university committee that was also investigating this but basically their report was based upon the report that the group, the professors [made]. There was an official committee but it was being fed information via the professors.

The crucial point here seems to be agreement that standards were low, and, after that, keenness to seize the initiative and control the design of a new, merged structure. The intervention of the Vice-Chancellor is acknowledged, but only as the trigger, not the driver, of change. The respondent goes on to say that the form of the restructuring was guided by scientific needs (as the professorial group saw them) but the external 'kick' was useful in achieving their ends: 'We thought we could do better science together than apart [...] we had a positive, it wasn't pure reaction. It was an intention to move to a system where there was more integration, more collaboration, better use of equipment, um, and in a sense the pressure from outside was used to achieve that end.'

In both these examples the leaders are rejecting the idea that they are acting as conscripted agents of senior management (or government policies), but rather emphasizing the common goals they share, and their own effective control of the terms of contract ostensibly set by the principal.

Whether these stories are defensive rationalizations or better approximations to how things are worked out in practice than the official record is of less interest than the readiness to embrace and assimilate change among these competitive scientific leaders. For these research leaders academic freedom does not extend to freedom to do second-rate research, and proactive management is needed to assure a physical and professional environment where good research flourishes. Another instance of this acceptance of management of research is provided by Henkel (2000) who notes that: 'Two heads of science departments made it clear that...they themselves kept the research performance of the members of their departments under regular scrutiny, Such was the level of competition for laboratory space, doctoral students and resources that [...] [m]embers would be challenged if their research seemed not to be progressing.'

It can of course be argued that this is an elitist view, not necessarily shared by all staff. Interviews with the staff of the centre and restructured department referred to above, however, mostly showed similar world-views—though, as a number of interviewees remarked, the early retirements of the 1980s and influx of new staff with different expectations would have had an effect on this. Wilson's fascinating account of the development of biological sciences at the University of Manchester in the 1980s (Wilson 2008) similarly notes the commitment to active management (and even research performance indicators) among the professorial staff, and also the substantial wastage of staff who did not 'fit' the new system. Wilson's study also shows that plans for restructuring biological sciences at Manchester were strongly driven by a scientific vision as well as resource considerations, and substantially predated the RAE (2008: 4–5).

Another area of reported central university intervention is strategic scientific planning, though the university role may be muted in life science departments. Three out of the four departments in my own study (Morris 2002) regarded themselves as having control of strategic matters. This included the RAE submission, but also, and perhaps more importantly, recruitment, fundraising, and areas of scientific concentration. In the fourth department, the hived-off centre appeared to be in full control of its research strategy, but the remainder of the department, seen as weak in research, had largely ceded authority to the dean and other senior managers. Lucas (2006), reporting on work undertaken in 1997–9, describes how authority for deciding the RAE strategy was hotly contested between a biology department head and the 'university research committee', since the head's strategy cut across a general university policy of submitting a full complement rather than a selection of staff to the RAE. Local knowledge and authority prevailed in the case of the biology department, in contrast with the reported experience of departmental heads in sociology and English of a strong university steer, curtailing the independence of departmental strategies (Lucas 2006: 78).

It seems reasonable to assume that such steering—with greater or lesser degrees of subtlety—would apply in all universities, and the extent of the role of departmental strategic planning has therefore to be continually negotiated within the university. Martin and Whitley (Chapter 2) cautiously conclude that on the basis of the available evidence the RAE (with other influences) has generally strengthened the authority of the institution over its constituent parts and prompted university management to be more interventionist in research matters. They note, however, that impacts on departments have varied across fields and institutions—a point reinforced by field-specific studies of impacts (see Chapter 10) and the examples quoted above. There is thus a space for 'impacts' to be co-produced by a process of continuing interaction and negotiation within and between different system levels. Those departments, like science departments, that were already accustomed to navigating the tricky waters of grant-funding schemes and conditions of award might be more likely to see such constraints as routine and negotiable hurdles rather than an infringement of academic autonomy.

Individuals and the RAE: Evidence from Life Science Departments

Empirical studies from the 1980s and 1990s demonstrate the multiplicity of pressures in science departments deriving from changes in the financial situation, related changes in the university labour market, and also from new developments within science (see, for example, Hackett 1987; Henkel 2000, 2005; Wilson 2008). Life scientists I interviewed in the late 1990s reported intensified pressures of keeping up with new scientific developments, increased global competition, and managing a career, which competed for attention with public funding policies (Morris and Rip 2006). Moreover, of the main sources of public funding available, research grant and contract funding (from research councils or government departments) was much more important to them than research infrastructure funding distributed via the RAE. Interestingly, Leišytė et al.'s (Chapter 9) more recent empirical data describe the RAE as dominating discussions in two 'biotechnology' departments in English universities in 2005, because RAE ratings were perceived as directly affecting not only internal university allocations but also (in contrast to my earlier respondents) research council grant award decisions. It remains the case, however, that the constraints associated with scientists' dependence on external project funding have inoculated them as a population against some of the perceived pathogenic features of the RAE. For example, McNay remarks on the criticism levelled at the RAE that the cycle of reviews and expected outputs encouraged short-term research[6] (HEFCE 1997: para. 115). The time straitjacket cited by life scientists, however, is more likely to be that of the three-year project grant favoured by the research councils and/or the length of their short-term contract.

Time-span is a particular constraint for more junior researchers who have not yet established a broad portfolio of research. For example, a young research fellow commented that, though she hoped to be able to take a longer term view in the future, now, with a young group, and a limited-term contract, she had to produce quickly to assure her next employment contract. They must bring out papers every year—some would be better than others. This was a planned element in her overall career planning. A research fellow at another university at a broadly similar career stage explained how he tried to reconcile long-term continuity and short-term management of his programme: 'We always have to look into what is the next thing: so it's usually things that I'm starting off now that will bear fruit in 2–3 years' time—the really interesting things—and so there is an undercurrent of [. . .] long-term planning.'

This was a strategy also used by a well-established group leader running a very large team: it could adapt itself to the time-constraints of the grants system to allow a coherent yet flexible development of the programme as a whole, while helping to assure continuity of funding. The culture engendered by exposure to the competitive grant application process or tendering for contracts could relatively easily accommodate the accountability demands of the RAE. Unsuccessful grant applications already constituted a kind of public audit of shortcomings. In short, the kind of discipline that required performance and deliverables within fixed timescales was in large part internalized by these researchers, but not, they claimed, allowed to stifle a longer term view.

A further area where the life scientists appear to deviate from the traditional view of academics is in their reaction to development of the management function—including the strategic management of the scientific programme—in research-intensive departments and centres. Management

---

[6] Intervals between RAEs have lengthened over the years. The first three—on which McNay's work was largely based—occurred in 1986, 1989, and 1992. Subsequent RAEs took place in 1997, 2001, and 2008.

comes to be seen as necessary to deal with the growing requirements in biological sciences for the larger groupings discussed earlier. As well as these science-driven needs, the new structuring of public funding and proliferation of separately funded initiatives required informed local management to coordinate quick responses. Two factors favoured departmental rather than central management: (i) traditional distrust of university central administrations remained while widening of functions of local administration was considered as supportive; and (ii) scientific strategy required a detailed, scientific knowledge of a department's strengths and where future opportunities lay. It also required the mediation of respected and trusted scientific colleagues who understood the importance of a light touch. The attraction of a departmental strategy for individuals and small groups is that it will take account of where funding is available and research strengths within the department, thus providing a stronger platform for individual grant applications. This complements individual strategies, of which there is a rich variety in science departments, all aiming to preserve the maximum freedom of manoeuvre for researchers (Morris 2003; Morris and Rip 2006). They too are relevant to authority relations, but relate to the authority of the project or programme funder to specify the content of the project or programme, rather than the more distant and collective form of control exerted via the RAE.

Researchers' Strategies

In describing how they went about their research, UK life scientists often emphasized the new strategic roles and activities they were developing to assure, or attempt to assure, continuity of funding. It is relevant to note that the opportunities for diversifying sources of funding and building alliances with particular funding sources are quite high in the UK, where there are 'multiple principals'—four or five research councils, government departments and agencies, large charities, industry—with slightly differing priorities. The biomedical sector in particular benefits from a number of well-funded charities and a strong research-oriented pharmaceutical industry. The UK is less pluralistic than the USA with its multiplicity of federal agencies, state support, and diversified sources of local and national philanthropic and industry funding, but still allows scientists (especially biomedical scientists) much more funding flexibility than does, for example, the Australian system (see Chapter 10). This flexibility lends some credence to the UK life scientists' portrayal of themselves as developing strategies to manage the effects of funding and evaluation regimes without serious deflection of research trails. They agreed that they sometimes had to trim their research (cf. Chapter 10)—for example, putting work 'on the back burner', using students as research assistants—when the funding strategies were not working well. Some reported having to abandon or curtail lines of research when essential facilities were closed down or normal funding channels dried up. Most, however, nonetheless claimed that they were still doing research that was of interest to them, and all who saw their future in research claimed that they could and would identify work that was of interest and importance both to them and to sponsors, and that would satisfy their long-term goals. Any sense of damage to an 'ideal' programme of research was (with a very few exceptions) missing—or well masked.

For example, a centre director said their approach was not 'to sit around, thinking, what's interesting, scientifically—not in an institute like this. What we do is we look for funding opportunities, and to do research that would interest us' According to his account, he does not see this pragmatic approach as resulting in work of less intellectual depth or rigour, or even implying a motivation or values any different from those of the traditional academic, as he illustrates by recounting an over-dinner conversation with a distinguished Fellow of the Royal Society (FRS): 'and he was saying, of course, you know, the only reason we do this is to find things out, to add to the general body of science. And I thought, well, what a wonderful thing to say, and I really agreed with that. But you have to be an FRS at the end of your career to be [*laughing*] able to turn around and say it.' The attraction of this other vision of research is not denied, but nor is it taken as a condemnation of his own institute's contribution to knowledge. (Such at least is the story as presented to the interviewer.)

A slightly different approach to reconciling the potential conflict between funding opportunities and research focus is taken by a biomedical scientist who describes himself as 'a mid-range lecturer' and a research manager. He incorporates fundability into his value scheme: 'My feeling is that if it's not fundable it's not going to be good science, and if it's good science it should be fundable from some source. So, I mean, I don't ever think first what is fundable, right, let's think about doing it. I think what we want to do, how do we get it funded?' He goes on to describe how he goes about 'selling an idea' and gives examples before concluding: 'So I think the idea comes first, decide what experiment you want to do and then you think how can this be funded? And if it can't be funded then fine, you've got to drop them.'

He accepts the need to 'sell' ideas without feeling compromised by it, and shows confidence in being able to generate 'good science'—*and* that the system will recognize it as such—that was shared by most established group leaders and also younger researchers. Both researchers quoted[7] expect to be able to produce 'good science' without compromise on quality, while simultaneously adapting to financial and policy frameworks.

Wilson reports on what is substantially the same phenomenon at institutional level in the University of Manchester, where 'biological science was self-consciously remodelled in line with political and economic conceptions of "good" science and "good management"'—the key to its success being recognition that 'visions of a "new biology" centred on molecular genetics' (with all its medical and commercial potential) formed the perfect match with 'political and policy changes [...] which stressed selectivity, competition, applicability and management' (Wilson 2008: 107, 109). In the USA, where the mix was slightly different, the commercialization of universities, rather than government policies, might take on the role of co-producer—along with academics themselves—of new institutional scientific strategies and personal strategies; these had similar implications for problem choice and shared understandings of what it means to do research (Cooper 2009; Jong 2008). In both the USA and the UK these processes were particularly evident in the life sciences where scientific advances and their potential for economic growth were in a phase of rapid development.

At the individual level, this ability to reinvent oneself, consciously or unconsciously, to fit with prevailing authority relations was reflected in UK life scientists across the spectrum. There remained a minority, mostly among senior academics and long-term serial contract researchers, who were unwilling or unable to change. The former absorbed themselves, with different degrees of declared satisfaction, in mentoring, teaching, or managerial/administrative activities; most of the latter were nevertheless looking for more contracts, but a few were already planning a career change. The majority at all levels, however, presented themselves as able to accommodate to changing research environments. Effects such as narrowing of research trails or moving to applied research were hardly present in the accounts of these UK researchers, or were referred to as problems that could and would be overcome, given time, by strategies they had at their disposal.

[7] Unpublished interview data, 1997–9.

255

4. Discussion

The overall thrust of these life scientists' presentation of their professional self is to demonstrate that in absorbing new structures and conditions concerning what it means to do science, they are enabling rather than constraining their pursuit of authentic academic goals. They may admit to making concessions, but deny abandoning values. One factor underlying these levels of confidence (or bravado) might relate to individual or departmental research performance—a cross-cutting factor highlighted by both Chapters 9 and 10. Systematic information on the performance of the UK samples is not available, but what is clear is that factors in the UK labour market make it difficult for poor performers—or those without a high level of confidence in their ability—to stay in university research. First is the competition inherent in the disproportion between numbers of untenured contract researchers employed on grants and permanent positions. In order to progress to some degree of independence researchers need to produce high-grade projects that can successfully compete for space and facilities in the laboratory, otherwise they are likely to drop out of the system. Secondly the population of tenured academics has been winnowed over the years by redundancy schemes following financial cutbacks and successive RAEs. Those in post whose interest in or capacity for research has waned take on administrative, management, teaching, or mentoring roles.

The UK data also suggest another cross-cutting factor, operating in a fashion similar to performance level and not totally separable from it: namely the degree to which a scientist is established in his/her career (see also Leišytė 2007; Chapter 9). Researchers have to make more compromises at the beginning of their careers because they lack the reputation to command funding, or attract good post-doctorates, and the small scale of their research portfolio gives them little flexibility. This may result in their taking on more applied research (because it is considered easier to get contract, as opposed to grant, funding), as may also be the case for low performers; but those trying to establish themselves construe any aberration as temporary and/or having the capacity to be a stepping stone to something else.

For those requiring external funding to progress their work, performance in the RAE plays an important but supporting role in the bigger question of mobilizing resources from project and programme grant awarding bodies. Here structural and operational features of the research system may come into play in adding or detracting from researchers' sense of having control of their research choices. A simple case is the matter of diversity of funding sources (where there are multiple grant-awarding bodies), so that no one funding body has the power to close down a career (see earlier comments in section on researchers' strategies). Thus to be solely dependent on the RAE risks limiting choice. A more complex question is how far the award and assessment system (whether RAE or research council) is trusted as reliably representative of professional and disciplinary values. Award systems in most advanced scientific societies work through intermediary bodies (such as funding councils and research councils in Europe and elsewhere; the National Science Foundation and National Institutes of Health in the USA: Rip 1994; Rip and van der Meulen 1996). These bodies are largely staffed and advised by academics and ex-academics, and have an important role in translating loosely formulated government policies into pro-grammes and standards capable of winning the cooperation and interest of the academic community (Braun 1993). Further to seal their legitimacy, their assessment systems are based on peer review (Rip 1994). The scientif-ic/academic viewpoints thus embedded and wielding influence in interme-diary bodies, however, are—or become by virtue of their location—those of disciplinary or other elites, thus adding to the authority of the elites and to fears in some quarters about the possible deleterious effects of elite dominance.

While natural scientists show reasonable confidence that their elites are serving the interests of science, and hence themselves (and are much to be preferred to politicians and bureaucrats), the greater individualism and comparative lack of disciplinary solidarity in some arts and social science communities leads to more questioning and contestation of the legitimacy of decisions and directions taken (Harley and Lee 1997; Chapter 2). The RAE's heavy reliance on peer review within a disciplinary structure has thus done it no favours with the latter group. By way of contrast, we may note the science lobby's protracted campaign against the all-metrics system proposed for the sciences in the Research Excellence Framework (REF): the replacement for the RAE. This campaign was crowned with success in 2009, when peer review (assisted by biometrics) was reinstated (Owens 2009). Natural scientists would agree with critics of the peer-review system that it encourages conservatism, and is likely to disfavour innovative, unfashionable, or interdisciplinary studies, but recognize a 'trade-off' be-tween its merits and demerits (Chubin 2002). In the context of the project/ programme grant award systems, the peer-review element encourages scientists to believe that when the cycle is complete and the results of their funding come up for evaluation, any deviations will be judged on their scientific merits rather than in a tick-box spirit, thus reinforcing their optimism about their ability to steer (to a large enough degree) their own course (Morris 2003: 364).

The proviso in the last sentence indicates another possible difference between natural scientists and their colleagues in other, more individualist, academic fields and sub-fields—the apparently low level of expectation of research freedom (except of course for some research 'stars') among the UK science community (Morris 2000: 433–4). Much of this can be attributed to the struggle for funding in a resource-intensive field, which has demanded a more flexible interpretation of what constitutes sufficient scientific freedom. There is also the question of how far freedom of choice has ever been the norm for an ordinary scientist. Take, for example, a problem often thought to be a product of strong research evaluation systems or excessively competitive project funding: namely, the potentially stifling and narrowing effects of trying to please the judges in choice of topics. Hagstrom (1965: 16–17) comments on how, even at that time, the scientist's freedom of choice is constrained because 'the desire for recognition [. . .] influences his selection of problems and methods. He will tend to select problems the solutions of which will result in greater recognition, and he will tend to select methods that will make his work acceptable to his colleagues.' Mulkay's (1972: 28) gloss on this is that 'originality is not valued unconditionally. It is valued and rewarded only in so far as it conforms to current research norms'. Whitley (2000: 19) confirms the continuation of this kind of disciplinary control that prevents work from being 'too original'.

These observations are consistent with the view reflected in the personal testimonies of some UK scientists that authority relations in science, as practised in universities, have changed very little since the mid-twentieth century. It is claimed that behind the façade of academic freedom there always existed a default authority system run by disciplinary elites. The culture of adaptation (but without losing sight of one's goals) may thus be said to be ingrained. Recent changes may have raised the stakes with regard to managing personal research programmes, but blend with an established culture. The hard school of competitive application for project funding has further shaped these resource-dependent scientists' views on what it means to do research. Management skills—to mobilize resources, manage uncertainty, and maintain flexibility—are an accepted component of doing research and the criteria for what are 'doable' projects have been widened to include (though not necessarily to regard) financial and political considerations. While arts and social sciences communities participate successfully in the RAE, and reportedly accept it as yet another competition one does one's best to win (Keenoy 2005: 310), its importance as a source of funding remains higher, its mode of operation more alien, and its potential to shift authority relationships[8] much higher than is the case for science departments. Far from seeing scope for adapting to work with the system while still retaining control, the views of academic staff in social science and business-related disciplines reported by Harley were that the result of the RAE was 'the mass production of research for a rating that was more important than what is produced, and a reorganisation of academic work in ways which violate traditional academic values' (Harley 2002: 187).

As well as trying to explain scientists' resilience in the face of the RAE and their ability to internalize its requirements without trauma or loss of self-belief, we need to consider what costs this may entail or alternatively whether their experience might justify some modification of the general verdict of the literature on the RAE. Martin and Whitley's analysis (Chapter 2) provides an authoritative summary of concerns about the RAE. The analysis bases itself on five expected consequences of a 'strong Research Evaluation System' (Whitley 2007): collegiality weakened by competition; publication inflation; homogenization of research; impediments to the development of new areas; and increased stratification and hierarchy. They note that shifts in authority relations at institutional levels attributable to the RAE contribute to these consequences. At the individual level also they find evidence of a number of effects, including: constraints on research agenda and growing pressure to perform; pressure to publish, sometimes prematurely; change in balance of research and teaching; a shift between individualism/collegiality, or competition/cooperation; game-playing to improve ratings; and overwork and stress. It is further noted that these shifts also have wider consequences in terms of changes in academic culture and rise of the 'audit society'.

It will be evident from the previous discussion that much of this ground was already familiar to scientists before the advent of the RAE, and while not disputing that significant changes have occurred in the way academic research is done, they would tend to be chary attributing these to the RAE,[9] and also to question whether some of the cited changes at individual level

---

[8] For an example, see Lee 2007.
[9] Except as an ally to force through changes in research focus and institutional structures already on the agenda of some forward-looking scientific elites: see earlier discussion of the scientific impetus behind the Merrison report and the first RAE, and of reconfiguration of biological sciences at Manchester. In relation to the latter, it should be noted that similar reconfigurations at Stanford and Berkeley took place at much the same period where the key power-shifting factor was not (other than indirectly) a government policy but an alliance with rising commercial biotechnology interests (Jong 2008).

were attributable to any shift in government or institutional authority or had not always been operative, in veiled form, under the old, informal academic arrangements. In relation to many of the cited effects, the natural science community deviates from the norm either in denying the effect is anything new or attributing it to a variety of causes rather than particularly to the RAE. Most characteristically they argue that the effects (whether of the RAE or of the competitive project/programme grant system) can be satisfactorily managed, with greater or lesser effort and adaptation. Developing a repertoire of personal and collective management strategies for this purpose becomes part of the job of the successful research scientist. By this means they create a negotiated personal, or departmental, space in which they can do their work, satisfy their paymasters, and also their peers by remaining loyal to traditional academic criteria of quality and creativity. If we accept, even in part, that high-quality work is still being done, does this then imply a need to reconsider some traditional assumptions about the necessary conditions for high quality and innovative research? While an effect of the RAE may, for example, be to weaken collegiality and the unity of research and teaching, what role these play in sustaining the university's research function is open to question (see, for example, Tapper and Salter 2002: 7). Nearer to the heart of concerns about the RAE and kindred systems is the question of autonomy. Though we may accept that 'the RAE [. . .] represented another stage in the erosion of university autonomy' (Tapper and Salter 2002: 30), and placed constraints on individual autonomy (Henkel 1999; Chapter 2), there is still room to consider what kind or degree of autonomy is required for successful research.[10] Scientists appear to be demonstrating that they can manage their work within a relatively small autonomous space and reconfigure their programmes to take account of changing research environments.

5. Conclusions

The general thrust of this chapter has been to show that the UK science community, particularly life scientists, had incorporated many of the principles embodied in the RAE into their way of doing science prior to its

---

[10] Rip, for example, in a discussion of different models for innovation and application in science, has questioned whether 'the autonomy claimed for pure science is really necessary to let it mature and bear fruit' (Rip 1997: 619).

eruption on the university scene. The same forces—competition, external scrutiny, collective priorities—that shaped the organization of successful science departments, shaped in greater or lesser degree the design and development of the RAE. Scientists' extreme resource-dependency, long-term exposure to research council competitive funding systems, and espousal of group-based work systems generated its own need for research management and monitoring. Thus the prevailing patterns of authority relations developed in science departments (which had little in common with the unchallenged individual autonomy once enjoyed by many academics in other parts of the university), provided structures and strategies ready for deployment in the face of perturbations produced by the RAE in, for example, relations between departments and central university management. Rather than discuss principles of academic or group autonomy they have tended to shift questions about authority relations to the more pragmatic grounds of who they believe is effectively in control of research programmes. The nub of my argument is that the major disruptive effects of the RAE experienced in some sections of academia were diminished—or swamped—by organizational features of natural science disciplines as they developed in the second half of the twentieth century, and prior acculturation of scientists to a competitive and increasingly selective funding system. Thus, whether the RAE reinforced or modified practices is an effect dependent on field. While the natural scientists I have quoted pride themselves on exerting a sufficient degree of control over the system, they have undoubtedly adapted their ways of working in order to do this, and their definition of what it means to do research, and what is required be a successful researcher. They have 'internalised relevance' (Rip 1997) and adapted to ways of working that disregard or substantially redefine many traditional values (individual autonomy, privacy, collegial governance) and accord them symbolic value rather than considering them conditions essential for the continued vitality of research. Where this places them in relation to the criticisms of institutionalized evaluation systems and other policies as ultimately devitalizing to research is obscured by the dual sets of values (politico-economic and academic-traditional) scientists claim to keep in balance. How this duality works, the authority relations it implies, the robustness of the boundary work entailed, and what it means for the content and conduct of research, are questions that have received some attention in other contexts (Calvert 2006; Felt *et al.* 2007) and would merit further exploration in relation to field-specific effects of strong policy interventions.

References

Berger, M., and Cooper, M. (1979) *Science*, 204 (4400) (29 June): 1369.

Beverton, R. and Findlay, G. (1982) 'Funding and Policy for Research in the Natural Sciences', in G. Oldham (ed.), *The Future of Research* (Guildford: Society for Research into Higher Education), 83–148.

Blume, S. (1980) 'The Finance of University Research in Western Europe', *European Journal of Education*, 15(4): 377–86.

—— (1982) 'A Framework for Analysis', in G. Oldham (ed.), *The Future of Research* (Guildford: Society for Research into Higher Education), 5–47.

Braun, D. (1993) 'Who Governs Intermediary Agencies? Principal–Agent Relations in Research Policy-Making', *Journal of Public Policy*, 13(2): 135–62.

Calvert, J. (2006) 'What's Special about Basic Research?' *Science, Technology and Human Values*, 31(2): 199–220.

Chubin, D. (2002) 'Much Ado about Peer Review, Part 2', *Science and Engineering Ethics*, 8: 109–12.

Cohen, A., and Ivins, L. (1967) *The Sophistication Factor in Science Expenditure* (Science Policy Studies, 1; London: HMSO).

Cooper, M. (2009) 'Commercialization of the University and Problem Choice by Academic Biological Scientists', *Science, Technology and Human Values*, 34(5): 629–53.

Cozzens, S. (2003) 'Science Policy: Two Views from Two Decades', *Prometheus*, 21(4): 509–21.

CSP (Council for Science Policy) (1967) *Second Report on Science Policy* (Cmnd 3420; London: HMSO).

Demeritt, D. (2000) 'The New Social Contract for Science: Accountability, Relevance, and Value in US and UK Science and Research Policy', *Antipode*, 32(3): 308–29.

DES (Department of Education and Science) (1987) *Higher Education: Meeting the Challenge* (Cmnd 114; London: HMSO).

Etzkowitz, H., and Kemelgor, C. (1998) 'The Role of Research Centres in the Collectivisation of Academic Science', *Minerva*, 36(3): 271–88.

Felt, U., Fochler, M., and Sigl, L. (2007) 'Re-thinking Biosciences as Culture and Practice: Tracing "Ethics" and "Society" in Genome Research. A Pilot Study' (GOLD II): http://www.univie.ac.at/virusss/projects/3/.

Georghiou, L. (2008) 'Forget Submission Rates: Funding is the Issue', *Research Fortnight* (18 Dec.): http://exquisitelife.researchresearch.com/exquisite_life/lukeg.html.

Hackett, E. (1987) 'Funding and Academic Research in the Life Sciences: Results of an Exploratory Study', *Science and Technology Studies*, 5(34): 134–47.

Hagstrom, W. (1964) 'Traditional and Modern Forms of Scientific Teamwork', *Administrative Science Quarterly*, 9(3): 241–63.

—— (1965) *The Scientific Community* (New York: Basic Books).

Harley, S. (2002) 'The Impact of Research Selectivity on Academic Work and Identity in UK Universities', *Studies in Higher Education*, 27(2): 187–205.

—— and Lee, F. S. (1997) 'Research Selectivity, Managerialism, and the Academic Labour Process: The Future of Nonmainstream Economics in U.K. Universities', *Human Relations*, 50: 1425–60.

HEFCE (Higher Education Funding Council for England) (1997) *The Impact of the 1992 Research Assessment Exercise on Higher Education Institutions in England, M6/97*: http://www.hefce.ac.uk/pubs/hefce/1997/m6_97.htm (accessed Apr. 2009).

—— *Circular Letter 13/2008. Annex A: summary of responses to consultation*: www.hefce.ac.uk/pubs/circlets/2008/cl13 08/.

Henkel, M. (1999) 'The Modernisation of Research Evaluation: The Case of the UK', *Higher Education*, 38: 105–22.

—— (2000) *Academic Identities and Policy Change in Higher Education* (London: Jessica Kingsley).

—— (2005) 'Academic Identity and Autonomy in a Changing Policy Environment', *Higher Education*, 49: 155–76.

Jong, S. (2008) 'Academic Organizations and New Industrial Fields: Berkeley and Stanford After the Rise of Biotechnology', *Research Policy*, 37: 1267–82.

Keenoy, T. (2005) 'Facing Inwards and Outwards at Once: The Liminal Temporalities of Academic Performativity', *Time and Society*, 14: 303–21.

Kogan, M., and Hanney, S. (2000) *Reforming Higher Education* (London: Jessica Kingsley).

Lee, F. (2007) 'The Research Assessment Exercise, the State and the Dominance of Mainstream Economics in British Universities', *Cambridge Journal of Economics*, 31: 309–25.

Leišytė, L. (2007) *University Governance and Academic Research*. Doctoral dissertation, University of Twente.

Lucas, L. (2006) *The Research Game in Academic Life* (Maidenhead: Society for Research in Higher Education and Open University Press).

Merrison, A. (1982) 'University Grants Committee/Advisory Board for the Research Councils', *Report of a Joint Working Party on the Support of University Scientific Research* (Merrison Report) (Cmnd 8567; London: HMSO).

Morley, L. (2005) 'The Micropolitics of Quality', *Critical Quarterly*, 47(1–2): 83–95.

Morris, N. (2000) 'Science Policy in Action', *Minerva*, 38: 425–51.

—— (2002) 'The Developing Role of Departments', *Research Policy*, 3: 817–33.

—— (2003) 'Academic Researchers as "Agents" of Science Policy', *Science and Public Policy*, 30(5): 359–70.

—— and Rip, A. (2006) 'Scientists' Coping Strategies in an Evolving Research System: The Case of Life Scientists in the UK', *Science and Public Policy*, 33(4): 253–63.

MRC (Medical Research Council) (1970) *Annual Report, 1969–1970* (London, HMSO).

Mulkay, M. (1972) *The Social Process of Innovation* (London and Basingstoke: Macmillan Press).

Oldham, G. (ed.) (1982) *The Future of Research* (Guildford: Society for Research into Higher Education).

OST (Office of Science and Technology) (1993) *Realising our Potential: A Strategy for Science, Engineering and Technology. Presented to Parliament by the Chancellor of the Duchy of Lancaster* (Cn 2250; London: HMSO).

Owens, B. (2009) 'Citations will Feed into Peer Review, HEFCE Says', *Research Fortnight*, 326 (17 June 2009).

Phillimore, A. (1989) 'University Research Performance Indicators in Practice: The UGC Evaluation of British Universities 1985–1986', *Research Policy*, 18: 255–71.

Power, M. (1999) 'Research Assessment Exercise: A Fatal Remedy?', *History of the Human Sciences*, 12: 135–7.

Price, D. de Solla (1963) *Little Science, Big Science* (New York: Columbia University Press).

Rip, A. (1994) 'The Republic of Science in the 1990s', *Higher Education Studies*, 28: 3–32.

—— (1997) 'A Cognitive Approach to Relevance of Science', *Social Science Information*, 36(4): 615–40.

—— and Van der Meulen, B. (1996) 'The Post-Modern Research System', *Science and Public Policy*, 23(6): 343–52.

Royal Society (2006) *Response to the DfES Consultation on Higher Education Research and Funding* (Policy document, 24/06; London: Royal Society).

Shore, C., and Wright, S. (1999) 'Audit Culture and Anthropology: Neo-Liberalism in British Higher Education', *Journal of the Royal Anthropological Institute*, NS 5: 557–75.

Strathern, M. (2000) 'The Tyranny of Transparency', *British Educational Research Journal*, 26(3): 309–21.

Tapper, T., and Salter, B. (2002) *The Politics of Governance in Higher Education: The Case of the Research Assessment Exercises*, OxCHEPS Occasional Paper, 6: http://oxcheps. new.ox.ac.uk/MainSitepercent20pages/Resources/OxCHEPS_OP6percent20doc.pdf (accessed Apr. 2009).

Teich, A. (1990) 'US Science Policy in the 1990s', in S. Cozzens, P. Healy, A. Rip, and J. Ziman (eds.), *The Research System in Transition* (Dordrecht: Kluwer Academic Publishers), 67–81.

Whitley, R. (2000) *The Intellectual and Social Organization of the Sciences*, 2nd edn. (Oxford: Oxford University Press).

—— (2007) 'Changing Governance of the Public Sciences', in R. Whitley and J. Gläser (eds.), *The Changing Governance of the Sciences: The Advent of Research Evaluation Systems* (Sociology of the Sciences Yearbook; Dordrecht: Springer), 3–27.

Wilson, D. (2008) *Reconfiguring Biological Sciences in the Late Twentieth Century: A Study of the University of Manchester* (Manchester: Faculty of Life Sciences, University of Manchester).

Ziman, J. (1983) 'The Collectivization of Science', *Proceedings of the Royal Society of London. Series B, Biological Sciences*, 219(1214): 1–19.

—— *Prometheus Bound: Science in a Dynamic Steady State* (Cambridge: Cambridge University Press).

9

Mediating Problem Choice

Academic Researchers' Responses to Changes in their Institutional Environment

Liudvika Leišytė, Jürgen Enders, and Harry de Boer

1. Introduction

New governance approaches are at the forefront of discussions of public-sector reforms, including academic research where a novel type of institutional environment has been created for most of the actors involved, and also the academic 'production unit', be it an individual researcher, a research group, a scientific community, or the academic profession as a whole. In this chapter we discuss how the interests and preferences of the academics, and the social norms that are considered important for academic research, mediate and influence the effects of these changes, focusing particularly on the consequences of changing authority relations for practices at the micro-level of academic research, the academic research groups, and researchers.

In the following we present, first, our conceptual considerations of understanding the complex interaction between the research groups and researchers and their changing institutional environment. Second, we present our case-study design investigating research groups in two countries (England and the Netherlands) and two fields of research (biotechnology and medieval history) and related methodological considerations. Third, we present and discuss interview data from eight case studies of research groups focusing on the issue of 'problem choice'. The final section summarizes the main conclusions that emerge from this analysis.

2. Conceptual Considerations

The focus of our chapter and empirical investigation is on changing authority relations in the governance of academic research, the responses of academic researchers to such changes in their institutional environment, and understanding what that means for their problem choice. When looking at problem choice, we focus on the reflections of academic researchers' on their formulation of research problems, and what considerations—which actors and institutions—researchers bear in mind in choosing their research problems. In other words, we look at the extent to which researchers' problem choice arises from the research process itself, mainly influenced by the feedback mechanisms incorporated into the practices of academic communities, and to what extent other considerations reflecting changing authority relations play a role in framing choices about what kind of research to pursue.

We assume that researchers in principle want to design their own research questions and keep their professional autonomy as much as they can. Attempts to influence problem choice touch upon the core of the academic profession, professional expertise, and academic freedom, and will be mediated by the researchers themselves being the only ones who can actually change their research. Nevertheless, academic freedom has always been constrained to a certain degree by 'material circumstances, historical opportunity, epistemic conviction, and above all, communal doctrine' (Ziman 2000: 204). Growing competition for external research funding, research priority setting by external funders or the management within universities, criteria set up for research assessment exercises of different kinds by academic elites and state bureaucrats, growing expectations as regards the socio-economic impact of academic research, all provide examples of institutional changes that might affect the rules of the academic research game at the shop-floor level (see Chapter 1). Such changes have modified the actor constellations and authority relations governing academic research, and contribute to changing behavioural expectations and resource dependencies that bear upon research groups and academic researchers.

Researchers are, however, by no means just passive recipients of such changes in their institutional environment. Research groups and individual researchers respond to such changes in their institutional environments in different ways, varying from passive acquiescence to active manipulation of external demands. In addition to active and passive compliance, another possible response is decoupling, where organizations seal off their core activities from the institutional environment in order to meet the inconsistent pressures for legitimacy and efficiency. In terms of problem choice, this strategy would mean, for example, writing grant proposals on the topics that fit the thematic area of the sponsor, but in fact carrying out the research of one's own choosing (see also Chapter 10).

A third strategy is manipulation, which can be seen as resistance to an institutional environment and the most active response to the institutional pressures, since it is intended to change the content of the institutional pressures and the actors that exert those (Oliver 1991). Examples of the manipulation strategy in terms of problem choice can be seen in influencing the research agenda in the intermediary funding bodies such as research councils or foundations, or promoting one's research topics within the institute/faculty so that they can become strategic research areas for the institutional management.

3. Methodological Considerations

We conducted a multiple case study that looked at similarities and differences in the responses of selected research groups and researchers to changing institutional environments at a given moment of time across two fields of research and two countries. Our selection of countries is based on shifts in governance in England and the Netherlands. Both countries were and are active in restructuring the academic research system. However, the starting point was different, since both systems were based on different higher education and public science systems: state-delegated and the state-shared (see Chapter 1). Shifts in governance also took different routes and led to somewhat different combinations of authority relations in higher education systems (Kehm and Lanzendorf 2006; Kickert 1997).

We chose two contrasting fields of research to address the variety of 'tribes and territories' in academia (Becher and Trowler 2001). Medieval history and biotechnology are selected as representatives of distinct research cultures of 'soft' and 'hard' sciences. Research groups in biotechnology and medieval history differ in the way they carry out research, how they are funded, and how they meet the expectations of different authorities they face. For bioscientists the ability to tap into a serious amount of (external) funding is much more vital than for medievalists. At the same time, bioscience is expected to benefit much more substantially from external expectations looking for socio-economic impact of academic research. Co-operation is perceived differently among bioscientists and historians. Team-work and extended cooperation is the name of the game for researchers in biotechnology. For medieval historians, work habits include much more working on the individual lines of research. Traditional publication cultures differ as well, given that bioscientists aim at peer-reviewed journal articles as the prime medium of communication while medievalists traditionally stress the role of (articles in) books.

The third aspect in the selection of our cases is the estimated quality of the research group itself. Here we want to distinguish between 'high achievers' and 'middle achievers'. The assumption is that (past) performance may have an impact on the responses towards changing circumstances. In England we followed the strategy of looking at the Research Assessment Exercise (RAE) 2001 results to identify 'high achievers' and 'middle achievers'; and in the Netherlands we also looked at the national evaluation results.

The study used different sources of data: documents, literature, and semi-structured interviews. The macro-level data were collected from 2003 to 2006, covering the period since the 1980s. This part of the data collection included major higher education and research policy documents in England and the Netherlands as well as policy studies of that period. During the same period of time the meso-level data were collected which included relevant university documents, such as strategic plans, financial reports, research policy development plans, and evaluation reports. During 2005, seventy-seven semi-structured interviews were conducted with researchers, university managers, and policy-makers in England and the Netherlands. Altogether six public research universities were visited.

Using the perspective of Marshall and Rossman (1997: 111), who see qualitative data analysis as 'a search for general statements among categories of data', and the procedures outlined by Holliday (2002), the raw data were brought together on the basis of their similarities into categories. We started with descriptive and topic coding, and ended with analytical coding. When the core categories were decided, we related all of them to each other and to the major core. To increase reliability we did coder-consistency tests for consistency over time and among colleagues. For the sake of parsimony, citations that best represented a category or opinion of the majority of the interviews were used.

4. Major Changes in Governance and Authority Relations in England and the Netherlands: A Bird's Eye View

Shifts in governance and related authority relations form part and parcel of an overall restructuring of the university sector in England as well as in the Netherlands. The point of departure for the reforms was, however, different for each country since the two systems were traditionally based on different higher education governance models. The historical-institutional context and the national styles of governance also influenced the way in which governance reforms are implemented in these countries and how they try to affect university governance and research practices (de Boer *et al.* 2007*b*). England is generally seen as an 'early adopter' of rather fundamental changes (see Chapter 2), and reforms occurred in a somewhat lighter fashion and later in the Netherlands (de Boer *et al.* 2007*a*; Meulen 2007). Both systems have undergone considerable change affecting the role and authority of the state, of intermediary bodies, universities as employers and managers of academics, of academic elites and individual academics as well as external stakeholders.[1]

State Regulation and Funding

In England, the traditionally limited regulation by the state in higher education and research has increased to a considerable extent. Several reforms, especially in funding and quality assurance, have strengthened the regulative authority of the state. The dominant motives of the reforms were to make universities more accountable and efficient, to call for 'value for money', and to make universities more responsive to the needs of the economy and the society. Since the time of Margaret Thatcher, the major instruments used to strengthen state regulation comprise the strengthening of funding councils and university leadership that act in the shadow of the state as well as the power of the purse by competitive resource allocation and research priority setting (Henkel and Little 1999).

In the Netherlands, state regulation of higher education and research via laws, decrees, and public funding was strong, as it typically is in a Continental model of governance. Since 1985, the concept of 'steering at a distance' has been at the forefront of higher education and research policies where the government is not the almighty system planner but instead

---

[1] For a systematic comparison, see Leišytė (2007).

serves the roles of catalyst, coordinator, and (financial) facilitator (Vijlder and Mertens 1990; Maassen and Van Vught 1988). According to this philosophy, government should try to keep its distance by taking the sector level as the point of departure for steering. Institutional autonomy should be enhanced, and universities were expected to become more adaptive to their environments. It was argued that all this would have positive effects on the quality of the primary processes.

In the same period, competition for resources among individuals and research groups has increased in both countries, partly due to the reductions in state funding for higher education and research, and partly due to the expansion of the systems. In England performance-based funding was introduced based on the results of the RAE that considerably strengthened the powers as well as the interventionist approach of the funding councils (see Chapter 2). The inclusion of former polytechnics into the university sector in 1992 contributed towards an even larger increase in external competition for the same pool of state funding. Further, the competitive allocation of research council funding has contributed to increasing competition for research money. As a result of competition, and especially due to the selectivity effects of the RAE, a stratification of universities appeared where most research funding is concentrated at the top research universities. Finally, more selective state funding increased internal competition within universities for scarce resources.

In the Netherlands, competition for resources also increased but was less pronounced. State funding for universities in the Netherlands is, to some extent, performance-based (and contributes around one-third of university budgets), but still includes the bulk of funding based on a stable historical component and student numbers. Regular quality evaluations are carried out, but are not automatically linked to funding and not organized by the funding council (de Boer *et al.* 2007b; Meulen 2007). University income from the state has remained relatively stable over time and much more widely distributed among Dutch universities compared to resource concentration in some English universities. The Dutch funding council has, however, gained in importance due to a variety of rounds and programmes established to set research priorities in collaboration with ministries and other stakeholders.

University Leadership and Management

Another important trend in the governance arrangements concerns the increased authority of the university as an employer and manager in both

271

systems. University top and middle management has gained powers through more centralized management structures in universities, performance monitoring, target-setting, and strategic management. Policies geared towards deregulation and efficiency were the key triggers for these developments.

In response to the Jarratt report of 1985, patterns of change spread across the English universities: mechanisms for strategic planning were put in place, there was more transparency in financial matters, and clearer management lines with fewer committees (Stephenson 1996). 'Committees were pruned, finance offices became larger and more powerful, central management teams were established, primitive computerized resource allocation models were developed' (Williams 2004: 235). Overall, universities moved strongly towards formal corporate management structures (Henkel 2000) that were facilitated by scarcity of resources as well as by quality assurance mechanisms such as the RAE. The role of deans and heads of departments has become more visible and prominent. This new responsibility of faculty and department leadership not only implies responsibility for procuring consumables and equipment expenditure, but also their greater authority in managing staff resources.

In the Dutch case, the devolution of state authority can also be illustrated by various policies meant to enhance the universities' institutional autonomy (de Boer *et al.* 2007*a*). Universities have more opportunities today to select their own students, make up their own internal financial allocation schemes, and have more authority in the area of personnel. They have owned their property since 1995. In the past, institutional management was almost absent. Today, some authority has moved down from the national government to institutional management. This process started somewhat later than in England and was more incremental. After the 1980s, financial and staffing matters devolved from the state to universities, which 'created' the possibilities for the universities' central management to increase their influence in strategic decision-making and budget allocations. Further developments in the Netherlands included the strengthening of the managerial responsibilities at the middle level (deans, head of schools, and research directors) and the 'verticalization' of authority lines (top–down appointments instead of elected posts).

Academic Self-Governance

Historically senior academics in both countries had strong powers in the internal governance of universities, and academic elites were influential in the external governance of the academic research system. As regards the internal governance of universities, the role of academic governing bodies as well as the powers of individual senior academics has been weakened. As the external governance of universities, the role of academic elites, especially via peer review and research priority exercises, has gained in importance.

Looking at the internal governance of English universities, at face value the position of leaders and managers has been strengthened at the expense of the powers of academic self-governance and individual chair-holders. Academics no longer have a decisive voice in non-academic matters, and their voice has even been restricted in terms of institutional and departmental research matters. The traditional academic committee structure is still in place, but academics do not make strategic decisions as these are in the hands of the university management. In such a context academics are increasingly confronted with decisions and activities framed by various layers of directives, strategies, and action plans devised by department, faculty, and top university management (Willmott 1995). At the same time, senior academics are still able to voice their interests through informal negotiations (Fulton 2003). When we look at external governance we can witness a growing role of the academic elites via quality assurance and evaluation procedures as well as via research programming. Representatives of the academic elites sit on the RAE panels and on the councils and boards of different governmental bodies.

Within the Dutch universities important powers have also been taken away from senior academics, although more recently than in England. Universities run more by academic leaders and managers than in the past. These new leaders and managers have, again in comparison with the past, gained more possibilities to run their institutions. However, they remain dependent on their professionals in many respects. Academics, and particularly the most prestigious ones, are still very capable of effectively voicing their opinions (de Boer *et al.* 2007*a*). In terms of external governance, peer systems are very important and are offering especially the academic elites the opportunity to influence strategic decision-making. Particularly in the area of quality control and research programming, professional expertise is providing members of the academic elites with a key role that is likely to be even stronger than in England.

Stakeholder Guidance

Nevertheless, the impact of academics tends to be modified by the inclusion of non-peers in internal and external decision-making bodies. In both countries, we witness attempts to incorporate industry's and society's needs in academic research planning. Within the universities, lay membership in governing and advisory bodies provides examples of stakeholder influence. Looking at the representation of lay members in the governance of academic research in England, we see external stakeholders represented in the boards of governors, in policy formulation, and in discussing assessment criteria for the RAE. In the Netherlands, a number of universities established so-called expert councils to advise on research matters. Institutional leadership is also influenced by external guidance as every university has a supervisory board entirely composed of external stakeholders. The authority of users as funders, such as businesses and industry, is also more important today, since the 'soft money' of contract research may be needed for research groups to survive in times of highly competitive funding from the public purse.

5. Choosing Research Problems: Views from the Shop Floor

In both countries, shifts in higher education and research governance have changed authority relations. But do these changes also affect the academic heartland of problem choice? In the following, we investigate how researchers in four biotechnology research groups and four medieval history groups in England and the Netherlands reflect on changes in their institutional environment, and what they mean for their problem choice.

English Biotechnology Groups

All senior and junior biotechnology researchers in the two English research groups report that they still have a lot of freedom to decide what and how they want to research provided they have funding for it, as Morris also indicates in Chapter 8. Therefore, both external funding bodies and university management are important for the capacity to carry out research. Funding bodies, such as research councils or charities, can in fact influence what area to research according to a professor in a high performing group:

You a have a certain amount of room for manoeuvre insofar as nobody actually comes in says: 'you will work on this protein'. But if you cannot get the funding to work on this protein, because the charities or the research councils or whoever it is doesn't provide you with a grant and you don't have the ground funding, then you've had it, see you don't have that, you don't have a massive room to manoeuvre. Nobody comes in and says go and work on protein X but if you are not working on protein Y, that someone will fund, then you won't work on anything. So in a way there is someone, the funding bodies guide it. I don't have a little pot of money that the university provides me with, so just go away and do your own thing.

The choice of research topics usually is related to the likelihood of funding. If there is no funding, there is a threat to that particular research line, and a researcher has to go where there is funding available. In this way, the freedom to choose research topics is restricted by external requirements. For English biotechnology researchers the main concern is thus how to fit the priority areas of external funding bodies to enhance the fundability of their projects. Strategies for increasing fundability and fitting project proposals into thematic priorities include the adjustment of topics, the strategic writing of proposals, and the repackaging of ideas as reported by the researchers.

They often play 'percentage games' and strategically decide which funding initiative is most suitable to meet the research unit's interests. One professor, for example, goes only for the highly probable funding, and diversifies the funding base:

When these initiatives come up, I think more so now we are taking a strategic view on looking at what the topic is, if we see that this is something that we are really involved in and we can really put together a good bid, then we go for it. You are just playing simple percentage games, you are looking at your likelihood for funding through different routes and that is partly down to just simply doing the numbers; how much money is there in this initiative, how many people are likely to apply, what's your percentage of getting it. But it's also looking at your belief in your ability to put together a really good proposal.

Applying for funding to him is about what is most efficient, weighing the pros and cons of the likelihood of funding of the research topic. He aims at less competitive grants due to a low application success rate in some programmes. Similarly, in the other group a senior researcher draws attention to the fundability and ability to pursue topics of interest: 'It was my agenda insofar as I started working in an area which I thought was interesting to me, and I've been fortunate on the whole until perhaps very recently, that it has also been fundable, so I carried on down that line until someone comes along and it gets worse.'

In his case, the area of interest was recently seen as 'outdated' and not fashionable, so the funding has been stopped from research councils in this particular area. If the research problem is not suitable for the funding priorities, then the solution for biotechnology groups is to be strategic and creative about how to maximize the chances of success by adjusting the theme. But this does not mean real change in their research direction, because 'you put a different spin on it, on what you are doing'. Thus, researchers in both groups strategically decide which research council theme fits their own research. They also follow certain strategies of using 'fire words such as relevance, innovation' in the proposals and, as reported by some researchers, repackage their ideas. In other words, in the grant proposals they emphasize what the reviewers of the grants are likely to want to hear in terms of 'excellence and relevance':

Actually what they are doing is that they are probably getting the same research in, but they are just getting the people to write it in a rather different way. And I have, just from a pure research funding policy point of view, I do have some problems with things they are saying; they are going entirely for things which are terribly innovative and you know, you get this lie a lot.

A researcher with a number of bad experiences in terms of reviews ending up with no funding is very concerned about this as it indicates that funding bodies and their reviewers did not regard the topic as fundable. His strategy to improve the situation is actually to change the topic area, which he thinks is a difficult thing to do and which he does not regard positively:

I am trying to look and see whether I can swap what might be regarded as a more attractive funding area. That's not all that straightforward to do. But yes, that is what happens whether you like it or not. And you have to try and shift the emphasis of what you are doing and trying to get my Ph.D. students in my group to move on a biochemical basis rather than a structural basis, to do more x-rays, to do more lab work that is not directly related to structures, synthesis, put value added into our grant publications and output. This is the way the funding bodies wanted to pay, and that's fine. It's public money.

Funding is crucially important while there is a question of how to balance between curiosity-driven research questions and fundable research questions. Researchers in both groups still predominantly think that the ideas are coming from unanswered questions from previous research, and only then do researchers try to fit into the theme of the funding initiative, as seen from a professor from a lower performing group: 'As a scientist I feel most comfortable with driving the next phase of research from what you have previously done. Now the danger of that is what people say, oh, it's incremental.' Similarly, a researcher from another research group reports that the process of selecting his topics is 'organic evolution'.

Such considerations are dominant in both research groups which preserve their problem choices as long as they find funding. They are not willing to change their research areas easily because 'it's hard to shift your thinking'. They find ways to adjust to the demands of external funding bodies by taking strategic steps in choosing the research topic.

At the same time, it is not only external funding opportunities and constraints that may influence problem choice in these research groups. In the case of the lower performing unit, university management tried to do this as well by centrally appointing a new group leader whose job was to change the direction of research within the group. This move was motivated by the end of a major long-term structural grant of the group and a perceived incapacity of renewal within the group.

When asked about the guidance in problem choice, the new group leader thinks that he is the facilitator for research themes, while the funding bodies are the ones that are directing problem choice. He admits that he could influence the research agenda of junior researchers but less so the research agenda of senior researchers:

Actually in research it's very difficult for people to turn the right hand corner. What happens is like a big ship, just turn slightly and to say well it's turning because of anything you've said or anything that is happening externally is hard to judge. What I think is probably correct, is to say that the research leaders are very clever and they have their own minds; they know what they are doing and they will adjust a bit but they are pretty mature individuals who need I'd say discussion but very limited advice. It's the more junior staff who are coming through the ranks who kind of witness what I say and how I say it and are potentially influencable [sic]. I think that's a key code: to develop and influence. Of course you do that in your own research groups but you also need to see to it with the other juniors.

Junior researchers admit that they contribute to discussions about research agendas, but the research topics are decided by the senior researchers—in this case, the group leader. The exception was the junior researcher who had an independent fellowship and could follow her own research line, but still had to fit into the overall research theme of the institute.

The question of the leader's role in research agenda setting and different decision power on problem choice between juniors and seniors was also a point of discussion in the high-performing group. This research unit is not due to reorganization as it has earned high credibility from its excellent performance in the RAE. In contrast to the lower performing unit, management has not become involved in major agenda setting.

Researchers in the higher performing unit share strategic considerations regarding external funding. The professor reports that a researcher can do something 'completely off the wall', in other words, completely follow his/her own research idea, only if he/she is established in the field and has funding for research. Both he and another researcher agree that the majority of researchers do not have this high credibility and no additional freedom to do what they want. A post-doctorate does not have any freedom to choose his research topic, since it is already funded by the project for which the PI applied and received funding. However, junior researchers can contribute to writing proposals.

Dutch Biotechnology Groups

Academics in both Dutch biotechnology research groups in principle have academic freedom to decide what and how they want to research. Their ideas usually come from unanswered questions from previous research; that is, organically following the developments in their research area. However, researchers indicate that they are conditioned by funding for their research, and therefore external sponsors can be influential in guiding their problem choice.

Due to the nature of biotechnology, academics in both research groups heavily rely on internal and external collaborations, and so problem choice may be a collective endeavour. Within the research groups a collaborative spirit is maintained by their leaders, where the open discussion of research topics is encouraged and in this respect problem choice may be regarded as a 'team effort'. For example, the high-achieving research group leader emphasized the democratic and strategic approach they undertake while choosing collective research directions:

Of course, we have strategic discussions within the group. And you rely also on your Ph.D. students and your post-docs and their skills; I value their opinion. So I certainly ask them also about their view to the future; where do you think the chances are in the future? So strategic discussions we have with the staff members, but also as a group. We even have sometimes a day [for this purpose].

The respondents in the Dutch biotechnology groups emphasize the importance of 'fitting' into the overall themes and strategic directions of the group that are, however, characterized as 'pretty vague'. This is not surprising given that they are reached through a discussion and consensus among the group members and a general willingness to be embedded in the research unit and collaborative work. This applies to both junior and senior academics, although their situation as regards freedom to choose their own research questions differs. Junior researchers tend to be influenced by their professors on what topics to pursue, especially if it comes to writing collaborative big project proposals for external funding bodies. More experienced post-doctoral researchers, however, add the name of the professor to the project proposal only at the end. There is no real steering of problem choice taking place, as noted by a post-doc from the high-achieving research unit: 'All the projects the post-docs have written are completely by themselves. The professor, if you need him or her for carriage then the name is there, the signatures at the end and that's the only thing the professor does.'

The attitude of maintaining own priorities in line with the overall research agenda of the group is supported by the faculty management. The managers do not impinge on the academic freedom of research groups. The group leaders think that as long as they play the game in the right way—that is, maintain the quality and attract external funding—the research groups can pursue problem choice of their own liking. As expressed by the research group leader the management understands that academics work best when they have freedom for manoeuvre in what and how to research:

You have to keep that door [the manager's] closed. If I went to every meeting where managers tell us how to get into running initiatives, I would never be here. It's probably for the best they do that, they push for new funds, new initiatives, but I just have to make sure we do good research and if there are any opportunities that pop up, you should be able to work on that. I think we get a lot of freedom, and rightly so. It also depends on which people you hired but if you trust that someone wants to do his job properly, you should give him the freedom to do so.

At the same time, external funding schemes can be prescribing certain thematic areas that are more likely to get funded. As substantial external funding is indispensable to the research groups, the academics tend to be careful in balancing their own research priorities with those of the external sponsors. Usually, they write proposals in a certain way so as to fit the external criteria and choosing a fashionable topic. Researchers are open about their strategic behaviour in this regard. For instance, a professor from the lower achieving research unit puts it succinctly: 'The theme we are working on is very popular. I mean there are many grants you can apply for, so there is a constant possibility to apply for grants. Of course, you try to fit in as good as you can in the theme they want. Try to write to some extent what they want to hear.'

When talking about EU funded research programmes, both Dutch bio-technology research groups are concerned about the constantly changing research priorities. They have to focus on specific areas and programmes, and that can be a problem as it may require a shifting of the research agenda. However, researchers do not indicate any examples of such beha-viour as they bid usually for funding that fits with their own research problem choice and try to be creative in choosing the right programme. This does not come without real concerns how to balance the two, as exemplified by a professor from the high-achieving research group:

EU priorities change so one has to be creative with themes in order to get proposals through: I mean EU is now focusing mainly on health, so it's all about health. So you have to refocus and if you are in a field where your bacteria are not really health related you have to find a way. Otherwise you cannot make proposals anymore. You have to be creative around the themes they choose.

In many cases this means framing the questions right, but in essence this does not change the key focus of their research agenda. The problem choice in biotechnology research groups is thus a bundle of decisions about a researcher's own research interests, collaboration opportunities, research unit strategic priorities, and external project grant requirements.

English Medieval History Groups

The selection of research topics among interviewed English medieval his-torians is predominantly driven by the dynamics of their own research inquiry, where the process of individual reflection and consultation with the wider academic community is central. Both senior and junior research-ers agree on this matter in the higher-performing departments. There is some pressure from the university in terms of the acquisition and manage-ment of research projects, but 'deciding what it is we want to do still lies with us as individuals'.

In the lower-performing department, the researchers strategically parti-cipated in the faculty multidisciplinary research themes to earn credibility in the eyes of the faculty management and to put their own topics on the faculty agenda. In other words, they proactively influenced the develop-ment of faculty themes. The department head still firmly believes that research should be idea-driven rather than guided by funding bodies' de-mands: 'There is an intellectual commitment in the department to move forward, to moving the university forward, but it's not simply chasing the money, that's what we are afraid of to some extent, that our research will, if we are not careful, be resource rather than idea-driven.'

Other respondents in this department note as well that they do not see a lack of room for manoeuvre in deciding what to research. A professor is able to look at a completely new area:

I can't find a green liberal metaphor for this, but it is rather like being the first whaling ship to enter the Arctic; all the whales are around you to pick young prey [...] or a less horrific image, walking into the new gold fields in Australia as the first person to walk there apart from natives and picking up lumps of gold everywhere, instead of having to refine the gold that has been recycled.

However, in both groups there are indications that researchers need to follow certain considerations when choosing topics for externally funded projects. Such strategic considerations have to do with the application to funding bodies for grants, where certain areas of research are more likely to be funded. Therefore, researchers have to make choices about how to fit into the priority area of the funder, such as a research council, a charity, or regional authorities, without compromising their own research interests too much. Research topics are thus not always driven by the academic agenda of the researchers, but are influenced by funding priorities and the perceived likelihood of getting funding for certain themes.

The higher-performing department collectively considered what to apply for and how to improve research proposals to secure external funding. Their considerations included not only the kinds of research questions that could be requested but also 'how we might package what we are proposing most effectively in a way that will attract the interest of outside funding bodies'. This careful consideration is not without reason. The experience of a senior respondent shows that the topics of the research projects funded by the funding bodies are related to the priorities of those bodies.

In the lower performing group there are indications of similar influences on problem choice when it comes to funding bodies. For example, while applying for project funding for three years, a junior researcher admits:

I wouldn't possibly immediately have chosen [the research topic] although it's actually very much connected; it's not central to my research. And so it's influencing. [...] Market funding has influenced what my future research expectations would be over a period extending about two to three year period. I would actually have to say, ok, I will be concentrating on something which I might not concentrate or probably would not concentrate otherwise.

This junior researcher regrets that he had to compromise his research topic: 'I'd like to have chosen to do something else which will interest me a bit more'. Similarly, a professor from this lower performing department is not satisfied with the restraints coming from the funding bodies. He reiterates that applying for external funding is 'exhausting and limiting', since the external funding bodies have many rules and restrictions: 'I've saddled myself with a research project with huge number of rules and restrictions attached that limit me and make my ability to research much less. And when these three years are over I am not going to do it again before another ten years, I should be free again, and have a much better research basis.'

Researchers from the lower-performing department emphasize that they follow practical considerations of the likelihood of funding while applying for the research projects; they employ the strategy of using specific topics that may fit the priorities of the funding bodies:

I think the funding bodies [. . .] influence enormously what actually gets done because ultimately any time you might have four or five equally kind of good projects which [. . .] selection you think, good projects which you could look at, and you'll say ok, those four give good projects, they've got lots of intellectual merit but I don't think a realistic way of getting funded—whereas this project will actually tap into and this has a good chance to get funded, so I'm going to go with that one. I think actually the funding bodies still actually have an effect on what's going to come out because ultimately [one has to] go after the money.

Obviously funding is very important to both research groups while they try to balance the demands of the funding bodies and their own research interests and agendas. This compliance with the rules of the game of project funding is visible among both junior and senior researchers.

An alternative response to the pressure for external funding from the university management is the diversification of funding sources. A professor from a lower performing group exclusively follows his own research interests while applying for external funding. His strategy is diversifying his funding base and being popular enough to have his own 'industry' which brings in money—that is, to participate in different TV shows, documentaries, and talks. This is a way to earn money for research that is not heavily taxed and provides some means to carry out research that he likes. He calls it 'entertainment business', as seen in this extended quote:

History and archaeology are hugely popular with the public. And there has been a tremendous growth of programs on them. It's starting to subside because they got over-funded, overstretched; there are too many bad programmes on archaeology and history, some of which I helped to make. But that is alright, my sources are very diverse, at ground level that means that village history society appears every year at this locality. If I wanted to, I could spend the year going from one local society to another speaking mostly to retired people. But people with grey hair have big wallets, because there is lots of spare time and cash. That's another way of increasing the income while entertaining people.

Dutch Medieval History Groups

Looking at the cases of medieval history in the Netherlands the autonomy of researchers is traditionally high in terms of deciding on their research agenda. Their research is not as resource-intensive as in biotechnology field, and external research funding is more a facilitator of research as it helps to buy time from teaching rather than a determiner of the research topics. The responses of the academics in medieval history show that their research agenda is predominantly driven by their own research interests, although they are concerned about the priorities of external funding bodies and the multidisciplinary themes of their research institutes.

Dutch medievalists indicate that the selection of research topics is a bottom–up activity where the most important considerations are the researcher's academic preferences, usually based on the consultations with the academic community and their own instincts. This is emphasized by professors as well as by junior academics. For example, a post-doctoral researcher from a high-achieving research group feels she has a lot of room for manoeuvre in deciding what and how to research:

There is a lot of freedom, really a lot of possibilities to find your own voice; do your own thing and that have led to the most wonderful results. For instance, an AiO[2] here who started two years ago or so, made a major discovery. She found manuscripts that were thought lost and people have been looking for them since the early nineteenth century. And she goes, reads her footnotes, thinks very deep and goes to the archives and finds them. That to us is something 'whaaa', to open your champagne for. In that sense, yes, there is, as long as there is no money involved. And that has not changed in my time.

The professor from a lower-achieving research group strictly follows the traditional academic view that research cannot be organized or programmed.

[2] In the Netherlands doctoral candidates can be staff members with an employment contract: they are then called AiOs. Doctoral candidates can also be funded via other sources, such as fellowships, and do not hold a status as staff.

Researchers in both Dutch medieval history cases mention certain factors that may, however, influence the selection of research topics, such as the overall themes of the research group or university institute, popularity of the topic and related likelihood of external research funding.

The multidisciplinary nature of the research programmes of the umbrella research institutes that the two research units belong to has been mentioned as a possible influence on their research agendas. In general, however, medieval historians are not that worried about them since they can easily find an area where their own research topics fit. Usually a rather broad research programme is drafted which 'is written in a way that there is plenty of possibilities for people', as shared by the professor of the high-achieving research group. A more important aspect for researchers in this unit is following the traditions of the research group, which has a specific research area and a specific medieval history period that it tackles.

When it comes to external funding bodies, both research groups emphasize the importance of 'wrapping' their ideas in the priorities and specific thematic areas of the external research funding body, which is usually the Dutch Research Council. The external sponsors may frame the research questions, but do not direct them. For instance, a post-doctoral researcher who secured external research funding shares his experience that is quite common among medieval historians in the Dutch cases: 'If I learn that NWO [Dutch Research Council] is starting a project, we are inclined to do this. Of course, you start thinking about well, what could I do with that, so it does influence your thoughts, but in the end, I guess, if you are really at the moment that you are writing a proposal, it's basically how can I sell this.'

The strategies for ensuring the likelihood of funding include making the topic look relevant for and attractive to the funding body. It can also lead to choosing a broad and interdisciplinary topic that would fit into the preferences of external research sponsor as vividly described by a professor in the lower-achieving research group:

I currently have a research proposal awaiting funding that involves urbanization and city culture. This is a non-recurring NWO funded programme. My colleague in history has submitted an application for three studies; one for an archaeologist, one for a literature historian, and one for a social economic researcher. They have a research proposal which uses all three research areas. This type of multidisciplinary research is usually very successful in getting funding.

Further, some researchers, especially post-doctoral researchers, think that the relevance of certain topics within the academic community can influence funding from the external funding bodies: 'One sees, for example, something about religion. Actuality of religion they think is something that scores at the ministry and may get funding. That is how the matters are.' Research hypes are thus mentioned as possibly influential. Junior researchers are particularly conscious of these fashionable topics, as exemplified by a post-doctoral researcher from the high-achieving research group:

In my case I grab everything I can get, simple as that. I need to keep a job, but in general when people write research proposals you just have to link up to international sexy research so to speak. Right now its ethnic identities and barbarians [. . .] I think that's one of the only ways to get subsidized. There are always these questions: 'How shall we write this?', 'Who might be the international referees?' 'Who might they choose?' There are five options for instance, not more than that. So, it is politics.

In fact, 'politics' influences the problem choice of junior researchers. But the actual implementation is left up to the researcher once the funding is secured, as seen by a post-doctoral researcher: 'As soon as the project is awarded, yeah, it's my project. I get all the freedom I want.'

Strategies of how to 'sell the topic', to balance between external sponsors' priorities and personal research interests, are common to both research groups. External funding bodies and research institutes do not seem to significantly influence the research agenda. Researchers still follow their own interests in choosing research topics, although they are conscious of 'relevant' and 'hot' topics while applying for external sponsorship.

Discussion

The evidence from all cases suggests that it is indeed not easy to influence academic problem choice. All groups and individual researchers struggle to make their own choices of research topics, and most of them succeed in maintaining their preferred lines of research. Researchers are highly unlikely to compromise their own problem choice. The data from our study suggested that this is 'the last thing on earth' they want to do, unless they are really forced into it. They are also aware of the potential costs of major changes in their problem choice in terms of losing part of their expertise and reputation. Their capacity for maintaining their problem choice is, however, dependent on their financial resource base, on their performance and reputation, and to some extent on researchers' seniority and the field of research.

The dominant theme throughout the interviews is 'funding' for either high-cost projects (biotechnology) or to buy out time from teaching (medieval history). Research groups also perceive increasing pressures to achieve external funding raised by respective behavioural expectations of external and internal research assessments, management expectations, and criteria for promotion. Research sponsors offer increasingly programmatic funding favouring certain topical areas and ask for 'relevant' research results not only in the sciences but also in the humanities. The need for research funding is not a new experience for researchers in biotechnology (see Chapter 8) but it has intensified due to the increasing lack of support from within the institution. In contrast, growing needs and expectations as regards external research funding are a more recent experience for medievalists.

In response to these changing conditions and expectations, bioscientists and medievalists act in specific ways. There are particular patterns across the groups in the use of certain strategies to pursue the research topics of their own liking. The most used strategy is to 'fit' into the research themes of external sponsors to increase the likelihood of getting funding. Many researchers say that, while following their own research lines and interests, they take account of the programmes and research priorities of their research institute and of external sponsors. Popularity of research topics and the chances of obtaining funding do influence many researchers' choice. They play the game by selling their own ideas in such a way that they fit research programmes and expectations that tend to be broadly and vaguely described. In some cases, these strategies are sustained by departmental policies supporting proposal writing and submission in an institutionalized way via internal advice and self-evaluation. In this way they try to enhance the likelihood of funding as well as the viability of research projects. The dominant response is thus symbolic compliance.

We also have one example of a proactive manipulation in the case of a lower-performing group in England. This group aims to gain legitimacy in the eyes of the university management by participating in the multidisciplinary faculty themes. The group proposed its own theme following the given criteria. In this way, researchers from this group could influence the faculty theme and receive a new professorial position.

The two higher achieving English groups as well as all the Dutch groups have managed to retain their preferred topics. They are to a large extent able to seal them off from internal and/or external thematic priorities (or internal reorganizations). They do so mainly by writing project proposals in a strategic way, formulating them according to the exigencies of the funding bodies, while following their own topics at the same time. Not all groups are fully successful in this respect. In England, two groups had to compromise their problem choice to some extent. These groups in England are in a highly uncertain environment; their ranking in the RAE is lower, and their dependence on resource providers is so high that they see no alternative to compromising their problem choice.

The lower performing medieval history unit was eager to strengthen its research capacities and reputation, and is encouraged to do so by the university management. This implies it has to 'play the game' of the university management and to become more active in getting external funding. The group and especially the junior researchers in the unit saw, however, less likelihood of receiving external funding with their traditional themes. Consequently, they compromise their problem choice to some extent. Researchers in the lower-performing English biotechnology group were forced to compromise on their problem choice as well. The group experienced the end of a long-standing basic research grant and a 'sudden' substantial need to obtain competitive external funding. In turn, this led to an internal reorganization, including the arrival of a new group leader who was been put in place by the university's management to change the research programme of the group. This example stands out as a case of multiple compliance, since researchers had to compromise their problem choice, taking into account the programmes, interests, and preferences of funding bodies nested with the intervention of the university's management decisions.

These examples indicate that resource dependence and apparent lower achievements are strong factors in changing the problem choices of research units. They also point in the direction of stronger nested effects of growing external resource dependence, hard research evaluations, and intrusive management in the English case, even though reorganizations are not unknown in Dutch universities. At the same time, the prevalence of symbolic compliance strategies indicates the persistence of routines and norms of academic problem choice. Particularly when the unit's performance and reputation is high, researchers tend to be successful in exploiting their professional autonomy also on the external market.

The biotechnology and medieval history groups also have to some extent different ways of choosing their research agenda. Although the bottom line in the biotechnology groups is that researchers pursue their own research interests in choosing research problems, individual freedom is bounded in several ways. Individual researchers do not only consider fundability as well as popularity of the topic in their selection process, but also the wider research agenda of the group. This agenda must be taken into account, but usually leaves sufficient space for individual selection. Although such restrictions affect individual choices, especially those of post-doctorals who usually work on the project of their seniors, they are all in the realm of the academic unit. There is at least 'collective' freedom of choice. Medieval historians, unlike bioscientists, have individual topics and do not negotiate the problem choice collaboratively. The department head/group leader does not have the same topic as other medieval historians in the group. Junior researchers determine their own research lines and topics of interest. All researchers claim that they have the room they need to select their own research topics within the group, given that they have funding and academic output.

6. Conclusion

The central research question of our chapter addresses the effects of shifts in governance and related changing authority relations on the problem choice of researchers in public universities. Shifts in governance in England and the Netherlands have resulted in changed institutional environments and actor constellations. Where the English government has intensified its interference with the university sector, in many instances the Dutch government has stepped back. Other important changes point into the same direction in the two countries: a stronger role of the funding councils and other research sponsors, less academic self-governance within universities and an increasing role of academic elites via peer review and evaluations, more strategic autonomy and management within universities, and increased competition for scarce resources. The timing and depth of these changes differ, however, between the two countries.

When we look at the responses on the shop floor of the investigated universities, struggles to maintain academic researchers' problem choice look largely similar in the two countries and in the two disciplines (biotechnology and medieval history). Our study shows that it is difficult to shift research agendas of individuals and research groups. Even in situations of growing dependencies, be it on internal or external stakeholders, researchers are keen to pursue largely their own agenda. One way of doing this is through symbolically complying with other agendas; another one is through participating in the agenda setting of other stakeholders. Researchers also make use of ambiguities in their research environment. Research councils or university managers may be active in steering research into certain fashionable directions, but research programmes tend to be designed so that quite a variety of themes can be 'fitted' into them.

The outcomes of our study also indicate that it would be wrong to suggest that research activities cannot be and have not been changed at all. 'Fundability' of research is the dominant theme for the researchers mediating the process of setting the academic research agenda. Many researchers, and especially junior researchers, choose in the light of topical fashions, research mainstreams considered important in peer reviews, and success rates in funding programmes. Vulnerability of groups' and individuals' research agendas is mainly due to lack of resources, lower scores in research assessments, and an interventionist management. There are examples in our study of English research groups and researchers that comply with the demands of external sponsors and their university management, or that are forced into thematic reorganizations. Thus, we observe symbolic compliance at work in many cases, and self-adaptations and enforced adaptations in some others.

References

Becher, T., and Trowler, P. R. (2001) *Academic Tribes and Territories*, 2nd edn. (Buckingham and Philadelphia: Society for Research on Higher Education/Open University Press).

Clark, B. R. (1983) *The Higher Education System* (Berkeley, Calif.: University of California Press).

de Boer, H., Enders, J., and Leišytė, L. (2007a), 'Public Sector Reform in Dutch Higher Education: The Organizational Transformation of the University', *Public Administration*, 85(1): 27–46.

de Boer, H., Enders, J., and Schimank, U. (2007b) 'On the Way Towards New Public Management? The Governance of University Systems in England, the Netherlands, Austria, and Germany', in D. Jansen (ed.), *New Forms of Governance in Research Organizations: Disciplinary Approaches, Interfaces and Integration* (Dordrecht: Springer), 137–54.

Gornitzka, A. (1999) 'Governmental Policies and Organisational Change in Higher Education', *Higher Education*, 38: 5–31.

Henkel, M. (2000) *Academic Identities and Policy Change in Higher Education* (London and Philadephia: Jessica Kingsley).

Henkel, M., and Little, B. (1999) *Changing Relationships between Higher Education and the State* (London: Jessica Kingsley).

Holliday, A. (2002) *Doing and Writing Qualitative Research* (London: Sage).

Kehm, B. M., and Lanzendorf, U. (2006) *Reforming University Governance: Changing Conditions for Research in Four European Countries* (Bonn: Lemmens).

Kickert, J. M. W. (1997) 'Public Governance in the Netherlands: An Alternative to Anglo-American "Managerialism"', *Public Administration*, 75(Winter): 731–52.

Leišytė, L. (2007) *University Governance and Academic Research* (Enschede: CHEPS, University of Twente).

Maassen, P., and Van Vught, F. A. (1988) 'An Intriguing Janus-Head: The Two Faces of the New Governmental Strategy for Higher Education in the Netherlands', *European Journal of Education*, 23(1/2): 65–76.

Marshall, C., and Rossman, G. (1997) *Designing Qualitative Research*, 2nd edn. (London: Sage).

Meulen, Barend van der (2007) 'Interfering Governance and Emerging Centres of Control: University Research Evaluation in the Netherlands', in Richard Whitley and Jochen Gläser (eds.), *The Changing Governance of the Sciences: The Advent of Research Evaluation Systems* (Dordrecht: Springer), 191–203.

Meyer, J. W., and Rowan, B. (1977) 'Institutionalized Organizations: Formal Structure as Myth and Ceremony', *American Journal of Sociology*, 83 September: 340–63.

Oliver, C. (1991) 'Strategic Responses to Institutional Processes', *Academy of Management Review*, 16(1): 145–79.

Stephenson, P. (1996) 'Decision-Making and Committees', in D. Warner and D. Palfreyman (eds.), *Higher Education Management* (Buckingham: Open University Press), 241–70.

Vijlder, F. J. de, and Mertens, F. J. H. (1990) 'Hoger onderwijs-arbeidsmarkt: Zorgenkind of betekenisvol perspectief?', *Tijdschrift voor Hoger Onderwijs*, 8(2): 42–54.

Williams, G. (2004) 'The Higher Education Market in the United Kingdom', in P. Texeira, B. Jongbloed, D. Dill, and A. Amaral (eds.), *Markets in Higher Education* (Dordrecht: Kluwer Academic Publishers), 241–70.

Willmott, Hugh (1995) 'Managing the Academics: Commodification and Control in the Development of University Education in the U.K.', *Human Relations*, 48 September: 993–1025.

Ziman, J. (2000) *Real Science: What it is, and What it Means* (Cambridge: Cambridge University Press).

10

The Limits of Universality

How Field-Specific Epistemic Conditions Affect Authority Relations and their Consequences

Jochen Gläser, Stefan Lange, Grit Laudel, and Uwe Schimank

1. The Field-Specific Nature of Authority Relations

Any account of the impact of changing authority relations on scientific research must consider differences between fields of research. Three kinds of variation can occur. First, authority relations themselves can vary between fields. For example, actors with commercial interests have little or no authority in fields that are remote from applications such as ancient history, pure mathematics, or high-energy physics. Ethics committees exercise considerable authority in fields researching human subjects but not in others. Second, field-specific instruments for exercising authority may be used. The proposal to assess science, engineering, and medical fields by using quantitative indicators but to keep peer review for the humanities and social sciences in future rounds of the British Research Assessment Exercise is a case in point. Third, authority relations may have different effects in different fields depending on the practices of a field's research, as for example demonstrated by Chapter 8.

Such field-specific modifications of authority relations and their effects have been common knowledge in science studies for a long time. The differences between research practices and social structures of scientific disciplines have been explicitly addressed in several influential theories (Whitley 1977, 2000; Rip 1982; Böhme *et al.* 1983). However, the role of these epistemic conditions is rarely taken into account when governance instruments are designed or investigated. Most instruments for the governance of science are applied across several if not all fields in the natural sciences, social sciences, and humanities, and are intended to have the same effects in those fields.

This logic underlies the research evaluation systems that have been established in many countries during recent decades (Whitley and Gläser 2007; see also Chapters 2, 5, and 8). A similar blind spot exists in studies investigating governance regimes, which ignore the impact of epistemic conditions on governance by (*a*) including only one field (for example, Meulen and Leydesdorff 1991; Morris 2000; Sousa and Hendriks 2007), (*b*) investigating several fields and only distilling results that are common to all of them (for example, Henkel 2005), or (*c*) listing variations in effects without integrating them into the study's causal explanations (for example, Henkel 2000; Leišytė 2007).

In this chapter we address this gap of policy design and analysis by providing preliminary answers to two questions. (1) Which properties of fields and research practices modify authority relations and their effects? (2) Which mechanisms produce the field-specific variations of authority relations and their effects?

2. The Empirical Investigation

The empirical material we use to answer these questions stems from a comparative investigation aimed at identifying the impact of a specific governance instrument—indicator-based block funding of university research—on the content of that research (Gläser and Laudel 2007; Lange 2007; Gläser *et al.* 2008). Australia has the oldest and most consequential indicator-based funding regime for university research, and the adaptation of the actors in the university system to that regime could be expected to have systematic effects on knowledge production.

Investigating these connections required the solution of three major methodological problems. First, changes in knowledge production must be identified, which requires comparisons between knowledge production processes and their outcomes—including comparisons with those research processes that could not be conducted because of the governance regime under investigation. Secondly, if field-specific effects of governance instruments are to be identified, the changes in knowledge production must be compared across fields. Thirdly, the observed effects must be causally attributed to a specific governance instrument whose impact is overlaid by influences of numerous other instruments and organizational procedures. We developed the following preliminary solutions to these problems.

1. For the identification of changes in the production of knowledge we utilised the concept of research trails and additionally derived a list of possible changes in knowledge from the literature. Following Chubin and Conolly (1982) we defined research trails as sequences of projects that are thematically connected because later projects use the theoretical, methodological or empirical knowledge that has been produced in previous projects. We used a variety of empirical strategies for identifying the initiation, abandonment, and topical change of individual research trails. A second approach consisted of studying general features of research that are discussed in the literature as possible candidates for changes under political pressure: namely, the type of research conducted (methodological, theoretical, experimental, or field research); the dominant research orientation (basic, strategic, or applied); the relationship to the community's majority opinion (non-conformist versus mainstream); time horizons (long-term versus short-term); the degree of interdisciplinarity; the degree of intellectual risk taken in the research; and the reliability of results (Gläser *et al.* 2002).

2. We conducted comparative case studies that used the general properties of research listed above, and selected six disciplines for which these properties varied.[1] We included mathematics as a non-empirical science, physics and biochemistry as experimental sciences, geology as discipline whose empirical work is largely based on field observations, political science as a social science that encompasses both theoretical and empirical research, and history as one of the humanities with yet another specific empirical programme. These disciplines also vary in another important property, namely their reliance on expensive equipment and additional manpower and thus their demand for external funding. In order to advance our understanding of field-specific properties and their modification of governance, we included two ethnographies: one in history and one in biochemistry.

3. A causal attribution of changes in the production of knowledge to the governance instrument under investigation can be achieved by specifying the social mechanisms that link the particular governance instrument to

---

[1] We could not draw on the theoretical approaches mentioned in the Introduction because they have not yet been (and probably cannot be) operationalized for empirical investigations. The approaches based on the constructivist sociology of scientific knowledge such as Actor-Network-Theory (Callon and Law 1982; Callon 1986; Law 1986; Latour 1987, 1988, 1996), the 'Mangle of Practice' (Pickering 1995), and 'Epistemic Cultures' (Knorr-Cetina 1999) are not of much help either because they combine idiosyncratic descriptions with 'grand theories' that do not support comparative frameworks.

these changes.[2] This we achieved by conducting nested case studies (Patton 2002) and comparing cases at all levels of our multi-level problem. At the level of national systems of university research funding we compared Australia to Germany, which at the time of the investigation (2003–7) was just beginning to introduce performance-based funding schemes for its universities, and was thus ideally suited as a 'reference measurement' of research not yet affected by performance-based block grant allocation. At the level of universities, we included seven universities from different strata of the highly differentiated Australian system and two German universities.[3] We also compared organizational sub-units (faculties and schools) in and across universities, and individual academics at different stages of their career from lecturer to full professor.

Data collection was conducted from 2004 to 2007. It combined analysis of documents and internet sites, bibliometric analyses, qualitative interviews as the core method of the case studies, and focused ethnographic observations. In Australia, a total of 179 interviews were conducted, including 61 interviews with managers at university, faculty, and school levels and 118 interviews with academics from the six disciplines and all career stages. German interviews included twelve interviews with managers and sixty interviews with academics (professors, their associates, and post-docs). Interviews with managers lasted 45 to 90 minutes and covered perceptions of funding conditions, the impact of the national and internal funding schemes on the core functions of the university (teaching and research), and university strategies for the internal governance of research, with special emphasis on performance evaluation schemes for organizational units and academic staff that are currently in place.

Interviews with academics lasted one to two hours. Following the theoretical considerations described above, the interviews addressed research trails, conditions for research, and university policies concerning these

---

[2] We define a social mechanism as *a sequence of causally linked events that occur repeatedly in reality if certain conditions are given and link specified initial conditions to a specific outcome* (see Mayntz 2004: 241; for similar but less precise definitions, Merton 1968: 42–3; Hedström 2005: 11). By identifying a social mechanism we demonstrate *how* a specific cause—in this case a governance instrument—produces changes in research, which in turn implies causally attributing these changes to the governance instrument. See Whitley (1972) for an application of the concept of mechanisms in science studies.

[3] Our sample included three members of the 'Group of 8' (research-intensive universities, most of which were founded before the Second World War), two 'Gumtree Universities' (post-Second World War foundations that are well established but less research-intensive), and two 'Universities of Technology' which are smaller and focus on applied research.

conditions. In order to go beyond the opinions of academics about the impact of their research conditions on the content of research, these two topics were clearly separated in the interviews. All interviews began with an extended exploration of interviewees' research trails and reasons for initiating, abandoning, or changing them. This part of the interview was prepared by bibliometric analyses of the interviewee's publication oeuvre, whose visualization was used in the discussion of continuity and reasons for changes of research trails. Only after the cognitive dynamics and the specific resource requirements of the interviewee's research were established, the actual conditions for research, including access to resources, behavioural expectations, and university policies, were explored. Ethnographic observations (in history and biochemistry) focused on field-specific epistemic practices and the impact of everyday activities on the conduct and content of research (see Gläser and Laudel 2007, 2009*a*, for a more detailed description of the methodology).

Data were analysed by qualitative content analysis (Gläser and Laudel 2009*b*). Cases were compared at all levels described above. The following analysis of selected results focuses on the Australian case. Results from the 'reference measurement' in Germany are included where appropriate.

From our empirical results we select those that can be used for a discussion of authority relationships and their effects in different fields. For this purpose we use the following aspects of our data analysis:

- For the discussion of field-specific uses of instruments for exercising authority, we used the categorization of governance tools according to their purposes (resource allocation, deciding on and introducing structural changes, distribution of workloads among academics, and individual performance management) and effects. We could not directly observe the role of field-specific instruments for exercising authority because the same instruments were applied in all fields. However, our analysis of the use of uniform instruments in all fields enables indirect conclusions about the role which field-specific instruments might play (section 3).

- The discussion of field-specific authority relations is based on the comparative analysis of academics' research conditions across the six disciplines. Since we had to embed our analysis of the effects of a specific governance instrument in the analysis of researchers' situations in their full complexity, we obtained data on authority relations concerning the formulation of goals and integration of results in the six disciplines under investigation (section 4).

- The identification of field-specific effects of authority relations is based on the analysis of academics' responses to their situation. We compared the content of and reasons for adaptation processes in the six disciplines. Particular attention was paid to the ways in which academics managed their 'research portfolios' (the bundle of actual and potential research trails they could follow) and the ways in which shortages of time and resources were responded to (section 5).[4]

The most interesting question of all is, of course, that about the causes for field-specific effects. Even though this was not the major focus of our investigation, we could draw some tentative conclusions about epistemic properties of fields (section 6) and their role in the mechanisms that produce field-specific effects (section 7).

3. Exercising Authority by Using Uniform Instruments

Authoritative agencies might use different instruments for exercising authority in different fields because instruments that are adapted to the specific conditions of fields are likely to be more effective. In our investigation we analysed instruments for distributing research funding that were applied to universities, within universities to faculties, and within faculties to schools. The different levels of aggregation represent different degrees of homogeneity of the (collections of) fields to which the instruments were applied.

Since funding instruments are likely to have different effects in different disciplines, one would expect both a variation between instruments applied at different levels of aggregation and a variation between the

[4] The comparisons of adaptation processes had to take into account a further intervening variable, namely the performance levels of the interviewed academics' research. Adaptation processes and their content are likely to depend on performance levels, which means that findings concerning field-specific effects are obscured by performance-dependent effects. In order to control for these influences, a variety of indicators and information from interviews was used to roughly categorize academics according to their performance levels. Since this is an inherently problematic exercise, we used as many indicators as possible and varied categorizations in order to test the robustness of our analysis. We included, where appropriate, numbers of publications, qualities of journals, book series, and publishers, citation counts, grant funding from competitive sources, reviews of books published by interviewees, and indicators of esteem such as memberships in editorial boards, invited lectures, and academic prizes and awards. We also included aspiration levels and research plans, which were covered by the interviews. On the basis of the categorization, we checked whether interview data from 'better' performing researchers were systematically different from those obtained from others (see Gläser and Laudel 2009a, for a detailed methodological discussion of the inclusion of informants' performance levels in qualitative science studies).

Table 10.1. Degree of disciplinary homogeneity of recipients of research block funding at different levels of aggregation

Authoritative agency	Recipients of funding	Degree of homogeneity
Government	Universities	Including all disciplines
Central university management	Faculties	Including similar disciplines
Faculty	Schools or departments	Including one discipline

instruments applied in different disciplinary environments; that is, by different faculties. Neither was the case in the seven investigated Australian universities. Instead, the national system of indicator-based block funding was copied by authoritative agencies at all levels.

National Level

In Australia, the block funding for university research has been transformed in an indicator-based system from the mid-1990s. Currently, a system of four indicators controls the allocation of the research block grants. In 2005, 1,135 million AU$ (7.9 per cent of the total income of universities) were allocated according to competitive peer-reviewed external grant funding (weight of the indicator 54.8 per cent), Masters and Ph.D. completions (29.1 per cent), numbers of publications (8.4 per cent), and research student load (7.7 per cent). The allocation is a zero sum game; that is, a university's share of the research block grants depends on its shares in the total numbers of publications, total amount of external funding, and total numbers of research student loads and completions of all universities.

These indicators are applied to all universities and thus to all disciplines. They obviously fit the research practices of some disciplines better than others. For example, in Australia only four types of publications are counted, namely peer-reviewed journal articles, book chapters, peer-reviewed conference contributions, and books (the last counting five times as much as the other publications). While these types cover the major output of many disciplines, they are by no means exhaustive. The Australian Government itself experimented with a significantly larger set of categories in order to accommodate, for example, the arts (by including works of art) and applied research (by including patents), but soon abandoned this approach for reasons of administrative convenience. As a result, the output of some fields is not appropriately covered, which in turn means that these fields do not 'earn research money' by publishing their results, regardless of the quality of their research. In our sample of disciplines, the most pronounced examples were edited books, which play an important role in political science, and book reviews, which are an important means of communication in history (for a general discussion of this problem see Laudel and Gläser 2006).

A similar point can be made for the most important indicator in both Australia and Germany: external funding. The indicator 'external funding' is popular in science policy because it is based on peer review (of grant applications) and because data can be easily collected. However, disciplines vary enormously in their need of external funding and in the size of typical grants. The weight given to this indicator in Australia and Germany makes some disciplines (especially engineering and the experimental natural sciences) 'breadwinners' for their universities, while others do not contribute much to the income, regardless of their performance.

Thus, at least two indicators (among them the most important one) return discipline-specific results. Their application in national resource allocation schemes does not pose too serious problems because most universities combine all or at least all types of disciplines, which effectively desensitizes them to the biases of the indicators.[5]

University Level

The biases of indicators do have effects within universities because the Australian universities responded to their funding environment by applying the same indicators in their internal resource allocation schemes. The logic behind this move is simple: the universities want to maximize their income and thus use the same 'incentives' for their faculties and schools that are applied to them. At the same time, the decision to distribute resources 'as earned' is the one that is easiest to legitimize in the internal struggles for resources. The important difference between the external and the internal application is that the latter is an application to disciplinary units. Within universities, the indicators are effectively used to distribute resources between disciplines. In this situation, the above-mentioned biases against certain fields do make a difference.

Being aware of that fact or using it to legitimize redistributions, some universities respond by slightly changing the weight of the indicators. The

---

[5] However, the use of these indicators has led to a certain devaluation of the social sciences, arts, and humanities in Australian universities because these disciplines contribute comparatively little to the university's 'research income'.

three most research-intensive universities in our sample (the members of the 'Group of 8') modified their internal application of the indicators in order to limit disadvantages for the humanities. They reduced the weight of the indicator 'external funding' in their internal funding formulae and increased the weight of the indicators 'research student load' or 'numbers of publications'. One 'University of Technology' also reduced the weight of the indicator 'external funding' and increased the weight of the indicator 'publications' in order to foster a culture of publication. The remaining three less research-intensive universities (the two 'Gumtrees' and one 'University of Technology') used the indicators unchanged. In one of these universities it was explicitly stated that the humanities do not receive much money this way, but also do not need much.

Faculties, Schools, and Departments

At the next lower level within universities, the isomorphism repeated itself. If there was money for research to distribute, the faculties used the same indicators and weightings that were applied to them. Only one faculty changed the weights by again increasing the weight of research student completions and publications and decreasing the weight of external grant income. This faculty included experimental natural sciences and mathematics, and wanted to account for field-specific differences.[6]

Interestingly enough, even the relatively homogeneous Faculties of Arts or Faculties of Social Sciences did not radically change the weighting of the indicators, although they could have accommodated the specifics of their fields. Like all other actors, deans considered it paramount that the 'right'—that is, income-maximizing—incentives were given to the sub-units. This overriding concern made the Associate Dean of Research of one humanities faculty devise a scheme for individual performance evaluation that included only the indicators as they are applied to the university. He was fully aware of the fact that this does damage to his own discipline (he was a historian) because encyclopedia entries and book reviews were not counted in that scheme. However, he felt that maximizing income was more important.

---

[6] The differences between mathematics and the experimental natural science in the same faculty also became important in one faculty in Germany. The faculty had just begun to design a performance-based resource allocation scheme that completely disregarded the specific nature of research, funding, and publication practices in mathematics.

These marginal adaptations indicate that Australian universities are aware of the tensions between universal governance tools and field-specific research practices (not least because disadvantaged disciplines kept complaining) but nevertheless felt compelled to adopt governance instruments regardless of unfavourable conditions for their functioning. Apart from the universities' wish to transmit the external signals they receive to their sub-units and researchers, the reluctance to modify indicators may also be due to the general scarcity of funding for universities. The income of most universities just covered salaries and basic infrastructure. Under these conditions, it is difficult for universities to legitimatize internal funding schemes that deviate from the 'received as earned' principle. The overwhelming importance of scarcity is indirectly confirmed by the fact that adaptations of indicators are concentrated in the older and more research-intensive universities. Apparently, only the 'richer' universities can address the problem of biased indicators at all.

4. Field-Specific Authority Relations and the Situation of Academics

Any impact of our focal governance instrument—indicator-based block funding of research—on the content of research is mediated by the actions of individual academics, which reflect general changes in their situation rather than particular shifts in governance. Such changes are created by a variety of actors exercising authority over three crucial elements of researchers' situations: (*a*) their access to resources, (*b*) their discretion over time for research, and (*c*) behavioural expectations concerning their research. We begin by characterizing the elements common to all academics' situations, and subsequently explore (*d*) the resulting authority relations and their variations between fields.

Resources for Research

The resource situation of academics in Australian universities is characterized by the near absence of any recurrent funding for research. Many heads of schools stated that they do not have any resources for research to distribute, but need all the money they receive from the faculty 'to keep their school in the black'. Australian researchers do not commonly receive recurrent funding for their research, except for some who may receive 'a few thousand dollars in a good year'. The little money that was available for research was centralized at the university and faculty levels, and was used either in the form of internal grants or for investments in infrastructure—both of which were aimed at improving the university's success in acquiring external funding.

This investment strategy was part of the response of Australian universities to the funding scheme described in the previous section. Internal resource distribution reflected the fact that the most effective means of increasing the block grant income of a university is to increase the external funding, which affects nearly 55 per cent of the government's block grants. Apart from copying the indicators and giving external funding a reduced, but still the highest, weight in their internal schemes, universities created sub-units that were likely to attract external funding and channelled the little internal funding they had to academics who were most likely to win external grants. The latter took the form of 'near-miss grants' and 'start-up grants'.

Australian funding councils informed universities about the relative success of failed grant applications. On this basis, universities gave internal grants to academics who came very close to being funded. These grants were not sufficient to conduct the actual research but enabled additional preparations that increased the likelihood of success in the following round of applications. The same motive made many universities offer 'start-up grants' to newly hired staff and to early career researchers. These grants should support the preparation of successful external grant applications. Recipients of these internal grants and the centres created by universities were envisaged as becoming 'self-sufficient' by entering a situation of continuous external grant funding.

None of the grants or infrastructure investments was sufficient for the planned research. Grants were limited to one year, were usually too small to hire staff, and did not include teaching relief. Thus, in many fields academics who received these grants found themselves in the same situation as the academics who did not receive any funding from their university: they crucially depended on external grants for conducting their research. This made the external funding landscape a very important condition for research.

The external funding environment of Australia is characterized by concentration and scarcity. The only major sources of funding are the two funding councils: the Australian Research Council (ARC) and the National Health and Medical Research Council (NHMRC). An applicant can hold no more than two funding council grants simultaneously at any time. There is little funding from other government agencies, charities, or industry.[7] The success rates of ARC grant applications vary between 20 and 30 per cent for investigator-driven 'Discovery grants' and between 40 and 50 per cent for 'Linkage grants' which are co-funded by industry partners.

Time for Research

The time available for research was constrained primarily by teaching and secondarily by administration tasks. Teaching was evenly distributed between academics in teaching and research positions. Actual teaching loads varied between years because of the necessity to teach the required courses with the available personnel, which in some cases included substitute teachers and tutors who were hired *ad hoc* if teaching funds were available. Since student numbers also varied between universities and within universities between subjects, the teaching loads of Australian academics varied enormously. Academics in our sample reported between four and fifteen contact hours per week.

These teaching loads could be reduced by three measures. One, which played a significant role in only one university, was a decision of a university to reduce teaching loads for selected academics who were designated directors of centres or otherwise important researchers who were likely to increase the university's grant income. A second measure was sabbaticals: the relief from teaching and administration tasks for one semester every three years. The third measure, which was important in many fields, was the opportunity for academics to 'buy themselves out of teaching' by requesting funds for a substitute teacher in their grant applications. An application for three years of funding could include up to three semesters with teaching buy-outs.

Behavioural Expectations

All academics in standard university positions (teaching and research positions) were expected to conduct research even though the university did not provide the necessary resources. Universities also expected high-quality research. However, quality measurement amounted to little more than applying the measures used in the indicator-based funding schemes. Thus, academics were expected to win grants, to publish, and to supervise

[7] Australia lacks the knowledge-based industries that fund university research in other developed countries. Probably for the same reason, there was little additional government funding for applied research (for example, by the various ministries). For an extensive analysis and comparison of Australian and German funding landscapes see Laudel (2006).

Ph.D. students. Owing to the weight of the grant indicator, the emphasis lay on grants. The expectations to publish were often reduced to the four types of publications that counted in the funding scheme: books, book chapters, peer-reviewed journal articles, and full peer-reviewed conference contributions. These expectations were communicated in annual performance reviews each academic had to undergo. These reviews were largely inconsequential. More importantly, the indicators were also applied in decisions about promotions.

The external funding environment unequivocally communicated the need to conduct useful research. This expectation was directly built into one of the major grant schemes—the 'Linkage grants' which required co-funding by an industry partner—and more indirectly institutionalized in the 'Discovery grants' which had to demonstrate a 'national benefit'. Even though 'workarounds' concerning this criterion were accepted in some fields, the expectation was clear and all academics knew that demonstrating a national benefit could make all the difference between being funded or not being funded.

Authority Relations

Following Weber (1947), a tradition within organizational sociology (Scott 1992), and the conceptual approach provided by Whitley (Chapter 1), we define authority as legitimate power. The authority structure of Australia public science system fits Whitley's type of a state-delegated competitive system, which is characterized by a high authority of research funding agencies and national academic elites and a medium authority of all other authoritative agencies. The Australian government used its authority to establish an indicator-based funding scheme for the block funding of universities. The universities reproduced this scheme internally, which reinforced the state's authority because its performance indicators became ubiquitous and were used to evaluate the research performance of each academic within the university.

More importantly, the money distributed by the block funding scheme was so scarce that universities used it primarily as investments in the ability of units and academics to win external grants. The absence of recurrent research funding made external grants a necessary condition for the conduct of research, and the clearly communicated expectation that academics should win external grants was addressed even to those academics who did not need them.

The strong pressure to win external grants dramatically increased the authority of the agencies that decided about grants: namely, the funding councils, the academic elites, the government, and collaborators from outside academia who could serve as 'industry partners'. The funding councils were by far the most important sources of grants. They reported to the government, which influenced their policy and had the authority to overthrow funding council decisions concerning the funding of individual projects (an authority which they actually used). Within these limits set by the state's retention of authority concerning decisions on grants, the members of the funding councils' disciplinary panels decided on the distribution of most of the Australian grant money. This gave the national academic elites, from whose ranks panel members were recruited, an exceedingly strong position.

Since the grants with the highest success rate depended on financial contributions from industry partners, the latter's authority over the research in question was also high. They were *de facto* in a veto position concerning the content of the research to be undertaken. Thus, by emphasizing external funding in its relations to the university and by creating a situation in which block funding is scarce, the government made university research dependent on competitive grants, thereby giving the authoritative agencies involved in grant funding the opportunity to use their monopoly on research funding for realizing their interests concerning the content of university research.

This general picture is incomplete unless another authoritative agency is included: namely, the academics who decide on research goals and approaches. Any exercise of authority over research goals is directed at the researchers who actually formulate these goals. The researcher thus is an 'obligatory point of passage' for authority relationships. The relative authority of researchers *vis-à-vis* the state, funding agencies, the epistemic elite, universities, and industry partners varied across fields and performance levels. Four groups of scientists were less affected by the authority relations described above. A very small group of top performers received all grants they wanted because of their high performance. These academics were sought after by both universities and funding councils. Their authority was high because they could threaten their exit by migration to another country, and because they belonged to one of the authoritative agencies: the elite. A second group of academics did not need many resources for their research, which in turn meant that the authority of all actors involved in the grant distribution process was diminished. A third group was those whose research interests coincided with the interests of the authoritative agencies. The actual authority of those academics might be low, and the authority of the external actors high. However, there was no need to exercise authority, because the 'right' decisions were made anyway. The fourth group was those with the lowest authority. Some academics in our sample conducted so little research that an exercise of authority concerning their goals (if they had any) was pointless.

These four groups represent two performance-dependent and two field-dependent variations. The highest and lowest performers were not susceptible to authority relations because the former's authority was high and the latter were of no interest to authoritative agencies. The authority of academics from fields with low resource dependency was relatively higher because the basis of external actors' authority—their control of resources—was of less importance to those academics. In our sample of disciplines, mathematicians, theoretical physicists, theoretical political scientists, political scientists relying on secondary (published) data, and some of the historians belonged to the group of academics with relatively high authority. Finally, in all disciplines except mathematics there were some academics whose interest in applied research coincided with that of the authoritative agencies, and whose 'felt authority' concerning their research goals was therefore high.

5. Adaptive Behaviour and its Effects

The situations described in the previous section triggered two kinds of adaptive behaviour, both of which had distinctive effects on the content of the produced knowledge. Academics responded to the accessibility of funding for their research and to the behavioural expectations by *adapting their research portfolios*—the set of (potential) research trails they could follow or were currently following. They also responded to the actual availability of funds and time for research by developing specific strategies for *coping with scarcity*. The two kinds of responses had different effects and can be analytically separated.

The Adaptation of Research Portfolios to Funding Conditions

The management of research portfolios included decisions

- to start specific research trails or projects because of opportunities provided by the environment;

- to abandon research trails, projects, or aspects of projects because of perceived constraints;
- to avoid (not to start) research trails or projects because of perceived constraints; and
- to change the directions of research trails or projects.

It must be noted that only very few researchers perceived themselves as actively managing a research portfolio, and that decisions about individual research trails were not usually made with a 'research portfolio' in mind. Furthermore, the adaptation of research portfolios varied between performance levels. We found most of the adaptive behaviour with the average performers in our sample (see above, note 4, for a short description of our procedure). Across all fields, the best academics started research trails or projects mainly out of personal interests. Funding opportunities were a motive for average rather than top researchers. Only very few of the top academics (all of them in the biosciences) abandoned research trails or projects because of a lack of funding, and avoided research they considered unlikely to attract funding. Abandoning research because of insufficient funding occurred mainly among the average performers.

Taking this modification into account, some patterns of field-specific responses can be identified. In biology, physics, and geology, and in a few cases in mathematics, average performers started research trails when funding opportunities arose. In political science, external stimuli (external demands, suggestions, or perceived opportunities) could also initiate new research trails. In history, these external stimuli were the only factor; funding opportunities did not play a role in the start of new research trails. The properties of newly started research trails did not show any clear pattern except for biology, where newly started research trails were more applied than the already existing ones.

In all disciplines, research trails or projects were abandoned due to lack of funding. Another reason for abandoning research, the lack of time, was more important than the lack of funding in mathematics and history, less important in political science and biology, and did not play a role at all in physics and mathematics. In four disciplines no pattern of abandonment of research was detectable. In biology, basic and non-mainstream research was abandoned, while abandonment in physics decelerated research and narrowed research portfolios.

Only biologists reported that they also avoided certain research trails because they considered them to be 'non-fundable' in Australia. Again, basic and non-mainstream research was characterized this way. Academics from other fields did not describe situations in which they avoided otherwise possible research trails.

In all fields except political science, academics also changed the directions of the research trails they were currently pursuing. This occurred because they followed funding opportunities (biology, history), responded to funding constraints (physics, geology), and complied with expectations of the university (mathematics). Across all five disciplines, the changes in directions of existing research trails consisted in a stronger emphasis on applied aspects of the research. While the adaptation of research portfolios showed a consistent pattern—a move to applied research and to the mainstream—its actual occurrence and the reasons given varied between disciplines. While funding for research played a major role in all disciplines, time for research was the primary reason for abandoning research trails in mathematics and history, and a secondary one in biology and political science. Only biologists abandoned lines of research because of their perceived 'non-fundability', and political scientists did not change their research trails by emphasizing applied aspects. We will further explore these differences when discussing the effects of adaptive behaviour.

Coping with Scarcity

Scarcity of time or resources occurred for a variety of reasons. The most important reason for insufficient resources was that academics were not successful with their grant applications and were forced to conduct their research with little or no money to spend. But even those who were successful often faced scarcity because grants were arbitrarily reduced. Time constraints were due to the competing tasks in the university: high teaching loads or administrative tasks. Coping with scarcity included the following decisions:

- to use supervised student projects (Honours, Masters, and Ph.D. students) as the main way of conducting research;
- to 'job': that is, to perform research projects or services for paying clients in order to maintain the resource base for research;
- to reduce the empirical basis of the research by choosing fewer or less suitable research objects or methods, or by conducting fewer experiments or observations; and
- to 'retard' research, either by temporarily abandoning it due to time or resource constraints or by slowing it.

Table 10.2. Academics' coping with scarcity

	Biology	Physics	Geology	Maths	Political science	History
Use of student projects	X	X	X			
'Jobbing'	X	X	X			
Reduce empirical basis	X		X		X	X
'Retard' research	X	X	X	X		X

Again, the decisions listed above were not in all cases conscious responses to a perceived situation of scarcity. For example, none of our interviewees described a decision to base his or her research exclusively on research student supervision. However, constructing the responses as decisions seems justified because they are clearly discernible behavioural patterns that are an aggregate effect of decisions, such as not to start an independent research project or whether to supervise Ph.D. students.

As is the case with the adaptation of research portfolios, coping with scarcity was not distributed evenly across performance levels. The top performers in our sample felt little need for such coping. Nor did the worst performers, because they did not conduct much research at all. Thus, there were only very few cases of reducing the empirical basis of research among the top performers. None of the top performers relied on student projects or 'jobbed'. However, instances of 'retarding' research were found across all performance levels except for the top biologists. If we take into account these modifications by performance levels, clear discipline-specific patterns of coping with scarcity emerge, summarized in Table 10.2. In three disciplines—biology, physics, and geology—some academics compensated for a lack of funding by continuing their research predominantly or exclusively on the basis of student projects they supervised or co-supervised.[8] Several of the low-performing academics gave the clear impression in the interviews that they rationalized their supervision of student projects as pursuing a research trail, even though they did not have a consistent research interest or strategy and just presented as their research what the students wanted to do. Interestingly, the academics in mathematics, political science, and history did not rely on student projects for their research.

---

[8] This was also a common practice in German universities.

'Jobbing' occurred in the same disciplines as the use of student projects. We found only very few instances of this practice (one each in biology and physics and two in geology). The absence of 'jobbing' in mathematics, political science, and history is most likely due to the absence of a demand for routine services.

Academics in four disciplines (unhappily) reduced the empirical basis of their research. These practices took field-specific forms. In biology, fewer objects were investigated, fewer methods applied, or fewer experiments conducted. In geology, unsuitable but cheaper sites for fieldwork were chosen, fewer sites included or fewer observations conducted. Political scientists investigated fewer cases or conducted fewer interviews. Historians cut the archival work; they did not go to particular archives, or spent less time there. There was no reduction of the empirical basis of research in mathematics, because mathematics is not an empirical discipline. Less trivial is the absence of this practice among physicists, which hints at the possibility that physicists have less leeway in designing their empirical research because of its more theory-driven and deterministic nature. Finally, 'retarding' research was a ubiquitous practice.

Variations in Effects

Four distinct changes in the production of knowledge could be observed as the result of the adaptation of research portfolios and of coping with scarcity. These effects varied between disciplines and between performance levels, as Table 10.3 suggests. A first widespread trend is the deceleration of otherwise unchanged research due to either funding or time constraints. This trend occurred across all disciplines except for the high performers in

Table 10.3. Changes in the knowledge production of the six disciplines

	Biology	Physics	Geology	Maths	Political science	History
Deceleration	AL	HAL	HAL	AL	AL	HAL
Narrowing of individual research portfolios	AL	HAL	HAL	HAL	HAL	HAL
Move to more applied research	HAL	AL	(HAL)			(HAL)
Reduced validity and reliability of research	HAL		HAL		HAL	HAL

HAL = effect occurs at high, average, and low performance levels; AL = effect shows only for average to low performers; parentheses = effect is weak.

biology and mathematics. A second general effect that could be found in all fields except for high performers in biology is the gradual narrowing of individual research portfolios. We found a significant imbalance between research trails and projects that are abandoned, avoided, or 'retarded' due to funding or time constraints, and research that is started because of new opportunities. Even high performers who win grants to fund their research, or do not need them, feel forced to leave out parts of their research because of felt restrictions in time or funding. The 'jobbing' also contributes to the narrowing of research portfolios because it consumes time—the most scarce of all resources for research. Jobbing and the imbalance between abandoned and started research trails reduced the breadth of academics' research portfolios. It is difficult to assess the consequences of this trend at the macro-level of fields. However, if breadth and diversity are a source of innovations in science, the innovativeness of Australian research is likely to suffer from this trend.

Third, in four of the six disciplines the content of research also changed in that research overall became more applied. Academics chose research problems or objects for their research which were relevant for the solution of societal problems, or conducted collaborative projects with partners from industry. While a good deal of window-dressing was involved, we also identified enough substantial changes in research content to state that the increasingly applied character of research was not just a matter of rhetoric. The trend towards more applied research can be clearly observed in biology and to a lesser extent in physics, history, and geology. No such trend could be observed in political science or mathematics. Mathematicians who have moved to or included applied research do so as a service to others—in competition to their research agendas rather than as a substitute.

Finally, a fourth change in knowledge production resulted from a surprisingly clear trend across several disciplines. The reductions in empirical work that occurred as a response to insufficient funding make the results in the affected empirical disciplines (biology, geology, political science, and history) less valid or reliable.[9]

---

[9] It should be noted here that our respondents did not like this at all. To the contrary, most of them were quite annoyed about being forced to work below the standards of their scientific community. But left with the choice between doing no research and doing it substandard, they felt forced to choose the latter.

6. Field-Specific Factors Modifying the Impact of Authority Relations

'Proximate Factors'

How can these field-specific effects of governance be explained? Our analysis of the factors that produce field-specific responses to authority relations suggests that it is useful to distinguish between two kinds of field-specific factors according to the degree to which a factor can be actively shaped by authoritative agencies. Analytically, these varying degrees of 'social shaping' correspond to two levels of abstraction. At a first level, 'proximate causes' of field-specific effects can be identified. These are factors that are properties of research that emerge from the interaction of a field with the environments it draws on in its knowledge production. These environments include neighbouring fields, sources of research technology, and the wider societal environment. Owing to their hybrid nature, these factors are highly dynamic and depend on local and historical circumstances. At the same time, this hybrid nature makes them compatible with the institutional and resource environments of a researcher. The proximate factors are 'authority-sensitive' properties of research processes, which enable researchers to negotiate compromises between the requirements of research processes and the requirements and opportunities of institutional and resource environments.

At a second, higher level of abstraction relatively stable 'remote epistemic factors' can be identified. These factors are produced by the relatively stable epistemic practices of a field. They create opportunities and constraints for producing knowledge which, under the specific conditions provided by the field's environment, manifest themselves as the proximate authority-sensitive properties of research. We will discuss them in the following section.[10]

Figure 10.1 lists the 'proximate' field-specific factors, tentatively ranks disciplines according to the strength of the properties, and links the factors to changes in the knowledge production by stating how it sensitizes or

---

[10] It is important to note here the many contingencies that must be taken into account in such an analysis. The epistemic factors are not universal timeless properties of research. 'Proximate' epistemic properties of a field are contingent on the development of that field as well as other fields and the social organization of research. 'Remote' epistemic factors are at least contingent on the current developmental stage of a field. This means that the following considerations—especially those on 'proximate' epistemic factors—are specific to the social context for which they were obtained: namely, Australian university research. We are also aware of the fact that our analysis is based on interview responses that lack a cross-disciplinary yardstick. However, we consider our exploration of these factors as an entry in their systematic *empirical* investigation, which requires comparisons between countries and over time.

'Proximate' field-specific factors	Tentative Ranking of disciplines according to the strength of the factor	Sensitizes research to	Desensitizes research to
Resource dependency	Biology Physics Geology Political Science History Mathematics	Scarcity of resources Expectations tied to the provision of resources	
Diversity of individual research portfolios	Physics, Biology Geology History Political science Mathematics	[Narrow portfolios-one trail]: Scarcity of resources External expectations	[Diverse portfolios-more than one trail]: Scarcity of resources External expectations
Correspondence to societal problems	Biology Physics Political Science Geology History Mathematics	Expectations to conduct applied research	Scarcity of resources (access to more sources, 'jobbing')
Competitiveness	Biology Physics Mathematics Geology Political Science History	Scarcity of time and funding	
Dependency on uninterrupted research time	History Mathematics Political Science Geology Biology, Physics	Scarcity of time	

Figure 10.1. 'Proximate' epistemic factors and their role in the mediation of governance

desensitizes research to specific conditions created by authoritative agencies.[11] A first factor is the resource dependence of research, which refers to all material resources (including the salaries of researchers) that are required for the production of new knowledge in a field. Research can be *existentially dependent* on resources because it requires equipment and consumables. Much empirical research belongs to this category, in particular the experimental sciences but also large-scale empirical policy research and geological fieldwork. Other research processes were *strongly dependent* on resources because they could be conducted without resources only at the cost of

[11] The 'proximate' factors do not only vary between disciplines but also between fields within disciplines. For example, theoretical physics and political theory belong in the same category of resource dependency as mathematics, and the resource dependency of history and empirical political science varies internally. Thus, any of the rankings of disciplines should only be taken as a rough estimate that also depends on our sample.

breadth, validity and reliability, or timeliness of results. This level of dependence was observed in empirical non-experimental disciplines such as small-scale empirical political science, geological fieldwork, and most sub-fields of history. Finally, some fields are only *weakly dependent* on resources because their knowledge can be produced by using only the resources that are provided as basic infrastructure independently of any specific research process. Weakly resource dependent fields in our sample were pure mathematics, sub-fields of history that do not require archival work, and political theory.[12]

Naturally, higher resource dependence means higher sensitivity to scarcity and a higher responsiveness to expectations tied to the provision of resources. This is why in the existentially resource-dependent disciplines of physics and biology only high performers who receive all the resources they need did not feel constrained in their research. In weakly resource-dependent disciplines (mathematics and history) there were more academics who did not feel constrained, and those included high to average performers. The other academics had to apply for money and therefore were forced either to follow the signals of their funding environment, or to cut down their research.

The second factor we could identify is the *diversity of individual research portfolios*. The disciplines in our sample markedly varied in the numbers of distinct research trails pursued by an individual academic. Again, these findings depend on the performance levels of academics. The high to average performers in biology and experimental physics and the average performers in history simultaneously pursued several distinct lines of research, while political scientists, mathematicians, geologists, high performers in history, and low performers in biology and experimental physics had only one research trail.

The diversity of individual research portfolios affected academics' opportunities to start, abandon, or avoid research in order to circumvent scarcity or external expectations concerning their research. In particular, academics who had narrow research portfolios consisting of only one research trail could not abandon this trail without either giving up research at all or

---

[12] This does not mean that additional resources would not accelerate or improve research in weakly resource-dependent fields. More money can always improve working conditions: for example, by increasing research time through teaching-buy-outs, by providing the opportunity to buy books rather than waiting until a library has acquired them, or by funding more travel and better offices. The main difference between strong and weak dependence is the difference between the indispensability of resources for the conduct of research according to the standards of the scientific community and the possibility of using resources to produce these results faster.

radically changing it. Thus, the pressure to respond to funding problems by emphasizing applied aspects, reducing the empirical basis of the research, or relying on student projects, was much stronger for researchers with single-stranded research portfolios than for those with multi-stranded research portfolios. On the other hand, researchers with multi-stranded portfolios could more easily satisfy external expectations or respond to scarcity by abandoning one of their research trails or starting additional ones.

The *correspondence to societal problems* affects research mainly through the existence of external users for the research. These users affect the availability of resources, either by paying for research themselves or by providing academics with the opportunity to claim an applied character or 'national benefit' of their research. A correspondence to societal problems could be found in all fields, albeit to varying extents. It is often missing in pure mathematics, and not all fields of the other disciplines have equal opportunities to claim such a correspondence.

The correspondence of research to societal problems sensitized that research to external expectations to conduct applied research. If applied research is clearly possible, the expectations to conduct such research are likely to be more articulated. At the same time, the correspondence to societal problems desensitizes research to the scarcity of resources because it increases the number of sources that can be accessed, and in some cases enables the acquisition of resources by 'jobbing'.

The *competitiveness* of fields in terms of being first to find a specific result was a factor that was explicitly mentioned only in biology but has the potential to affect other fields as well. Our tentative ranking of fields in Table 10.3 is based on our findings but confirmed for some fields by the literature (on mathematics: Hagstrom 1965; Heintz 2000; on the biosciences: Latour and Woolgar 1986; Cambrosio and Keating 1995; Knorr-Cetina 1999). The competitiveness of a field sensitizes research to the scarcity of time and funding. Thus, the high competitiveness of biology made some biologists in our sample avoid specific research trails because they felt unable to compete under the Australian funding conditions. It also explains why the high performers in biology did not retard their research—they would abandon insufficiently funded research rather than conduct it at a slower pace. Historians, on the other hand, were not concerned at all about somebody else anticipating their work.

Finally, the disciplines in our sample varied according to their *dependency on uninterrupted research time*. This is a quite specific factor, because it does not refer to the absolute time available for research but to the necessity to work on a problem without interruptions. This necessity was highest in history, closely followed by mathematics. Biology and physics appear to be far less dependent on uninterrupted research time. The collective nature of much of the research in these disciplines appears to correspond to the possibility for the lead investigators to make plans for others, to let them work, and to attend to the research process only occasionally and *ad hoc*.

Dependence on uninterrupted research time sensitizes research to the scarcity of time. If there is less time, it is very likely that there is also less uninterrupted time. This is why historians predominantly applied for external grants in order to buy themselves out of teaching. Time problems were given as reasons for a retardation of research and for reducing the empirical basis of research mainly in history and mathematics, and to a lesser degree in political science and geology.

'Remote Epistemic Factors'

The 'proximate' field-specific factors discussed in the previous section are obviously co-produced by the research practices of a field itself and the field's environment. For example, the resource dependence of a field is affected not only by the nature of the research but by the development of research technologies; the competitiveness of a field partly depends on the number of researchers financed in that field, and the dependence on uninterrupted research time partly depends on research technologies and the organizational design of research projects.

However, while the proximate field-specific factors depend partly on such environmental conditions, they also depend on inherent epistemic properties of fields; that is, on properties of the practices by which a field produces new knowledge. We call these properties 'remote epistemic factors' because they cannot be affected by authoritative agencies. The production of knowledge about a certain object requires actions that are specific to this particular object, which means that some of the properties of these actions cannot be changed without relinquishing the opportunity to obtain the knowledge.[13] In our investigation we inductively derived a tentative list of

---

[13] This does not mean that these properties are not subject to social influences. The nature of the knowledge produced by a field over time and the practices by which that knowledge is produced evolve, and they evolve in the human practices of knowledge production. For example, what is thought about in mathematics as 'proof' has changed quite dramatically over the last two centuries (Heintz 2000). However, these properties are 'hard' at any given time in that 'they cannot be wished away' (Berger and Luckmann 1967).

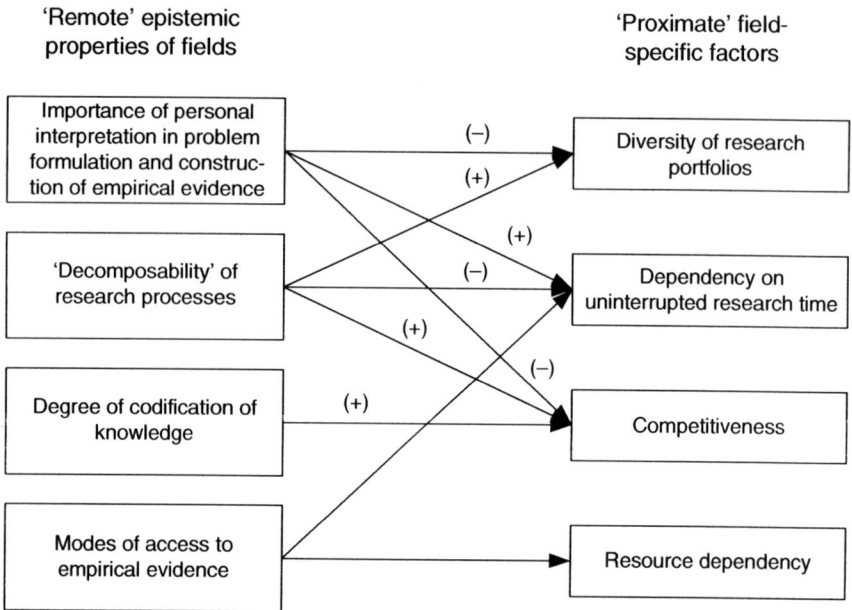

Figure 10.2. 'Remote' epistemic properties and their contribution to 'proximate' field-specific factors (the signs indicate a positive respectively negative influence of a stronger 'remote' epistemic property where strength is a relevant dimension)

some of these properties and their contribution to the authority-sensitive 'proximate factors' discussed above, as shown in Figure 10.2. This list is incomplete and awaits extension and modification in further research. The 'remote' epistemic properties are likely to interact with each other in ways that are not yet known. Some of them have been mentioned in the literature before.

Fields of research vary significantly in the extent to which problem formulation and construction of empirical evidence rely on the personal perspectives of researchers. In mathematics there is little room for personal interpretation since there is little disagreement about the truth of a mathematical statement. This is why sociologists of science have struggled with the unclear role of 'the social' in mathematics for decades—without having found a satisfying solution (Heintz 2000). At the other end of the spectrum, history (and other humanities as well) rely to a large extent on individual interpretation. In these fields, what constitutes empirical evidence to support a claim, what a problem is, and how it should be addressed, depends on the perspectives of individual researchers.[14]

The *role of personal interpretation in problem formulation and construction of empirical evidence* affects the degree to which research tasks can be delegated to colleagues or assistants. Most historians who commented on the use of grants in history felt that there is no point in sending research assistants to the archives because they have to see the evidence for themselves. Furthermore, none of the interviewed historians considered Ph.D. students as a resource in their own research. Ph.D. students in history must find their own perspectives on problems, the literature, and evidence. They thus cannot contribute to the research of their supervisors. The same holds true to some extent for political theory.

The resulting necessity for the academics to do most of the research themselves sets limits to the diversity of research portfolios and produces great dependence on uninterrupted research time. It also limits the competitiveness of the field. It is entirely possible for two biographies of the same person to be published at the same time, and for both to be equally valuable contributions because they provide different perspectives.

The *degree of codification of knowledge*—the degree to which knowledge is represented by formal symbols with agreed-upon meanings—was first introduced as a property of fields by Zuckerman and Merton (1972; see also Cole 1983). It affects the competitiveness of fields because high degrees of codification limit the interpretive flexibility of the knowledge, which in turn increases the likelihood that researchers formulate similar problems and arrive at similar solutions to these problems. History and mathematics are again at opposite ends of the spectrum, with the codification lowest in history and highest in mathematics. It is possible that the degree of codification is the inverse of the role of interpretation. However, more research is needed in order to establish the co-variation of these two properties.

Another feature that is closely linked to the previous two but needs to be treated separately is the *'decomposability' of research processes*. We have already mentioned that this is low in history, where the low decomposability is due to the importance of personal interpretations, and noted some of the consequences. According to our interviews, the 'decomposability' of

---

[14] Which does not mean that 'anything goes' in the field of history. Personal perspectives are shaped in the interaction with other's research, and there are strong conventions about legitimate problems and fact constructions. However, these conventions only provide frames within which the individual historian decides about interesting and legitimate perspectives on historical processes as well as the suitability of empirical information to confirm or disprove a claim. See Lamont (2009: 79–87) on history as a consensual field with respect to methods.

research processes is also low in many fields of pure mathematics, although the reasons are likely to be different. Owing to their low 'decomposability', these fields are similar to history in the diversity of research portfolios, the separation of Ph.D. student supervision from the supervisor's research, and the sensitivity to time constraints. The similar consequences of different properties point to the need for further research to establish the relationships between the two. High 'decomposability' enables a division of labour in the research process and a delegation of tasks. Thus, additional resources can be used for hiring additional personnel and conducting more research under the direction of one group leader, which is a common practice in many fields of the biosciences. Obviously, this affects the resource dependence of fields.

A fourth property that plays a role as a 'remote epistemic cause' of authority-sensitive properties of research processes refers to the modes of access to empirical evidence. There is of course the obvious distinction between those disciplines which use empirical evidence in their knowledge production and those that do not. However, empirical evidence can be obtained in a variety of ways. It can be produced and thus be enhanced, or it can be inherently limited (for example, for ancient history). Furthermore, empirical evidence can be produced by observation or experiment. Experiments can be conducted as single events or as series providing data for statistical analysis. Observations can use more or less intrusive methods. This wide variety of modes of access to empirical evidence affects the resource dependence of fields. The extreme cases are fields of pure mathematics and high-energy physics. The former do not access empirical evidence at all, which contributes to their relatively low resource dependence, while high-energy physics accesses some of its empirical evidence by using supercolliders, which can only be funded by a collaborative effort of several countries.

This tentative list and discussion of relatively stable epistemic properties of fields that co-produce 'proximate' authority-sensitive properties indicates that it is possible and worthwhile to look beyond the immediate causes of adaptations. Since 'proximate' factors are co-produced by social contexts and epistemic properties of fields, they are difficult to compare across countries, organizations and time. Furthermore, the impact of changing authority relations in different disciplines is difficult to assess if disciplines are only compared in terms of their inevitably changing proximate epistemic properties. The identification of relatively stable epistemic properties of fields enables the construction of comparative frameworks that solve these problems.

7. Conclusions

In this chapter we have demonstrated that the exercise of authority had field-specific consequences for the production of knowledge. We have provided a tentative list of epistemic properties of fields that are responsible for the variation of effects between fields, and proposed mechanisms that produce field-specific governance instruments, field-specific authority relations, and field-specific effects of both. As a conclusion to this chapter we attempt a further generalization by asking how the exercise of authority by authoritative agencies other than the researcher changes the content of research.

The most general answer to this question is that authoritative agencies change the environment of researchers, who adapt their decisions about research to the changed situations. The translation of authority into changes in knowledge occurs at the individual and group level: namely, in the strategic decisions of academics about their research including the formulation of research problems, the selection of objects and methods, negotiations, collaborations, and the selection of communication channels. These decisions are partly made on specific occasions (for example, when project proposals are written or when experiments are designed). However, they are also woven into the everyday research activities of academics, and are often made implicitly (Knorr-Cetina 1981).

Like all actors, researchers make decisions in response to their perceptions of their situations. Authoritative agencies can change these situations by providing time or resources for research, increasing or decreasing a researcher's reputation in a specific social context, or formulating behavioural expectations. These elements of situations overlap with a researcher's epistemic room of manoeuvre, thereby creating opportunities to conduct research as well as constraints for that research. When academics make decisions about their research they must satisfy two conditions, both of which are necessary for continuing research. First, they need to design their research so that behavioural expectations are met to an extent that guarantees approval by their organizational and political environment. This approval is necessary for access to resources and to time for research, and possibly for achieving other goals such as promotion. Secondly, the research design must enable the production of a contribution that is accepted by the scientific community—a contribution that meets the community's expectations concerning relevance, originality, validity, and reliability. These expectations are tied to a field's specific research practices and epistemic properties. The latter produce 'technological' opportunities and constraints which are largely but not entirely the result of an historically contingent consensus of the scientific community.

The simultaneous adaptation of decisions to both authority-induced necessities and 'technological' necessities of the research process makes researchers respond to authority relations in a way that is specific to the field to which the research belongs. This is why field-specific effects are not only created by field-specific governance instruments or field-specific authority relations, but also by uniform governance instruments that are applied in different fields, or by uniform authority relations.[15]

Since all channels through which authority relations can influence scientific innovation go 'through' individual researchers and their decisions, where they are adjusted to epistemic conditions of research, these epistemic conditions modify the exercise of authority.[16] At the meso- and macro-levels, the extent of such modifications is reflected as a varying compatibility of epistemic conditions of research with interests of authoritative agencies, which makes some directions of research possible or easier to follow than others.

Our findings on the field-specific nature of (responses to) authority relations and their causes are explorative and preliminary in nature—not the least because, in the context of our investigation, identifying epistemic factors meant dealing with intervening variables in a research design that was aimed at identifying impacts of governance. Further empirical research and theoretical discussion are necessary in order to find the best way of constructing remote epistemic properties. The results of this study also enable conclusions concerning the methodology of studying field-specific effects of governance tools. We have demonstrated that a careful operationalization of research questions makes it possible for science studies to identify links between governance and changes in knowledge production of which researchers are not aware—to go beyond collecting researchers' *opinions* about the impact of governance on the content of research. Our study also shows that it is important to take research performance levels of interviewees into account. These performance levels are an important

---

[15] For example, the authority relationships in biology and physics were similar, but the effects on knowledge production still differed because of the different epistemic properties of the two fields.

[16] Examples of unmitigated exercises of authority over research in socialist countries demonstrate that attempts to 'override' epistemic conditions lead to situations where no knowledge is produced at all, which is exactly why there were relatively few such interventions (Gläser and Meske 1996).

intervening variable because they modify responses to governance and thus the effects of governance. Since the moderation of effects of governance by performance levels varies between fields, it is necessary to include both dimensions when the impact of governance on the content of research is investigated.

A third methodological contribution of our study is the identification of 'remote epistemic causes' of field-specific effects. These variables appear to enable the construction of a stable framework for the comparative analysis of fields and their responses to governance. It seems difficult to construct such frameworks from the 'proximate' factors because these factors are co-shaped by authority relations, which makes them a composite of both independent and dependent factors and altogether too fluid for comparative analyses.

Two more important methodological lessons need to be taken into account in further research. The discipline is far too large a unit for the analysis of specific consequences of governance. Field-specific effects and their 'remote epistemic causes' must be studied at the level of epistemically homogeneous knowledge production processes: that is, scientific specialties. Furthermore, our data show that even carefully prepared and conducted interviews are of limited use for investigating field-specific conditions and effects of governance. Owing to the inevitable time constraints of qualitative interviews, they cannot go deep enough to produce the necessary data. Ethnographic observations appear to be the only method that can provide these data.

Our results enable tentative generalizations with regard to the channels through which authority is exercised. First, the results strongly suggest that the 'governance by money'—creating scarcity and tying behavioural expectations to the resources that are made available—cannot be successful in all fields. It can be used to change the directions of research in existentially and in strongly resource-dependent fields, but is unsuitable for achieving aims related to all fields such as higher research quality. Secondly, in the fields where the 'governance by money' is most effective it is also likely to produce a strong unintended side effect: namely, a diminishing research quality of average researchers who cannot secure adequate resources for their research.

From our project it also follows that the frequent implicit assumption that a governance regime affects all fields in the same way (or affects some 'target fields' in the intended way and leaves the others unaffected) is likely to be wrong in most cases. Whenever governance instruments change the content of research by affecting the practices of knowledge production, it is very likely that the enormous variation of research practices between fields causes variations in both the ways in which governance instruments are shaped within fields and the actual outcomes of governance. This is why science policy studies must address the problem of 'implicit uniform models of science' that inform policy decisions, and must systematically discuss field-specific consequences of authority relations.

References

Berger, Peter L., and Luckmann, Thomas (1967) *The Social Construction of Reality* (Harmondsworth: Penguin).

Böhme, Gernot, van den Daele, Wolfgang, Hohlfeld, Rainer, Krohn, Wolfgang, and Schäfer, Wolf (1983) *Finalization in Science: The Social Orientation of Scientific Progress* (Dordrecht: Reidel).

Callon, Michel (1986) 'Some Elements of a Sociology of Translation: Domestication of the Scallops and the Fishermen of St Brieuc Bay', in John Law (ed.), *Power, Action and Belief* (London: Routledge), 196–233.

——and Law, John (1982) 'On Interests and their Transformation: Enrolment and Counter-Enrolment', *Social Studies of Science*, 12: 615–25.

Cambrosio, Alberto, and Keating, Peter (1995) *Exquisite Specificity: The Monoclonal Antibody Revolution* (New York: Oxford University Press).

Chubin, Daryl E., and Connolly, Terence (1982) 'Research Trails and Science Policies', in Norbert Elias, Herminio Martins, and Richard Whitley (eds.), *Scientific Establishments and Hierarchies* (Dordrecht: Reidel), 293–311.

Cole, Stephen (1983) 'The Hierarchy of the Sciences?', *American Journal of Sociology*, 89: 111–39.

Gläser, Jochen, and Laudel, Grit (2004) *The Sociological Description of Non-Social Conditions of Research* (REPP Discussion Paper, 04/2; Canberra: Australian National University).

——and ——(2007) 'Evaluation without Evaluators: The Impact of Funding Formulae on Australian University Research', in Richard Whitley and Jochen Gläser (eds.), *The Changing Governance of the Sciences: The Advent of Research Evaluation Systems* (Dordrecht: Springer), 127–51.

——and ——(2009a) 'On Interviewing "Good" and "Bad" Experts', in Alexander Bogner, Beate Littig, and Wolfgang Menz (eds.), *Interviewing Experts* (Basingstoke: Palgrave), 117–37.

——and ——(2009b) 'Creating Competing Constructions by Reanalysing Qualitative Data', *Historical Social Research*, 33: 115–47.

——and Meske, Werner (1996) *Anwendungsorientierung von Grundlagenforschung? Erfahrungen der Akademie der Wissenschaften der DDR* (Frankfurt a.M.: Campus).

——Laudel, Grit, Hinze, Sybille, and Butler, Linda (2002) 'Impact of Evaluation-Based Funding on the Production of Scientific Knowledge: What to Worry about, and How to Find out', Report to the German Ministry for Education and Research: http://www.sciencepolicystudies.de/dok/expertise-glae-lau-hin-but.pdf (accessed Oct. 2009).

——Lange, Stefan, Laudel, Grit, and Schimank, Uwe (2008) 'Evaluationsbasierte Forschungsfinanzierung und ihre Folgen', in Friedhelm Neidhardt, Renate Mayntz, Peter Weingart, and Ulrich Wengenroth (eds.), *Wissensproduktion und Wisenstransfer* (Bielefeld: transcript), 145–70.

Hagstrom, Warren O. (1965) *The Scientific Community* (Carbondale, Ill.: Southern Illinois University Press).

Hedström, Peter (2005) *Dissecting the Social: On the Principles of Analytical Sociology* (Cambridge: Cambridge University Press).

Heintz, Bettina (2000) *Die Innenwelt der Mathematik: Zur Kultur und Praxis einer beweisenden Disziplin* (Vienna: Springer).

Henkel, Mary (2000) *Academic Identities and Policy Change in Higher Education* (London: Jessica Kingsley).

——(2005) 'Academic Identity and Autonomy in a Changing Policy Environment', *Higher Education*, 49: 155–76.

Knorr-Cetina, Karin (1981) *The Manufacture of Knowledge: An Essay on the Constructivist and Contextual Nature of Science* (Oxford: Pergamon Press).

——(1999) *Epistemic Cultures: How the Sciences Make Knowledge* (Cambridge, Mass.: Harvard University Press).

Lamont, Michele (2009) *How Professors Think: Inside the Curious World of Academic Judgement* (Cambridge, Mass.: Harvard University Press).

Lange, Stefan (2007) 'The Basic State of Research in Germany: Conditions of Knowledge Production Pre-Evaluation', in Richard Whitley and Jochen Gläser (eds.), *The Changing Governance of the Sciences: The Advent of Research Evaluation Systems* (Dordrecht: Springer), 153–70.

Latour, Bruno (1987) *Science in Action* (Cambridge, Mass.: Harvard University Press).

——(1988) *The Pasteurization of France* (Cambridge, Mass.: Harvard University Press).

——(1996) 'On Actor-Network Theory: A Few Clarifications', *Soziale Welt*, 47: 369–81.

——and Woolgar, Steve (1986 [1979]) *Laboratory Life: The Construction of Scientific Facts* (Princeton: Princeton University Press).

Laudel, Grit (2006) 'The Art of Getting Funded: How Scientists Adapt to their Funding Conditions', *Science and Public Policy*, 33: 489–504.

——and Gläser, Jochen (2006) 'Tensions between Evaluations and Communication Practices', *Journal of Higher Education Policy and Management*, 28: 289–95.

Law, John (1986) 'Power/Knowledge and the Dissolution of the Sociology of Knowledge', in John Law (ed.), *Power, Action and Belief* (London: Routledge & Kegan Paul), 1–19.

Leišytė, Liudvika (2007) *University Governance and Academic Research*, Ph.D. dissertation (Enschede: University of Twente, CHEPS).

Lynch, Michael (1985) *Art and Artifact in Laboratory Science: A Study of Shop Work and Shop Talk in a Research Laboratory* (London: Routledge & Kegan Paul).

Mayntz, Renate (2004) 'Mechanisms in the Analysis of Social Macro-Phenomena', *Philosophy of the Social Sciences*, 34: 237–59.

Merton, Robert K. (1968) 'On Sociological Theories of the Middle Range', Robert K. Merton (ed.), *Social Theory and Social Structure* (London: Free Press), 39–72.

Meulen, Barend van der, and Loet Leydesdorff, Loet (1991) 'Has the Study of Philosophy at Dutch Universities Changed under Economic and Political Pressures?', *Science, Technology, and Human Values*, 16: 288–321.

Morris, Norma (2000) 'Science Policy in Action: Policy and the Researcher', *Minerva*, 38: 425–51.

Patton, Michael Quinn (2002) *Qualitative Evaluation and Research Methods* (Newbury Park, Calif.: Sage).

Pickering, Andrew (1995) *The Mangle of Practice: Time, Agency and Science* (Chicago: University of Chicago Press).

Pinch, Trevor (1986) *Confronting Nature: The Sociology of Solar Neutrino Detection* (Dordrecht: Reidel).

Rip, Arie (1982) 'The Development of Restrictedness in the Sciences', in Norbert Elias, Herminio Martins, and Richard Whitley (eds.), *Scientific Establishments and Hierarchies* (Dordrecht: Reidel), 219–38.

Scott, W. Richard (1992) *Organizations: Rational, Natural, and Open Systems* (Englewood Cliffs, NJ: Prentice-Hall).

Sousa, Celio A. A., and Hendriks, Paul H. J. (2007) 'That Obscure Object of Desire: The Management of Academic Knowledge', *Minerva*, 45: 259–74.

Weber, Max (1947) *The Theory of Social and Economic Organizations* (New York: Free Press).

Whitley, Richard (1972) 'Black Boxism and the Sociology of Science: A Discussion of the Major Developments in the Field', in Paul Halmos (ed.), *The Sociology of Science* (Sociological Review Monograph, 18; Keele: University of Keele), 61–92.

——(1977) 'Changes in the Social and Intellectual Organisation of the Sciences: Professionalisation and the Arithmetic Ideal', in E. Mendelsohn, P. Weingart, and R. Whitley (eds.), *The Social Production of Scientific Knowledge* (Dordrecht: Reidel), 143–69.

——(2000 [1984]) *The Intellectual and Social Organization of the Sciences* (Oxford: Clarendon Press).

——and Gläser, Jochen (eds.) (2007) *The Changing Governance of the Sciences: The Advent of Research Evaluation Systems* (Dordrecht: Springer).

Zuckerman, Harriet, and Merton, Robert K. (1973 [1972]) 'Age, Aging, and Age Structure in Science', in Robert K. Merton (ed.), *The Sociology of Science* (Chicago: University of Chicago Press), 497–559.

11

Public Science Systems, Higher Education, and the Trajectory of Academic Disciplines

Business Studies in the United States and Europe

Lars Engwall, Matthias Kipping, and Behlül Üsdiken

1. Introduction

The twentieth century has seen a strong growth in higher education and public research organizations worldwide. Within this development there has been a successive addition of new disciplines to systems of higher education and research. What once were primarily institutions for theology have with the passage of time developed into organizations teaching and researching in a large number of disciplines. The inclusion of these new disciplines has had, and continues to have, an important influence on the production of new scientific knowledge and its dissemination, since academic status confers legitimacy and access to human and financial resources. And it confers this legitimacy not only on those admitted into academia, who thereby can have the opportunity to join the scientific elites, nationally and, possibly, internationally; it also confers an increased status on those graduating from these academic institutions, helping them gain entry into national elites—depending on the status of the discipline and the educational institution.

Yet, this development has not been without conflicts regarding the appropriateness of new fields of study and research. Those inside higher education institutions are often reluctant to grant academic status to new disciplines, since this means increased competition for students and financial resources. Moreover, insiders will likely resist admitting what they consider inappropriate disciplines, fearing that they might reflect negatively on the institution and reduce the legitimacy of existing scientific elites. At the same time, admission into institutions of higher education and research not only confers advantages on the new disciplines, but also exposes them to the scrutiny of and authority relations within extant higher education and public science systems (see Chapter 1). It is actually these existing systems which determine whether and in which way new disciplines are included. Since there are, as shown by many chapters in this book, significant national differences among these systems of education and research, the outcome of the process is likely to vary significantly from one country to the next. They will also change over time since the systems themselves are subject to change—and possibly convergence.

Drawing upon neo-institutional theory in organizational analysis and utilizing in particular the notion of 'field', the central idea on which this chapter builds is that institutional fields are nested in other fields, often with 'hierarchical' relationships between them (Lounsbury and Pollack 2001). In this case, disciplinary fields are nested within broader higher education and public science systems. Thus, variations in the super-ordinate fields of higher education and research as well as the changes in them are likely to influence the ways in which a new discipline becomes accommodated and develops over time (see Clark 1995; Lounsbury and Pollack 2001). To illustrate this broad proposition about the incorporation and development of disciplinary fields in nationally based institutional frameworks of education and research and the resulting differences in outcomes in terms of organizational forms and knowledge production, we examine the institutionalization of business studies as an academic field in the USA and several European countries since the late nineteenth century.

As an academic discipline, business studies pertains to education and research in general management, organization, accounting and finance, marketing, production, and purchasing with a primary, though not exclusive, focus on business firms (for a further discussion see Engwall 2009: ch. 1). This particular discipline provides an opportune case for an examination of the above proposition for at least three reasons. First of all, business studies emerged roughly at the same time in the USA and a number of European countries. However, it was differentially accommodated within the respective higher education systems. Whereas business studies became a part of many universities in the USA as a professional school, in much of Europe it was originally established in separate organizations outside the university system.

Second, irrespective of organizational form there was significant initial resistance to its inclusion in both contexts. Thus, at the turn of the last century the Rector of the University of Würzburg spoke of the *Handels-hochschulen* (commercial schools) that were at the time emerging in Germany as 'pathological creations' which should not exist at all (Üsdiken *et al*. 2004: 385). A similar position was also taken by the famous sociologist Thorstein Veblen in the USA, who just after the First World War wrote that '[a college of commerce] belongs in the corporation of learning no more than a department of athletics' (Veblen 1918: 210; see also Ewing 1990: 267–8; and Sass 1982).

Third, even after business studies became a part of higher education, tensions did not necessarily subside and have persisted until the present day. In the relevant literature, this has been conceptualized most frequently as a tension between scientific rigour and practical relevance (for a long-term overview see Daniel 1998; and for the most recent debate Fincham and Clark 2009). For business studies, the achievement of scientific status proved particularly difficult, since it could not rely on a single 'science' as the basis for its claim to such a position (see below for German exceptionalism in this respect). Moreover, and strongly related, in most countries it took a long time until business graduates could actually aspire to the leading positions in organizations, which had previously been held by graduates of other disciplines, namely engineering and law (see, for example, Engwall *et al*. 1996), and to elite status at the national level.

The chapter is in two main parts. In the first part we provide a brief overview of the evolution of the field of business studies in the USA and several European countries. This part is subdivided into three chronologically organized sections. The first period, from the late nineteenth century until the Second World War, saw the emergence of two rather distinct organizational models: (1) the university-based business school in the USA; and (2) the stand-alone institutions, such as the commercial schools in France and the *Handelshochschulen* in Germany—the latter turning later in this period into university faculties. The second period, following the Second World War until the 1990s, saw both a transformation of the US model in terms of a shift towards graduate education and increased scientization and its active expansion around the globe. Its actual influence remained limited, however, with the persistence of the original organizational models, in particular in the core European countries. Only the final period, since the 1990s, saw some increase in convergence between the USA and Europe, though more so with respect to the accent on, and the assessment of, research output.

On the basis of this empirical account, the second part discusses the influence of the broader fields of higher education and academic research—and their evolution—on variations in organizational patterns and knowledge production in the nested field of business studies. To account for these variations (and their persistence over time), we highlight two main factors. One of these relates to the way that national higher education fields are structured and thus the organizational templates that they provide for the new disciplines, such as the other professional schools in the USA or the engineering *grandes écoles* in France. Secondly, we consider the nature of nationally based public science systems. This concerns the extent of state guidance and regulation, the relative significance of state-based or private funding, and the internal governance of higher education and research organizations—specifically, the distribution of power between administrators and the academic elites (see also Chapter 1).

2. The Development of Business Education and Research

The Founding Years: From the Late Nineteenth to the Mid-Twentieth Century

In the USA as well as in Europe, the first institutions explicitly offering business education at the post-secondary level were founded more or less concurrently around the mid-nineteenth century, with the bulk of the development occurring towards the turn of the twentieth century. However, as the following more detailed account will show, the entry of the new discipline into the higher education field differed significantly between the USA and Europe. In the former, business education was incorporated into existing universities, whereas in the latter it was formed as stand-alone establishments outside the university system, at least initially. What should be noted here is the almost exclusive focus in both the USA and Europe on education rather than research—something that changed only towards the end of this first period.

THE UNITED STATES

From the very beginning, business studies found a place within universities in the USA. Among the frontrunners were the University of Louisiana (1851), the University of Wisconsin (1852), and Washington and Lee University (1869). Then in 1881 came the Wharton School of Finance and Commerce at the University of Pennsylvania, which would eventually become one of the top US institutions. It was followed by a row of establishments at prestigious universities: namely, California (1898), Chicago (1898), Dartmouth (1900), New York (1900), Columbia (1908), and Harvard (1908) (Engwall 2007). Although private business colleges had sprung up in the USA too, a definitive turn had taken place towards what in US parlance was referred to as collegiate business education (Lyon 1922). Thus, the new discipline achieved acceptance within the university, even if this was by no means an easy process (as the quotations in the introductory section indicate). Part of the explanation can be found in the fact that the emergence of business education coincided with the making of the US university in the form we know today (Thelin 2004). Put differently, its integration within the overarching field of higher education was facilitated by the fact that the latter was itself in a process of being organized.

Business education in the USA during this period exhibited two major additional characteristics, which differentiated it from similar efforts in Europe. First, it was infused with a so-called liberalizing mission, as exemplified in the founding of the Wharton School (Sass 1982). At the undergraduate level this manifested itself through the inclusion of a general (or 'liberal') education component into the curriculum, comprising subjects such as literature, humanities, social and natural sciences, as well as mathematics (Üsdiken 2007). The vocational-cum-liberalizing logic prevalent at the time stemmed, on the one hand, from the liberal arts tradition coming from the early colleges and, on the other hand, from the aspiration to generate a more cultured and, in that sense, more professionalized business class (Khurana 2007; Lyon 1922; Thelin 2004).

This incorporation of liberal studies into undergraduate education paved the way for the emergence of 'graduate schools' for advanced training in the professions. The professional graduate school in universities served as a vehicle in raising admission standards for the 'learned professions', notably, in fields such as law and medicine (Clark 1995). These were then to provide a model for the appearance of the 'graduate school of business' epitomized by the one established at Harvard in 1908 (Daniel 1998), which constituted the second major distinction from simultaneous developments in Europe. For business studies the graduate school served as the organizational base for the emergence of the MBA as an advanced degree and the ensuing claims towards professionalism (Khurana 2007). The graduate business school and the MBA degree did, however, have a slow up-take until the mid-twentieth century. Before the 1950s, college departments and the undergraduate degree in business continued to prevail (Hutchins 1960). Nor was there any notable expansion in the research base of the knowledge imparted in undergraduate or MBA programmes. Nevertheless, the 'university-based graduate school' had made some headway before the middle of the twentieth century as a uniquely American form of organizing business education.

EUROPE

For Europe, existing studies on the development of 'academic' education in business usually mention the foundation in 1852 of two institutions: namely, *l'Institut Supérieur de Commerce de l'État* and *l'Institut Supérieur de Commerce Saint Ignace* in Anvers, Belgium, as a starting point (for example, Engwall and Zamagni 1998). Two other institutions founded in the 1850s were *l'École Supérieur de Commerce* in Paris (1854) and *Wiener Handelsakademie* in Vienna (1856)—both established again as separate institutions outside universities. In Italy *Scuola Superiore di Commercio* opened in Venice in 1868, to be followed by similar schools in Genoa, Bari and Milan, founded respectively in 1884, 1886, and 1902 (Fauri 1998).

Despite these early foundations in various parts of Europe, two countries in particular—France and Germany—have been regarded as the 'first movers' in European higher education for business (for example, Amdam 2008). Although there had been some cross-national influence in the founding of the pioneering institutions, such as the first Italian school in Venice taking the one in Anvers as its model (Longobardi 1927), it was essentially the initiatives in these two countries that served as exemplars for others in Europe and elsewhere (Amdam 2008; Kipping and Üsdiken 2009). In France the chambers of commerce took the initiative in one commercial centre after another in the establishment of 'lower-level' schools (Whitley *et al.* 1981: 62), such as in Mulhouse (1866), Le Havre (1871), Lyon and Marseille (1872), and Bordeaux (1874). More notable, however, was the founding of the École des Hautes Études Commerciales (HEC) in Paris in 1881, with a stronger claim towards offering education at the 'higher level' (Locke 1984). For HEC and similar institutions that followed in France, the engineering *grandes écoles*, constituting a category distinct from the universities, served as the model (Locke 1984).

In Germany it was also the chambers of commerce and local business communities that initiated the creation of higher commercial schools (the *Handelshochschulen*) around the turn of the twentieth century: Aachen and Leipzig (both in 1898), Cologne and Frankfurt (1901), Berlin (1906), Mannheim (1908), and Munich (1910) (Meyer 1998). Like in France, the *Handelshochschulen* were founded as separate institutions. For them too, the technical schools in the country supplied an already existing model for practically oriented higher education. In turn, both the French and the German institutions served as models or at least as a source of inspiration for similar initiatives in other countries. Thus, the *écoles de commerce* proved very influential in Spain and Turkey, for example (Kipping *et al.* 2004), whereas the German institutions served as role models for *handelshögskolor* in Scandinavia, which were established in Stockholm (1909), Helsinki (1911), Copenhagen (1917), and finally in Bergen in 1936 (Engwall 2000).

Thus, the emergence and early development of higher-level business education in Europe in the late nineteenth and early twentieth centuries differed significantly from the pattern observed in the USA. It was only in the UK that this new field of study, as in the USA, found a home in universities. Commerce degrees had begun to be offered within universities in Birmingham (1902) and Manchester (1904) (Engwall 2007). However, during this formative period, the most prestigious universities, such as Oxford and Cambridge, did not espouse the new discipline—quite unlike their US equivalents. And nor did business education make significant progress elsewhere in the country after the early foundations (Tiratsoo 1998).

In Continental Europe, on the other hand, business became a subject for higher education outside universities through newly established standalone institutions usually called 'commercial schools'. As much as with the resistance of universities, this had to do with the aspirations of their initiators who came from outside the scientific establishment. As noted, in the case of the first movers in Europe, in France and Germany, these initiators were chambers of commerce, wealthy individuals, or groups of businessmen (Engwall 2007; Üsdiken *et al.* 2004). The same was the case for followers like Sweden, Finland, and Denmark. Only in countries such as Italy and Spain were such initiatives led by both private actors and governments, with the former also including religious institutions (Durand and Dameron 2008; Engwall 2000; Kipping *et al.* 2004).

The motivations of the founders were similar to the ones in the USA: namely (1) to address the perceived functional needs for educated staff due to the growing size of business firms and increasing competition; (2) more importantly perhaps, raise the social status of people in business; and (3) possibly to a lesser degree, strengthen the position of salaried personnel *vis-à-vis* the owners (Locke 1984; Üsdiken *et al.* 2004). But unlike the USA, where the new degree programmes were influenced by the broader vision of the liberal arts, in Europe the curricula of the new schools were imbued almost exclusively with a practice-orientated vocational logic. Similarly, the selection of the first instructors also valued predominantly their practice-based competence. Not surprisingly, most of them had limited experience of research (Engwall 2009: ch. 4).

However, the vocational orientation of the emerging discipline did not fit well with the existing fields of higher education and research in the first-mover countries. In Germany it did not align with either the humanistic education or the strengthening research orientation of specialized academic fields in the universities (Clark 1995). Nor did it fit into what was becoming the standard faculty structure within French universities, comprising law, letters, science, medicine and, somewhat later, pharmacy (Clark 1995). Although the universities were unwelcoming in both cases, the higher education fields in the two countries differed in the extent to which they accommodated a powerful alternative form of organization. This in turn served to shape the fate of the commercial schools.

In Germany, the technical schools had also developed outside the universities, despite similar negative reactions. They were granted, through state intervention, similar internal structures and rights at the end of the nineteenth century (Clark 1995). So, in order to become legitimate members of the higher education field in that country, the commercial schools had little choice other than to struggle for the right to grant doctoral degrees, extending the duration of studies and making their discipline science-based (Üsdiken *et al.* 2004). These efforts led to the creation of *Betriebswirtschaftslehre* (or business economics) as a separate discipline with the claim of being a 'science' of the business enterprise (Locke 1984). Not only did this enable the gaining of an academic identity, but also led the way towards becoming part of the public science system in the country. The conversion of the *Handelshochschulen* in Frankfurt and Cologne in 1914 and 1919 into universities marked the beginnings of a major process of organizational transformation whereby the commercial schools were eventually integrated into the university system as faculties of business economics (Meyer 1998).

This is particularly noteworthy, since the followers of the German *Handelshochschulen*, the Scandinavian *handelshögskolor*, in most cases remained independent. And nor did the same happen in France. This had to do with the presence of the engineering *grandes écoles* as a separate category and the reputation that they enjoyed relative to the universities. They thus served as a source of legitimacy and aspiration for the commercial schools, which also remained separate from the universities, although clearly as second-tier with respect to the *grandes écoles* of engineering (Locke 1984).

CONCLUSION

Overall therefore, the period up until the Second World War was significant for the entry of business studies into higher education systems. This entry occurred in parallel on both sides of the Atlantic and was made possible through strong support from the business communities on both continents. However, the way it happened was significantly different. In the USA it was organized within universities, whereas in Europe it required the creation of separate commercial schools independent of the university. This early period also saw the first influential attempt to turn business studies into a science, though as a uniquely German phenomenon.

Scientization and Expansion: The Era after Second World War

The four decades or so following the Second World War saw fairly dramatic changes in the relationship between business studies and the broader field of higher education and research, particularly in the USA and to some degree in Europe, together with the role (and reputation) of graduates within business firms as well as society at large. The changes that originated in the USA spread to other parts of the world, albeit gradually and often superficially. The USA became a global reference for business education and research, but the established organizational forms and public science systems in other countries remained largely unchanged—with a few exceptions in isolated pockets of 'Americanization'.

THE UNITED STATES

In the USA there were several reasons for the changes in the institutional logics governing business studies and for its positioning within the field of higher education and public sciences after the Second World War. These included a shift in the kind of background required for managerial roles towards a more generalist education—itself a reflection of the changing strategies and structures of business firms (Fligstein 1990). Equally, if not more importantly, these managers and in particular the graduates of both undergraduate departments and business schools, clearly aspired to a more prominent role in both their own organizations and society at large (Drucker 1954), thus promoting a process of at least superficial 'professionalization' of business. Last but not least, at a systemic level, the support for and belief in 'big science', funded by both big business and big government, and its practical benefits, reached unprecedented heights following its

333

undeniable success during the Second World War and in the immediate post-war decades (Galison and Hevly 1992).

Within this context a major trigger and/or catalyst regarding business studies in the USA was the publication of the two so-called 'foundation reports'—one by the Carnegie Corporation (Pierson 1959) and the other by the Ford Foundation (Gordon and Howell 1959). These reports pointed both to the weak science basis of business education and the drift at the college level away from the liberalizing mission towards vocationalism. With hindsight, the most lasting influence of these reports turned out to be in their claim that business education had to become more based on scientific research. Together with sizeable funds provided notably by the Ford Foundation, US business schools came to focus more on quantitative studies, particularly operations research, as well as the so-called behavioural sciences (Locke 1989). In order to handle this reorientation these schools became multidisciplinary by recruiting mathematicians, political scientists, psychologists, sociologists, and statisticians. Business studies turned into a discipline increasingly governed by the norms of science. A number of academic journals emerged, and faculty members in leading universities were expected to be, and did become, active in publishing (Engwall 2007).

A munificent resource environment shaped by demand for graduate study in business and public funding of research also helped this reorientation (Augier *et al.* 2005). Attention was turning towards graduate education, and the vision was to retain the practical orientation inherent in the MBA together with expanding the use of science in solving managerial problems. Along with the claim that management was a profession, the new professional logic involved a blending of managerialist and academic logics (Üsdiken 2007). Although remaining strong at the discursive level, actual practice in US business schools began to deviate, coming under the increasing influence, throughout the 1970s and the 1980s, of the 'academic drift' within the blend that the professional logic implied. This had to do with aspirations for status within the university as well as the incentive structures that were employed and the competition among universities for reputation (D'Aveni 1996).

The radical shift in logics and the ensuing academic drift, however, did not lead to the emergence of new forms of organizing within the US business education field. What it instead did was to alter the balance among the existing forms. The strengthening of the professional discourse and then the academic logic led to a shift in balance towards the graduate version of the university business school (Locke 1989). Moreover, the more prominent US universities tended to withdraw from undergraduate education and to concentrate on the MBA and its variants (D'Aveni 1996). Nevertheless, the undergraduate version did not disappear, and three forms of organization came to be established in the USA: namely, (1) undergraduate only, (2) undergraduate and graduate combined, and (3) graduate only. Yet the graduate school and the MBA increasingly became the flagship, making the US business education field more structured and homogenized.

Another major outcome of this shift in logics was to help the USA rise to a position of the world's centre in business education as well as research. Together with the country's post-war international supremacy, business education and research in the USA was becoming a superordinate field for national fields in other parts of the world. As a result, the USA started to influence education and research for business and management elsewhere. Probably most obvious was the spread of US content—for example, in the adoption of American textbooks, sometimes translated into national languages (Engwall 2000). In addition, the USA also became a 'model' for the way in which business education and research was to be organized. At the national level the effect was more indirect, with the perceived US model used to enact national reforms. More significantly, at the organizational level the USA provided advice and funding through government-led 'technical assistance' projects or, more importantly, through the private Ford Foundation (Gemelli 1998). In this process, some US business schools—in particular, Harvard—helped create a number of 'clones' in several European countries, as well as in Canada and Latin America. But the overall influence was broader and affected to some degree the overall structure of the field of business education and, to a lesser extent, research.

EUROPE

As for Europe, a first significant change was that governments in countries such as France and Sweden, where business education traditionally had been outside universities, began to introduce the discipline into public universities (Engwall and Zamagni 1998). In the case of France this occurred first through the establishment of the so-called IAEs (*Instituts d'Administration des Enterprises*) by graduate institutes attached to faculties of economics or law, which initially offered one-year 'certificate' programmes to university graduates from other fields (Kipping and Nioche 1998). This was followed by the introduction of first-level degrees—most notably in the case of the *Université Paris Dauphine*, founded in 1968. The same had, already happened in Sweden a decade earlier, when chairs in business studies were established first in 1958 at the universities in Lund and Uppsala, followed by others in the 1960s (Engwall 2004). This had to wait until the 1970s in countries such as Spain and the Netherlands, where in the former case the extant commercial schools were also integrated into public universities (Kipping *et al.* 2004; Noorderhaven and van Oijen 2008).

A second change was brought about by the opportunity for the creation of private stand-alone schools or American-type business schools in private universities. Notable among these were the three 'international' ones created in the 1950s: namely, INSEAD in France as well as IMEDE and IMI in Switzerland (which were merged in 1990 to become IMD). Spain provided another notable case with the founding of ICADE as a separate school, and of ESADE as well as IESE as a part of private religious universities (Kipping *et al.* 2004). Similar to the latter case was the establishment of a business school in 1971 (SDA) within Italy's private Bocconi University (Kipping *et al.* 2004).

A third change was exemplified by the UK, where in the initial instance in the mid-1960s an adoption of the American university-based graduate business school model occurred. This was, however, limited to two such schools: namely, the London Business School and the Manchester Business School (Whitley *et al.* 1981). Academic business studies' more widespread diffusion had to wait for some two decades, and this was dominated by undergraduate education, particularly in the less prestigious universities. Eventually, a broader adoption of the American undergraduate and MBA model became established in most British universities, though with an alteration in the duration of the MBA which was mostly turned into a twelve-month programme (Tiratsoo 1998).

While the 'Americanization' of business education in Europe during this period seems undeniable, the result was far from a Europe-wide dominance of the US model that rested on the university-based graduate business school. On the contrary, in terms of the overall field structure these developments had, by the late 1980s, led to the consolidation of a tripartite organizational panorama in Europe consisting of university departments, university business schools, and stand-alone business schools. Universities that housed business departments were typically public institutions. They only offered first-degree programmes together with doctoral studies, though, as in the case of the IAEs in France, some form of postgraduate training could also exist. Unlike the USA where the undergraduate degree also incorporated a significant element of liberal arts, these first degrees were more likely to involve discipline-based education.

The ownership and programme portfolio of the stand-alone schools, on the other hand, were more varied. The international ones, such as INSEAD in particular, were almost entirely independent of the higher-education field of the countries in which they operated, and tended to concentrate on postgraduate and executive education. Some of the more local ones, such as those in France, though part of the higher-education system in the country and offering similar types of degrees, still remained in private governance and outside the university system. Those in Scandinavian countries remained separate too, increasingly becoming specialized universities. Indeed, there was also a case where a stand-alone institution (Gothenburg) became integrated in 1971 into a public university due to financial difficulties. Countries in Europe varied quite significantly in the degree to which their business education field followed this tripartite organizational scheme or were dominated by a single organizational model. Germany, the Netherlands, and Italy constituted examples of homogeneous fields dominated almost entirely by faculties or departments in public universities, although Italy also had a business school patterned after the US model within a private university. The business education field in the UK was also quite homogeneous, with business schools in public universities as the dominant form. Scandinavian countries, France, and Spain, on the other hand, had relatively more heterogeneous fields as both university departments and stand-alone schools coexisted.

'Americanization' of business studies as a research field was even less advanced. Influence did occur, as from the 1960s onwards Europeans increasingly made study trips to the USA, often supported by the Ford Foundation (Gemelli 1998). From the 1970s the European Institute for Advanced Studies in Management (EIASM) not only connected European management scholars among themselves but also to those in the USA (Engwall 2009: ch. 7). These links with the USA began to breed tendencies towards publishing in US journals, though not with great success (Engwall 1996, 1998). More notably perhaps, starting in the 1970s, academic associations similar to the ones in the USA began to emerge across Europe, accompanied by the launching of subdiscipline-based scholarly journals again patterned after those in the USA (Engwall 2007). The direction of influence was thus essentially one way, although in some fields there were significant theoretical and methodological differences between research coming out of the USA and Europe in the latter part of this period (Üsdiken and Pasadeos 1995). Moreover, publishing in USA or pan-European outlets remained confined to scholars in a small number of countries, as academic careers could be developed through publishing in nationally based journals and books.

CONCLUSION

To recapitulate, the development in the USA after the Second World War was characterized by expansion and a strong shift towards scientization of business education, accompanied by widespread international influence. Serving as a reference and source of inspiration or through the support of aid agencies and private foundations, the USA also had an impact on European business education. This involved, variably across countries, the creation of new organizations, alterations in the curricula of existing ones, and the introduction of business studies into public universities. The pressures for a stronger research orientation in the USA, however, were late-coming and only gradually appearing in European business departments and business schools.

More Convergence: Business Education and Research after 1990

From the 1990s onwards the now firmly established field of business studies underwent another round of externally driven change. Together with increasing strains overall in public funding of higher education and scientific research, it became subject to increased scrutiny both from public and private funding agencies and from the press, in the form of rankings. These developments also affected the super-ordinate fields of higher education and public sciences, as discussed in many chapters in this book, but, unlike the previous period, business studies was exposed to these pressures early on. While many of these developments originated in the USA, they reached many European countries almost concurrently and seem to have initiated a more profound process of change, leading to a somewhat stronger convergence of institutional logics and internal governance among clusters of organizations in different national settings.

THE UNITED STATES

The university business schools in the USA were the first to encounter rankings as a novel institutional element (Wedlin 2006). The advent of the rankings in the late 1980s accompanied reductions in public funding of education and research and therefore an increased dependence on student tuitions, executive education, and corporate funding (Trank and Rynes 2003). Some degree of challenge also appeared from private providers of business education (Pfeffer and Fong 2002).

Together, these new sets of institutional and material conditions have not, as yet, led to marked changes in the organizational panorama of business education in the USA, other than perhaps making business schools more autonomous and loosely coupled with their universities. They have, nevertheless, made the already market-based business education field in the USA even more competitive (Spender 2008). Indeed, there are strong indications that the discourse around US business schools is becoming imbued with business terminology. The market logic appears to have gained ground, increasingly shaping the actions of business schools (Zell 2001). With greater competition they have altered their domains by expanding the diversity of the programmes that are on offer. Another notable move has been increased internationalization that has involved more active recruitment of foreign students and entering into various kinds of linkages with similar institutions in other parts of the world.

Such institutional changes and greater competition seem to have intensified business schools' responsiveness to the demands of major resource providers, students as well as corporations (Morgeson and Nahrgang 2008). On the one hand, there is an increasing concern with bridging what has conventionally been labelled as the theory–practice divide by driving research more towards practically useful ends. There has also been a growing attention to teaching and teaching performance, as research output turns out to be only weakly linked to rankings (Trieschmann *et al.* 2000). Often this has led to the hiring of less research oriented, non-permanent teaching staff (Zell 2001). Altogether these changes appear to be culminating in reducing 'the power of research-oriented faculty in business schools' (James March, quoted in Zell 2001: 337).

EUROPE

In Europe, changes since 1990 have concerned primarily the research side, with mounting pressures for business studies to obtain outside funding and increased scrutiny of research outputs. But there has also been some more convergence in the way business education is organized. The latter has occurred quite dramatically in the UK, where MBA degrees and business schools, usually within the confines of public universities, have flourished (Tiratsoo 2004). Moreover, the MBA has now gained legitimacy for access to high-level positions, at least in certain sectors (Thompson and Wilson 2006)—a legitimacy it had acquired much earlier in the USA, but which had escaped it so far in the UK. Organizational changes also occurred in some of the other European countries, albeit more slowly and partially.

Paramount in driving these changes, as in the case of the expansion of university business schools in the UK, have been reductions in public funding (Tiratsoo 2004). This has been coupled with the continuation of minimal and regulated student fees in most European countries for public universities (Durand and Dameron 2008). In addition to more adverse material conditions, organizations providing education in and doing research on business in Europe have also come under novel institutional pressures. These have been primarily of three kinds. First, rankings have appeared on the European scene too, a decade after the ones in the USA (Wedlin 2006). Second, and differently from the USA, whilst state funding has been decreasing, accountability pressures have increased and been accompanied by the introduction in many countries of national evaluation systems (Whitley and Gläser 2007; Chapters 5 and 9). Finally, the Bologna process has served as another set of institutional pressures, though its effects appear to be limited to bringing especially Continental European institutions closer to those in the USA, not least because of the pressures to adopt the three-tier system of bachelor's, master's and doctoral degrees (Hedmo and Wedlin 2008).

The rankings and indeed resource constraints too have been more forceful in their effects on institutions offering graduate programmes and executive education. The primary concern here has been the quest to obtain international recognition and visibility—seeking accreditation from one or more of the accrediting agencies, such as the American AACSB, the European EQUIS and the British AMBA, and, more importantly, a place in the business school rankings. Accreditation has been obtained by a large number of institutions in Europe, including faculties of some public universities. Achieving a listing in the rankings, however, has been largely confined to university-based and stand-alone business schools. Some of these have also been able to obtain a place in the so-called 'global' rankings that also include the US business schools. It is in these institutions and those aspiring for similar achievements that increasing commonalities with the latter can be observed. This is partly apparent in the similarities in the portfolio of programmes that are offered, which often incorporate an emphasis on the MBA as well as executive education. It can also be observed in their engagement in international operations and networks. Given these similarities and the more pronounced manner in which leading US business schools not only serve as their main reference but also major competitors, they are also likely to be influenced by the emerging market logic (see, for example, Tyson 2005).

That the university-based and stand-alone business schools have been more adept in entering into rankings and even finding a place in global listings appears to have led to mimetic reactions on the part of departments or institutes in public universities. This becomes apparent in the increasing adoption of the 'business school' label. The Department of Business Administration at Stockholm University, for instance, has adopted the name 'School of Business'. Similarly, the Université de Lyon now refers to its IAE as 'The University of Lyon Business and Management School'. That these examples come from settings where stand-alone business schools in particular have been prominent actors may be indicative of the institutional pressures that they generate on public universities.

More notable are perhaps the effects that the US graduate business school model and the rankings may be having on the emergence of novel organizational forms. This is perhaps best exemplified by the creation in 2004 of the Mannheim Business School in Germany (abbreviated as MBS) as a separate legal entity and with a governing board made up of executives from industry. It offers MBAs and Executive MBAs, company programmes as well as international joint ventures with university-based and stand-alone schools in the rankings. Noteworthy is also what looks like the revival of the initiatives that characterized the foundation of commercial schools in the late nineteenth and early twentieth centuries. An important example is the *Wissenschaftliche Hochschule für Unternehmensführung* (WHU) founded in Germany in 1984, involving a chamber of commerce and industry. The WHU offers bachelor's, MBA and EMBA programmes, but as the name suggests has also strong research aspirations. Its establishment spearheaded the re-emergence of stand-alone business schools in a country where they had vanished a long time before (for other examples see Kipping and Üsdiken 2009).

Rankings and the introduction of new evaluation regimes, in particular, appear to have even more profound effects on business studies as a public science. To the extent to which such national evaluation systems have been imposed on business studies and the degree to which they have been tied to funding, as in the UK for example, they have generated pressures towards, on the one hand, greater internationalization and, on the other, the strengthening of the academic logic. More specifically, this has meant attempting to find ways for increasing publications in internationally prominent outlets as well as making European and national journals more international. A companion implication has been turning more to external markets to obtain faculty members with good or promising publication records. Overall, it has also meant becoming more like business schools in the USA, where publication norms have increasingly, since the 1960s, guided recruitment and promotion decisions (Crainer and Dearlove 1999).

CONCLUSION

In sum, the post-1990 era can be considered as a period during which higher education and research organizations have come under greater institutional pressures, often at the same time as declining public funds. The outcome has been a more competitive environment especially for US business schools and those in Europe trying to match them, as well as the ones aspiring to get greater international recognition. Emergent conditions appear to force US university business schools to become more 'market' oriented and to search for ways of making their research more 'relevant'. In Europe, on the other hand, especially to the degree that national evaluation systems have been introduced and tied to funding, as has been the case in some countries, the turn has been towards internationalization—not only by seeking accreditation, but also by an increased accent on, particularly, publications in international outlets (see Chapters 2 and 5).

3. Higher Education Fields, Public Science Systems, and Academic Organizations

The preceding historical account has indicated that the emergence of business education has been largely co-terminous in the USA and Europe. It has also shown that the motives of the initiators, at least in the first-mover countries, were also largely similar. So was the accompanying voca-tional logic as well as the practice-driven knowledge base that shaped the content of the curricula. These similarities notwithstanding, the form of organization in which business studies was to be located in the first-mover countries turned out to be different, marked as it was by a distinction between the USA and much of Europe. From the outset, business studies found a place in the USA in the form of *university business schools*, whereas in Europe, with the exception of the UK, this was to happen through the creation of *stand-alone commercial schools* outside the university—particu-larly in the two first-mover countries, France and Germany.

A transformation not only in the way business studies was organized but also in the claim towards a scientific knowledge base first occurred in Germany in the inter-war period, as the commercial schools were turned into *university faculties or departments*. A second major transformation, similar in intent but different in the alteration in organizational form, took place in the aftermath of the Second World War, this time in the USA. The shift was towards the graduate version of the university business school, especially in the leading universities, and an accompanying strong drive towards becoming a separate scientific discipline.

Following these profound transformations, the university-based graduate business school became institutionalized as the dominant model in the USA in the period up to the 1990s, with a further strengthening of the academic logic despite espoused concerns with solving managerial problems. Field structures in Europe remained and indeed became more heterogeneous, partly due to US influence, accommodating in nationally variant ways stand-alone business schools, university faculties or departments, and university business schools.

The developments in material and institutional conditions in the last couple of decades both in the USA and Europe, though not identical, appear to have had relatively limited impact on the way national fields of business education and research had come to be structured in the previous period. As the preceding account has shown, however, significant internal organizational changes seem to be occurring, as business schools in the USA are increasingly becoming driven by market logics, whereas those in Europe are coming under greater pressure to become research-driven and to internationalize. The divergent outcomes across the USA and Europe, as well as the transformations that have taken place and the more recent changes that are occurring, we argue, can be accounted for by two interrelated factors: namely, the way higher education fields have come to be structured, and the nature of the national science systems.

Educational Field Structures and Organizational Templates for New Disciplines

A major factor shaping the differences regarding the place of business studies as a new discipline towards the end of the nineteenth century was the available organizational templates within existing higher-education fields in different national settings. In both the USA and Europe the entry of business programmes into higher education required external forces. Entry required from these outsiders both lobbying for the need of vocational training and financial backing. In the USA, as the university was taking shape at the turn of the century it also 'brought professional schools into its structure and organized them with sequencing and connection with academic units' (Thelin 2004: 129). Thus as a potential 'professional school' business studies could also find a place within the American university and in a way which could provide, as its early benefactor Joseph Wharton wanted, 'a liberal course of study' (Thelin 2004: 86). Indeed, the graduate school alternative for purportedly advanced training was also available, which Harvard took up in the early 1900s (Daniel 1998).

This, however, could not happen either in France or Germany, as in both cases professional studies other than medicine and law had no home within existing universities. Engineering or technical schools had already been set up as separate organizations. The initiators and benefactors of commercial schools, therefore, also had to build them outside the university. Nor was there the idea of graduate education, as undergraduate study could be more advanced and academic, as prospective students at the time were likely to be better educated relative to their counterparts in the USA during high school (cf. Trow 1989).

The different routes that the commercial schools took in France and Germany in the first half of the twentieth century also depended on the status of different institutions of post-secondary education in the two countries (Becher 1993; Henkel and Kogan 1993; Neave 1993; Neave and Edelstein 1993). First, the pre-eminence of the universities varied. Universities enjoyed higher status in Germany, whereas this was not necessarily the case in France. And in Germany, technical schools which had also emerged outside the university system had with state support become like universities, whereas this did not happen in France as the *grandes écoles* of engineering there were attached to government ministries. In Germany, for the commercial schools to command status they had to become integrated into the university system. This pushed them towards becoming more academic and inventing a discipline of their own which enabled them to gain a 'scientific character' and thus become university faculties. In France the commercial schools could try to emulate the engineering schools—which they actually did quite successfully, later entering into the status of *grandes écoles*.

Evolution and the Role of Public Science Systems at the National Level

Later development of business studies into a scientific discipline was in turn shaped by the respective national science systems in which it was embedded, together with international influences at play as well as the more recent changes in material and institutional conditions impinging upon education and knowledge production.

THE UNITED STATES

Based on the ideal types of public science systems introduced by Whitley in Chapter 1, the USA in the post-war era could be characterized as a competitive version of employer-dominated public science systems, where the authority of the state and of organizational academic elites are low, whereas administrators, private interests, research funding foundations, and national scientific elites enjoy relatively higher levels of influence.

As Clark (1995), for example, has observed, in the USA at the early stages of development the federal government played a very limited role in shaping and regulating the higher education field. Both private and public universities emerged, and funding for both types came from a variety of sources, including, in particular, private philanthropy. In the absence of rule-making by governmental authorities, and due to the existence of private as well as public institutions, the higher-education field in the USA became much more market-driven than in Europe. As pointed out by Trow (1989: 378), this had to do with 'the absence of a federal ministry of education with the power to charter new institutions, or a single pre-eminent university that could influence them in other ways'.

Another important element of the employer-dominated science system in the USA has been the internal governance of universities in which significant powers have been accorded to administrators of universities. Historically, US colleges and then the universities have come to be characterized by a strong central administration. As Thelin (2004: 12) observed, 'the creation and refinement of [the] structure [based on an] external board combined with a strong college president is a legacy of the colonial colleges that has defined and shaped higher education in the United States to this day'. This stronger position of US university leaders in combination with funding through private sources and the market has implied, as Whitley suggests (Chapter 1), relatively less influence on the part of the academics within universities, especially when they have had to operate in more competitive conditions.

So the academic elites within the universities, despite their reluctance to embrace business studies, were not able to block its entry into the 'corporation of learning', as Veblen (1918) put it. Even the most prestigious universities relatively quickly espoused the new discipline—in particular when it came to the establishment of the 'professional' business schools, often with substantial donations from private sponsors. Those private interests shaped the new institutions at the outset. Even more important during the subsequent development of business studies in the USA was the role of business interests and the research funding foundations. As discussed above, the Carnegie Corporation and the Ford Foundation acted as catalysts after the Second World War in turning what were primarily institutions of vocational training into more research-oriented schools, subsequently funding much of this research. An important consequence of this process for the discipline was the gaining of a scientific identity and status as well as the creation of a national scientific elite, which has since then exercised a significant degree of authority over the field.

That US universities at large as well as the business schools within them could operate in a relatively resource-abundant environment in the period until well into the 1980s (Augier *et al.* 2005) has allowed the latter to drift away from the post-war practical science aspirations towards a strong academic logic, despite occasional calls from within or from the outside for greater relevance (Daniel 1998; Porter and Kibbin 1988). Changing resource conditions since, as well as the rankings, which focus much more on employment-related criteria than academic achievement, appear to be altering the character of business schools so that they are increasingly operating more like businesses in a market (Pfeffer and Fong 2004). Moreover, the new resource dependencies not only seem to be generating further pressure for producing managerially relevant knowledge, but also possibly weakening further the power of research faculty within business school environments.

EUROPE

Europe presented a rather different picture from the USA. Here, the state exercised a much more significant authority over the public science system, albeit with some variation across countries. As Whitley (Chapter 1) suggests, many state-dominated systems are characterized by considerable authority granted to academic elites within the universities, and variably to those at the national level, with other external actors wielding much less influence. In the words of Trow (1989: 385), this implied that the European university leaders faced 'a situation in which power is already held jealously by the professoriate, or by the academic staff broadly, or by government ministries, or student organizations, or trade unions'. More specifically, in contrast to the employer-dominated system in the USA, the state-dominated European systems, such as those in France and Germany, have come to be characterized by a weak central administration and strong chair-holding professors, as was also the case in post-war Japan (Chapter 4).

So in France, as an example of the state-centred version of what Whitley calls the state-dominated system, the early commercial schools, backed as they were by business interests, could only be established (and did remain) as private organizations, although their exemplars, the engineering *grandes écoles*, were public institutions attached to different government ministries.

Cut off from the public science system, these business schools remained dominated by a vocational logic in France and in most other countries which adopted them. It was only after the Second World War, that under US influence, governments, and possibly some university academic elites, were convinced that business studies could find a place within universities and be acknowledged as a part of the science system in the country. In Germany, where the state was also quite powerful, but nevertheless shared its authority not only with organizational but also national academic elites, the initial result was the same. Nevertheless, given the much stronger scientific ethos at that time in this country, gaining status, and indeed even legitimacy, necessitated recognition as a 'science' and becoming a part of the university system, which, though not without tensions, did occur after about two decades or so (Üsdiken *et al.* 2004).

The largely state-dominated nature of public science systems in European countries not only affected the way business studies became initially organized but also their subsequent development. In the period after the Second World War, running through to the end of the 1980s, there was some limited borrowing of educational models from the USA, with a much stronger orientation towards importing curricula and content for teaching. On the other hand, the penetration of the US model of a strong emphasis on publication performance was very limited in this same period, especially in leading universities. This had to await the material and institutional challenges arising after the 1990s. But even then it turned out that there were two types of organizational formations for business studies in Europe that were more adept both in responding to these challenges and adopting the US model for publications to a greater degree (see, for example, Baden-Fuller *et al.* 2000). Some of the university-based business schools burgeoning in what Whitley calls the state-delegated system of the UK, and similar ones in private universities within a few other European countries, constitute the first category. The second includes the stand-alone schools in various parts of Europe, which essentially operate outside the state-dominated science system in those countries.

4. Conclusions

The main theme of this chapter is that the developmental trajectory of scientific disciplines is highly dependent on the nationally based higher education and science systems in which they are embedded, despite possibly increasing flows of international influence. By comparative examination of the emergence and evolution of business studies in the USA and Europe, we have shown that not only the ways in which this particular discipline came to be organized in the first place, but its later development into a scientific field, differed in these two settings.

Differences in the pre-existing structure of higher education fields mattered, as they defined the organizational templates available to actors trying to make business studies a subject for academic study. So did the variations in the way and the degree of development of public science systems in this early stage in the two contexts, and particularly in the period after the Second World War. With relatively weak links to the state, more powerful administrators, greater openness to the influence of business interests, and through more diverse funding sources, US universities took the lead in the scientization of business studies in this latter period. Later development was also largely shaped and consolidated by US universities as they operated under the same governance structures as well as generous support from private sources of funding. The way business studies had come to be organized during the formative period in the first-mover countries and their imitators in Europe continued to the post-war period, with some alterations, however, as US-inspired new organizations were established, though again in the form of private stand-alone schools dedicated largely to MBA and executive education. The US model had an impact in public universities too, as new faculties, departments, and institutes were set up to house business studies. Nevertheless, the flow of influence from the USA remained confined to teaching. Although new publication outlets also followed in the later stages of this period, within the largely state-dominated context in Europe there was little in the way of moving towards a US-style accent on publication performance, at least until more recently.

We have shown that the different nature of public science systems in the two settings has also served to moderate the impact of the somewhat similar recent changes in material and institutional conditions. This has been the case despite a movement in European countries in the direction of less state regulation. In the specific case of business studies the new challenges appear to have led US university business schools further towards market-oriented logics, which may possibly be working against the strong academic logic that has been dominant throughout the second half of the twentieth century. In Europe, on the other hand, the turn has been largely towards a greater accent on publications and international impact.

The latter developments bring to the fore the question regarding international convergence versus diversity. There is some convergence around the 'business school' label, although even this does not appear to be complete, as not all organizations teaching and researching business in Europe carry this name (Wedlin 2006). In fact, diversity of forms still exists in Europe even under this common label. There are not only schools attached to universities but also the often private, but sometimes public, stand-alone schools, as well as the novel versions that are private but loosely linked to a public institution. Despite these variations in form however, they are often similar with respect to non-academic involvement in their governance and a strong internal administrative structure. They are likely to be operating, increasingly perhaps, as Pfeffer and Fong (2004) have it, 'more like business than school', focusing primarily on fee-paying graduate, post-experience, and executive education, though some may also include undergraduate and indeed doctoral studies. Finally, they are also likely to be concerned with their positions in rankings and therefore highly similar to their US counterparts in their recruitment and promotion practices, with expectations of publication records in prominent outlets, although perhaps more so now with a greater promise of managerial relevance.

University faculties, institutes, or departments in many European countries have not been entirely immune to the same kinds of institutional pressures, as some have, for example, begun to adopt the 'business school' label. These are likely to be coupled with the more recent governance changes influencing public universities at large (Chapter 1). Greater pressures for international publications as well as closer industry ties and managerially relevant research are therefore likely to ensue, especially in national contexts where other organizational forms may be seen as outperforming the universities. Nevertheless, to the extent that the state-dominated nature of higher education and public science systems persists, complete convergence around the US-type 'business school' phenomenon does not appear to be imminent.

References

Amdam, R. P. (2008) 'Business Education', in G. Jones and J. Zeitlin (eds.), *The Oxford Handbook of Business History* (Oxford: Oxford University Press), 96–119.

Augier, M., March, J. G., and Sullivan, B. N. (2005) 'Notes on the Evolution of a Research Community: Organization Studies in Anglophone North America, 1945–2000', *Organization Science*, 16(1): 85–95.

Baden-Fuller, C., Ravazzalo, F., and Schweizer, T. (2000) 'Making and Measuring Reputations: The Research Ranking of European Business Schools', *Long Range Planning* 33(5): 621–50.

Becher, T. (1993) 'Graduate Education in Britain: The View from the Ground', in B. R. Clark (ed.), *The Research Foundations of Graduate Education: Germany, Britain, France, the United States, Japan* (Berkeley, Calif.: University of California Press), 115–53.

Clark, B. R. (1995) *Places of Inquiry: Research and Education in Modern Universities* (Berkeley, Calif.: University of California Press).

Colasse, B., and Pavé, F. (1995) 'Claude Riveline: Une pédagogie médiévale pour enseigner la gestion', *Annales des mines*, 14–32.

Crainer, S., and Dearlove, D. (1999) *Gravy Training: Inside the Business of Business Schools* (San Francisco, Calif.: Jossey-Bass).

D'Aveni, R. A. (1996) 'A Multiple Constituency, Status-Based Approach to Interorganizational Mobility of Faculty and Input–Output Competition among Top Business Schools', *Organization Science*, 7(2): 166–89.

Daniel, C. A. (1998) *MBA: The First Century* (Lewisburg, Penn.: Bucknell University Press).

Drucker, P. F. (1954) *The Practice of Management* (New York: Harper & Row).

Durand, T., and Dameron, S. (eds.) (2008) *The Future of Business Schools: Scenarios and Strategies for 2020* (Basingstoke: Palgrave Macmillan).

Engwall, L. (1994) 'Anders Berch's Followers: The Development of Modern Academic Business Studies in Sweden', in L. Engwall and E. Gunnarsson, E. (eds.), *Management Studies in an Academic Context* (Acta Universitatis Upsaliensis, Studia Oeconomiae Negotiorum; Uppsala: Almqvist & Wiksell), 45–65.

—— (1996) 'The Vikings versus the World: An Examination of Nordic Business Research', *Scandinavian Journal of Management*, 12(4): 425–36.

—— (1998) 'Mercury and Minerva: A Modern Multinational. Academic Business Studies on a Global Scale', in J. L. Alvarez (ed.), *The Diffusion and Consumption of Business Knowledge* (London: Macmillan), 81–109.

—— (2000) 'Foreign Role Models and Standardisation in Nordic Business Education', *Scandinavian Journal of Management*, 15(1): 1–24.

—— (2004) 'The Americanization of Nordic Management Education', *Journal of Management Inquiry*, 13(2): 109–17.

—— (2007) 'The Anatomy of Management Education', *Scandinavian Journal of Management*, 23(1): 4–35.

—— (2009) *Mercury Meets Minerva* (Stockholm: EFI; 1st edn. Oxford: Pergamon Press, 1992).

—— and Zamagni, V. (eds.) (1998) *Management Education in Historical Perspective* (Manchester: Manchester University Press).

—— Gunnarsson, E., and Wallerstedt, E. (1996) 'Mercury's Messengers: Swedish Business Graduates in Practice', in R. P. Amdam (ed.), *Management Education and Competitiveness: Europe, Japan and the United States* (London: Routledge), 194–211.

Ewing, D. W. (1990) *Inside the Harvard Business School. Strategies and Lessons of America's Leading School of Business* (New York: Times Books).

Fauri, F. (1998) 'British and Italian Management Education before the Second World War', in L. Engwall and V. Zamagni (eds.), *Management Education in Historical Perspective* (Manchester: Manchester University Press), 34–49.

Fincham, R., and Clark, T. (2009) 'Introduction: Can we Bridge the Rigour–Relevance Gap?', *Journal of Management Studies*, 46(3): 510–15.

Fligstein, N. (1990) *The Transformation of Corporate Control* (Cambridge, Mass.: Harvard University Press).

Galison, P., and Hevly, B. (eds.) (1992) *Big Science: The Growth of Large-Scale Research* (Stanford, Calif.: Stanford University Press).

Gemelli, G. (1998) 'From Imitation to Competitive Cooperation: The Ford Foundation and Management Education in Western and Eastern Europe (1950s–1970s)', in G. Gemelli (ed.), *The Ford Foundation and Europe (1950s–1970s)* (Brussels: European Interuniversity Press), 167–304.

Gordon, R. A., and Howell, J. E. (1959) *Higher Education for Business* (New York: Columbia University Press).

Hedmo, T., and Wedlin, L. (2008) 'New Modes of Governance: The Re-regulation of European Higher Education and Research', in C. Mazza, P. Quattrone, and A. Riccaboni (eds.), *European Universities in Transition: Issues, Models and Cases* (Cheltenham: Edgar Elgar), 113–32.

Henkel, M., and Kogan, M. (1993) 'Research Training and Graduate Education: The British Macro Structure', in B. R. Clark (ed.), *The Research Foundations of Graduate Education: Germany, Britain, France, the United States, Japan* (Berkeley, Calif.: University of California Press), 71–114.

Hutchins, J. G. B. (1960) 'Education for Business Administration', *Administrative Science Quarterly*, 5(2): 279–95.

Khurana, R. (2007) *From Higher Aims to Hired Hands: The Social Transformation of American Business Schools and the Unfulfilled Promise of Management as a Profession* (Princeton: Princeton University Press).

Kieser, A. (2004) 'The Americanization of Academic Management Education in Germany', *Journal of Management Inquiry*, 13(2): 90–7.

Kipping, M. (1998) 'The Hidden Business Schools: Management Training in Germany since 1945', in L. Engwall and V. Zamagni (eds.), *Management Education in Historical Perspective* (Manchester: Manchester University Press), 95–110.

—— and Nioche, J.-P. (1998) 'Much Ado about Nothing? The US Productivity Drive and Management Training in France, 1945–60', in T. Gourvish and N. Tiratsoo (eds.), *Missionaries and Managers: American Influences on European Management Education, 1945–1960* (Manchester: Manchester University Press), 50–76.

—— and Üsdiken, B. (2009) 'Beyond Isomorphism and Translation: The Role of Foreign Models in the Development of Management Education in Germany, Sweden and Turkey', in L. Wedlin, K. Sahlin, and M. Grafström (eds.), *Exploring the Worlds of Mercury and Minerva: Essays for Lars Engwall* (Acta Universitatis Upsaliensis, Studia Oeconomiae Negotiorum, 51; Uppsala: Uppsala University), 45–68.

—— Üsdiken, B., and Puig, N. (2004) 'Imitation, Tension and Hybridization: Multiple "Americanizations" of Management Education in Mediterranean Europe', *Journal of Management Inquiry*, 13(2): 98–108.

Locke, R. R. (1984) *The End of the Practical Man* (Greenwich, Conn.: JAI Press).

—— (1989) *Management and Higher Education since 1940* (Cambridge: Cambridge University Press).

Longobardi, E. C. (1927) 'Higher Commercial Education in Italy', *Journal of Political Economy*, 35(1): 39–90.

Lounsbury, M., and Pollack, S. (2001) 'Institutionalizing Civic Engagement: Shifting Logics and the Cultural Repackaging of Service Learning in US Higher Education', *Organization*, 8(2): 319–39.

Lyon, L. L. (1922) *Education for Business* (Chicago: University of Chicago Press).

Neave, G. (1993) 'Séparation de Corps: The Training of Advanced Students and the Organization of Research in France', in B. R. Clark (ed.), *The Research Foundations of Graduate Education: Germany, Britain, France, the United States, Japan* (Berkeley, Calif.: University of California Press), 159–91.

—— and Edelstein, R. (1993) 'The Research Training System in France: A Microstudy of Three Academic Disciplines', in B. R. Clark (ed.), *The Research Foundations of Graduate Education: Germany, Britain, France, the United States, Japan* (Berkeley, Calif.: University of California Press), 192–220.

Meyer, H.-D. (1998) 'The German Handelshochschulen, 1898–1933: A New Departure in Management Education and Why it Failed', in L. Engwall and V. Zamagni (eds.), *Management Education in Historical Perspective* (Manchester: Manchester University Press), 19–33.

Morgeson, F. P., and Nahrgang, J. D. (2008) 'Same as it Ever Was: Recognizing Stability in *Business Week* Rankings', *Academy of Management Learning and Education*, 7(1): 26–41.

Noorderhaven, N., and van Oijen, A. (2008) 'Higher Management Education in the Netherlands', in T. Durand and S. Dameron (eds.), *The Future of Business Schools: Scenarios and Strategies for 2020* (Basingstoke: Palgrave Macmillan), 242–54.

Pfeffer, J., and Fong, C. T. (2002) 'The End of Business Schools: Less Success than Meets the Eye', *Academy of Management Learning and Education*, 1: 78–95.

—— and —— (2004) 'The Business School "Business": Some Lessons from the US Experience', *Journal of Management Studies*, 41 (Dec.): 1501–20.

Pierson, F. C. (1959) *The Education of American Businessmen* (New York: McGraw-Hill).

Porter, L. W., and McKibbin, L. E. (1988) *Management Education and Development: Drift of Thrust into the 21st Century?* (New York: McGraw-Hill).

Sass, S. A. (1982) *The Pragmatic Imagination: A History of the Wharton School 1881–1981* (Philadelphia: University of Pennsylvania Press).

Spender, J. C. (2008) 'The Business School in America: A Century Goes by', in T. Durand and S. Dameron (eds.), *The Future of Business Schools: Scenarios and Strategies for 2020* (Basingstoke: Palgrave Macmillan), 9–18.

Starkey, K., and Tiratsoo, N. (2007) *The Business School and the Bottom Line* (Cambridge: Cambridge University Press).

Thelin, J. R. (2004) *A History of American Higher Education* (Baltimore: Johns Hopkins University Press).

Thomas, H. (2008) 'UK Business Schools', in T. Durand and S. Dameron (eds.), *The Future of Business Schools: Scenarios and Strategies for 2020* (Basingstoke: Palgrave Macmillan), 117–32.

Thompson, A. W., and Wilson, J. F. (2006) *The Making of Modern Management: British Management in Historical Perspective* (Oxford: Oxford University Press).

Tiratsoo, N. (1998) 'Management Education in Postwar Britain', in L. Engwall and V. Zamagni (eds.), *Management Education in Historical Perspective* (Manchester: Manchester University Press), 111–26.

—— (2004) 'The "Americanization" of Management Education in Britain', *Journal of Management Inquiry*, 13(2): 118–26.

Trank, C. Q., and Rynes, S. L. (2003) 'Who Moved our Cheese? Reclaiming Professionalism in Business Education', *Academy of Management Learning and Education*, 2(2): 189–205.

Trieschmann, J. S., Dennis, A. R., Northcraft, G. B., and Nieme, A. W. (2000) 'Serving Multiple Constituencies in the Business School: M.B.A. Program versus Research Performance', *Academy of Management Journal*, 43(6): 1130–41.

Trow, M. (1989) 'American Higher Education: Past, Present, Future', in T. Nybom (ed.), *Universitet och samhälle: Om forskningspolitik och vetenskapens samhälleliga roll* (Stockholm: Tiden), 369–96.

Tyson, L. D. (2005) 'On Managers Not MBAs', *Academy of Management Learning and Education*, 4(2): 235–6.

Üsdiken, B. (2004) 'Americanization of European Management Education in Historical and Comparative Perspective: A Symposium', *Journal of Management Inquiry*, 13(2): 87–9.

—— (2007) 'Commentary: Management Education between Logics and Locations', *Scandinavian Journal of Management*, 23(1): 84–94.

—— and Pasadeos, Y. (1995) 'Organizational Analysis in North America and Europe: A Comparison of Co-citation Networks', *Organization Studies*, 16(3): 503–26.

—— Kieser, A., and Kjaer, P. (2004) 'Academy, Economy and Polity: *Betriebswirtschaftslehre* in Germany, Denmark and Turkey before 1945', *Business History*, 46: 381–406.

Veblen, T. (1918) *The Higher Education in America* (New York: B. W. Huebsch).

Wedlin, L. (2006) *Ranking Business Schools* (Cheltenham: Edward Elgar).

Whitley, R. and Gläser, J. (eds.) (2007) *The Changing Governance of the Sciences: The Advent of Research Evaluation Systems* (Berlin: Springer).

—— Thomas, A., and Marceau, J. (1981) *Masters of Business? Business Schools and Business Graduates in Britain and France* (London: Tavistock).

Wilson, J. F. (1992) *The Manchester Experiment: A History of Manchester Business School* (London: Paul Chapman).

Zell, D. (2001) 'The Market-Driven Business School: Has the Pendulum Swung Too Far?', *Journal of Management Inquiry*, 10(4): 324–38.

Concluding Reflections

12

From Governance to Authority Relations?

Jochen Gläser

1. Experimental Reinterpretation

One of the practices by which the social sciences and humanities advance their knowledge is the constant exploration of new perspectives on the knowledge they have produced so far and can currently produce. The loose coupling between interpretation and empirical methods in these fields makes experimental reinterpretation a worthwhile activity that may lead to novel perspectives combining existing research in a fruitful manner. Edited volumes are often both documents of, and devices for, such exercises in experimental reinterpretation. In many cases emerging from dedicated workshops or conferences, they represent attempts by authors to reinterpret their research in a new context, to link it to research they have not previously considered to be connected to their own, and to derive new avenues of research from these new perspectives. This volume is a case in point.

The aim of this concluding chapter is to consider how these chapters reinterpret research on governance in the sciences by focusing on changing patterns of authority. Thus, rather than summarizing and integrating the findings of the chapters, I use some of the findings to discuss the utility of analysing the impact of governance changes on scientific research through the study of shifting authority relations.

This requires linking the concept of authority relations to the dominant framework of current science policy studies: namely, the governance perspective. Governance is a concept that is so general and diffuse that it inevitably includes authority relations. However, less is sometimes more. The major argument of this chapter is that 'authority relations' can be a conceptual device that provides governance research with a specific focus and with a specific tool for conceptual integration.

2. Authority Relations as Part of the Governance Structure

If one searches for a definition of governance, the usual result is not a single definition but a set of various meanings (Rhodes 1997; van Kersbergen and van Waarden 2004; von Blumenthal 2005). These sets are not even exhaustive; more general definitions of governance refer to the intentional adjustment of actions among interdependent actors and even include the unintentional order (spontaneous order, Hayek 1991) of the market. The rise of the concept 'governance' reflects both the growth of non-hierarchical practices of political coordination and the accompanying shifts in social science perspectives, which departed from their original focus on 'government' as hierarchical control by the state and arrived at the investigation of the relationships between conditions, processes, and outcomes of a wide variety of forms of governance (Benz *et al.* 2007).

The governance of science and technology is no exception here. The extremely wide and diffuse meaning of 'governance' is probably the reason why both empirical research and attempts to develop theories of the governance of science are devoted to much more specific subjects such as the governance of higher education, of academy–industry relations, of innovation, or of specific fields of research and technological development such as nanotechnology. Empirical investigations of governance are focused on specific instruments or sets of instruments that deal with a particular problem. Theoretical attempts try to understand the dynamics of particular governance regimes, governance modes, or governance mechanisms.

Thus, one of the reasons why turning to authority relations might be advantageous is that this concept provides a clear focus. The concept of authority as used by the contributions to this book refers to *legitimate power of actors*. Authority relations, then, mean the relative authority of a set of interdependent actors. If authority relations are used as an analytical perspective, they can be specified in relation to a specific decision process. This notion is both more specific than concepts of governance and more inclusive than much empirical research on governance. It is more specific insofar as it focuses on actors (authoritative agencies) and uses institutional structures and processes of governance as 'background information' on how authority is produced and exercised. At the same time, it is more inclusive because it always includes all actors who have authority concerning a specific decision process regardless of their inclusion in particular governance instruments. Collective actors such as academic elites and individual actors such as researchers are included in the study of authority relations.

While authority relations are produced, maintained, and realized by processes of governance, they represent a conceptual device that focuses our attention on certain aspects of governance that are not as easily caught by approaches that emphasize institutional or process aspects (Mayntz 2004; Benz *et al.* 2007). In the remainder of this chapter I discuss some of the uses to which the 'authority relations perspective' (ARP) can be put in governance research by relating Whitley's typology of public science systems to some of the typologies of governance that have been suggested in the higher education literature, and by highlighting some of the ideas that result from the application of the ARP to the empirical studies we have presented.

3. 'Authority Relations' as Integrative Theory and Comparative Framework

One of the enduring challenges in research on the governance of science is the theoretical integration of, and the provision of a comparative framework for, the investigation of the numerous governance structures and processes. Over the last three decades, higher education research has responded to that challenge with a variety of frameworks including

- Clark's (1983) famous triangle of state, market, and academic oligarchy coordination of higher education, which he later enriched by a fourth mode of organization (Clark 1998);
- Curry's and Fischer's (1986) distinction of state-agency, state-controlled, state-aided, and corporate models of state–university relationships;
- Van Vught's (1994) distinction of state control and state supervision;
- Braun's and Merrien's (1999) typology that uses the dimensions of substantial autonomy of universities, procedural dimension of universities, and belief systems of governments concerning higher education; and
- Schimank's (2005; Lange and Schimank, 2007) 'equalizer model' that characterizes governance according to the strength of five modes: financial support and regulation by the state, external guidance whether by the state or by delegated authority, institutional competition, hierarchical management, and academic self-governance.

Most of these and other similar suggestions expand on Clark's original proposal. They all have in common that they are based on either a governance structure—the relationship between the state and higher education institutions—or modes of governance; that is, modes of coordination in higher education systems. Consequently, they also implicitly refer to authority relations. The typology of public science systems proposed by Whitley in the Introduction nevertheless differs from these typologies by its focus on research, which leads to the inclusion of research-oriented actors such as funding agencies and national scientific elites.

It also avoids two particular problems of frameworks based on governance modes. First, by using the relative authority of actors rather than governance modes, it is not forced to combine governance modes from different levels of aggregation. The more complex frameworks listed above are based on a combination of state–university coordination, inter-university relationships, and intra-university modes of coordination. This can become problematic since, secondly, these modes are not independent of each other.

In particular, participation in decision processes and the authority of actors concerning specific decisions is often a zero-sum game—one actor's gain in authority implies other actors' losses. Even though Whitley's framework is not completely balanced (there seems to be more authority in state-delegated competitive systems than in others), its focus on the distribution of authority among actors could tighten comparative approaches to governance. By bringing into the limelight the actors in whose situations and actions governance modes 'meet', authority relations may help answering the question how governance modes are linked, and how many 'degrees of freedom' a governance system actually has.

4. Authority Relations as an Analytical Perspective: Integrating Findings on Governance

The preceding contributions highlight two particular advantages of the ARP. First, the ARP easily deals with multi-level problems because it includes authoritative agencies relevant to the investigated phenomenon regardless of levels of aggregation. Such a perspective is especially important for the study of the governance of science because of the latter's specific authority structure, which results from the overlap of a formal authority system and highly fluid communities with their informal authority structure (Gläser and Lange 2007). Owing to the multiple overlaps of formal governance structures and scientific communities, the macro- and meso-levels of the public sciences host very different authoritative agencies: namely, corporate actors (state ministries, intermediary organizations such as research councils, public research organizations), collective actors in form of the elites of the various scientific communities, which compete with each other for resources and adequate governance, and individual actors who translate governance into changed research.

Second, the ARP supports an analytical focus on changes in research which, albeit partly addressed by various governance instruments, are not the primary target of any of them. Research on the governance of science is likely to choose empirical objects that are part of the governance system: governance instruments, institutions, or corporate actors. This makes inevitable a focus on the purposes of instruments and institutions and the interests of corporate actors. However, if one is interested in a specific aspect of research such as the conditions for intellectual innovation, it seems advantageous to turn the question around and ask how processes of governance jointly contribute to changes of that particular aspect. The ARP is again a useful integrative tool for this kind of research, because it enables the capture of the 'resultant force' of qualitatively different and asynchronous changes in governance. Treating authoritative agencies as 'crystallization points' where changes in governance are translated into interests and action capabilities thus supports the investigation of 'emergent unintended effects'—of effects of governance that are not side effects of any specific instrument but result from the interaction of changes in authority relations which are an inevitable side effect of governance.

Varieties of Scientific Elites

The multifaceted nature of scientific elites and the variety of their roles in the governance of the public sciences is one of the interesting outcomes of this collection of studies. In Chapter 1, Whitley introduces the distinction between national and organizational elites and demonstrates that their relative authority differs in all types of public science systems. The authority of national elites is highest in competitive public science systems because there is no other arbitrator for the competitions (as has been demonstrated by Gläser *et al.*'s comparison of research evaluation systems in Chapter 5). The authority of organizational scientific elites is highest when there is little competition for resources, because these are then largely controlled by the organizational scientific elites themselves.

Shifts of elite authority due to changes in the governance of science are addressed by many studies in this volume. Taken together, these contributions demonstrate the complex and heterogeneous nature of the authoritative agency 'elite'. Instead of one unitary actor, we find several elites, which are integrated in formal governance structures through several interfaces, each of which enables a specific way of exercising authority.

Several contributions highlight differences and frictions between elites. In her discussion of the role of national elites in the conception of the RAE, Morris shows that the natural science elites were an important supporter of its institutionalization. Being aware of the difficulties of maintaining an internationally competitive resource base across all universities, and having already incorporated a culture of competition for funding, the natural science elites were quite open to the idea of a competition for block funding. The elites of the social sciences and humanities, on the other hand, feared a loss of authority if they did not submit themselves to a competition that was designed, and after more than twenty years is still conducted, around blueprints from the natural sciences.

Meier and Schimank even report a split between national and organizational elites. In the evaluation of Lower Saxony's university research, the national elites of all disciplines appeared to agree on the importance of creating 'critical mass' and coherent research profiles. The organizational scientific elites, however, show the expected difference between laboratory sciences and other disciplines: While the former agree with their national elites because of the collaborative and resource-intensive nature of their research, the organizational elites of fields featuring an individualised research culture do not take easily to the collective approach inherent in profile-building.

The importance of treating elites as a multitude of collective actors with different interests also becomes clear from Benninghoff and Braun's account of the evolution of the Swiss National Science Foundation. The frequent struggles about the responsibility for application-oriented funding programmes reflect not only the attempts of the foundation to keep its organizational domain but also the competition between national scientific and technological elites for the control of resources—a competition which was exacerbated because the corresponding organizational elites belong two different types of organizations: the scientific elites are concentrated in cantonal universities, and the technological elites in the two federal universities of technology.

Having disentangled the complex relationships between scientific elites, we can also try to assess how the previous and current changes in governance affect the authority of the various elites. Chapters 2–5 and 7 support the thesis in Chapter 1 that the increasing competition for block funding and grant funding and the evaluations that are built into these competitions increase the authority of national elites. Kneller's observation that the highly skewed distribution of grants towards the Japanese elite universities does not correspond to the representation of the elite universities in the peer-review system sheds light on the specific kind of authority of elites. Adopting French and Raven's (1959) discussion of bases of power, we can describe the authority of elites as 'expert authority' which is based on the perception of superior knowledge. Kneller observes that across several different funding programmes a near-representative sample of members of scientific communities and external experts still arrives at decisions on grants that channel a large proportion of the funding towards academics at elite universities.

The findings on the authority of organizational scientific elites are more complex. Their authority is highlighted by several chapters in this book. A particularly instructive example is provided in Chapter 11, where Engwall *et al.* trace the different paths of institutionalization of business studies in the USA and Europe. The subject of business studies was relatively easily integrated in the emerging universities in the USA, while the already established organizational scientific elites of universities in continental Europe and of the elite universities in the UK 'repelled' business studies.

The recently increasing autonomy of universities and the strengthening of university management in Germany, the Netherlands, and Switzerland weaken the influence of organizational scientific elites. At the same time, the transition from block funding to grant funding in all countries increases both the independence of grant winners from their universities and the dependence of universities on their researchers. This shift in authority relations affects all researchers who can secure a steady stream of grant income: predominantly the organizational scientific elites. Finally, as shown by Chapters 8 and 9, the organizational elites dominate research groups and often define the intellectual domains in which others have to find their topics—a process that is supported by the increasing authority of university management.

As a result of these processes, authority over the formulation of research goals and the integration of results has become more concentrated in the various academic elites, and what once was considered to be a functional imperative for all competent researchers—autonomous problem choice—is gradually turning into the privilege of academic elites. This is happening at different speeds in the various fields depending on their governance-relevant epistemic properties, especially resource-dependence (see Chapter 10). However, since several authoritative agencies contribute to this trend via different channels, it seems likely that it remains stable and will occur in all fields.

Researchers as 'Obligatory Points of Passage' for Authority over Research

As with academic elites, individual researchers constitute a category that is not easily dealt with by governance research. They must be included be- cause they mediate the effects of changing governance without being part of any formal governance structure. Researchers are always implicitly in- cluded in the analysis of authority relations over the content of research because this content is ultimately shaped by the decisions of researchers. In other words, researchers are an 'obligatory point of passage' (Latour 1988: 43–4) for the exercise of authority over research content.

Authority relations concerning the content of research can thus be thought of as the 'authority ego-centric networks' of individual researchers. Several contributions address this role of researchers in the exercise of authority over research. Martin and Whitley's account of the evolution of the RAE can be read as an account of the redistribution of authority over problem choice from the academics at British universities to their univer- sity management, the state, the funding councils, and the academic elites. The only agency that has not gained authority due to the RAE appears to be the users of research (but those have been amply compensated by other governance instruments).

Four contributions demonstrate the advantages of moving from the study of specific governance instruments to the study of authority relations when the level of individual researchers is addressed. In Chapter 10, Gläser *et al.* demonstrate how their search for the effects of a particular governance instrument made them study the situations of researchers, which in turn required including the authority relations concerning the strategic deci- sions of researchers. In a similar vein, Leišytė *et al.* include authority rela- tions concerning external funding in their analysis of the impact of governance changes on the problem choice of researchers in British and Dutch biotechnology and medieval history research. These authority rela- tions contribute much to the explanation of differences between problem choices in both fields and countries.

Morris demonstrates the compatibility of the governance instrument 'RAE' with the (much more complex) authority relations in the biosciences and many other natural science fields, which explains the relative ease with which this instrument has been adopted by these fields. Finally, Louvel shows how the complex interplay of changes in authority relations in the French public science system and resulting changes in French research laboratories affects researchers' opportunities to choose research problems. Her investigation most clearly describes a possible dilemma that is inherent to the situations of researchers in many science systems and to the impact of authority relations in general. The emergence and diffusion of intellectual innovations appears to require both 'protected spaces' in which they can be developed regardless of the forces of quasi-markets favouring the mainstream and short-termism, and the flexibility of approaches and careers supported by these very quasi-markets of evaluation-based block funding and grant funding.

These findings underline the importance and usefulness of the ARP. In order to understand the impact of any given governance instrument on the content of research, one has to understand how this instrument modifies the situation of researchers. Thus, the whole situation of researchers that is relevant to the effects of interest must be included in the analysis, and the ARP provides a tool for doing just that.

Researchers are not only obligatory points of passage for authority concerning research. They also have a special position in the authority relations because they are the only authoritative agency that makes *positive* decisions on the content of research. Several other authoritative agencies have veto powers concerning research projects or whole lines of research. Elites, funding agencies, and managers of public research organizations can decide that a particular project is not to be conducted—mainly by not granting resources or by taking away research time. In many cases the combined veto powers of authoritative agencies may even amount to the authority to direct research—simply by cutting off all but one opportunity available to a researcher. However, the researchers have not only that same veto power but also are the only actors who decide what will be done. This fundamental imbalance in the authority to make decisions on research content is produced by the 'self-selection' mode of task definition in research (Benkler 2002; Gläser 2006).

It is nicely illustrated by a case of 'authority failure' analysed by Benninghoff and Braun. The Swiss National Science Foundation's funding programme for 'Priority Research Programmes' failed because not enough researchers were interested in the topics defined by the administration, parliament, and stakeholders. Even though Benninghoff and Braun rightly argue that this failure was partly due to the limited authority of the actors

365

attempting to define the topics, since researchers did not depend on their funding, the case illustrates the principle: authoritative agencies can prevent certain research from being done but can rarely enforce specific research. (See also Chapter 9 on the problem choice of researchers and Chapter 10 on research-level effects of authority relations.)

Looking from a Particular Effect to its Causes

The contributions to this volume have in common (and share with most investigations of the governance of science) that they start with governance structures or processes and try to establish their effects. The ARP makes it possible to turn this around by starting with a particular effect (a changed condition for or property of research) and asking how this effect is produced by the numerous processes of governance in which every research process is embedded. This approach underlies Whitley's question about the consequences of changing authority relations for intellectual innovations. It builds on one of the strengths of the ARP: by focusing on authority relations it cuts across all processes and instruments of governance, opportunistically 'harvests' their impact on the relative authority of actors concerning a specific aspect of research, and integrates these partial definitions of authority.[1] It thereby enables a high specificity of studying effects of governance and a high inclusiveness concerning instruments and processes of governance. Rather than asking what changes in research are produced by a specific governance instrument or process, one can use the ARP perspective for asking how all the various governance processes impact upon one specific aspect of research.

What could such an approach produce? Chapter 6 indicates the opportunities. Changes in the governance of the French public sciences—particularly changes in funding structures, personnel recruitment, and evaluations—led to a change in the authority structures of French joint CNRS/university laboratories, which entailed a change of conditions for intellectual innovation. Even though this is a small-scale approach, it demonstrates the feasibility of identifying 'resultant forces' and discovering the synergies of seemingly unrelated governance structures and processes. By including all actors who have authority concerning a particular aspect of research, the ARP is able to capture long-range effects that are produced by indirect changes of authority relations. This way it also avoids what has been termed by Mayntz

[1] This short description already indicates the enormous methodological difficulties involved in doing this empirically; see below, section 5.

(2004: 74) the 'problem solving bias': namely, the overemphasis on the purposes for which a governance instrument under study was (apparently) built, and on the assumption that the major aim of all the actors involved is to solve the problem at hand.

5. Conclusion: The 'Authority Relations' Handle on Governance

I hope to have demonstrated that the ARP is not some upstart alternative to the study of governance but a specific perspective on governance that has some specific advantages. Comparatively describing authority relations makes it conceptually easier to integrate collective and individual actors on an equal footing and provides a means of studying emergent effects of overlapping governance instruments at various levels of aggregation. In particular, it enables us to study the emergent impact of a multitude of governance instruments on specific aspects of research such as problem choice or intellectual innovation. It is thus an interesting perspective for comparative research on governance—both for the integration of results and for designing empirical investigations.[2]

Having said this, I would like to emphasize some of the problems of the ARP—problems which remain hidden as long as it is only used for developing theoretical perspectives or reinterpreting empirical results that were obtained without applying the perspective. First of all, the ARP does not escape the problem of comparative research. In order to explore its full potential we have to 'comparatively measure' authority across different authoritative agencies, across fields, and across countries. This is by no means an easy task. Using the various bases of authority—resources, rewards and sanctions, intellectual influence—might help solve this problem. However, an empirical methodology for the 'measurement' of authority does not yet exist.

Secondly, the ARP must not neglect the complexity of conditions of action for the actors involved. Chapter 7 by Meier and Schimank indirectly highlights this problem by identifying several conditions that are not

---

[2] The ARP is currently applied in an empirical project on 'Restructuring Higher Education and Scientific Innovation'. The project investigates how changes in the governance of the public sciences have modified authority relations in the UK, the Netherlands, Sweden, Germany, and Switzerland, and how these changes affect conditions for scientific innovations in four fields. See http://www.esf.org/activities/eurocores/running-programmes/eurohesc/eurohesc-projects. html.

directly related to the authority relations concerning research profiles, but to other processes such as teaching, university reforms, and changes in university leadership. These conditions, while not affecting the authority of the agencies involved, nevertheless affect their willingness and ability to act, and therefore the exercise of authority and its outcome.

This leads to the third and most important point that needs to be taken into account. The ARP remains static as long as the exercise of authority is not included in the picture. We are, after all, interested in the social mechanisms through which authority relations affect research. The identification of these mechanisms requires integrating the various means by which actors exercise their authority concerning specific actions in science. The key to this integration appears to be the situation of a researcher which is partly shaped by actors who provide access to resources, communicate behavioural expectations, and affect the researcher's conditions for gaining reputation. Integrating these influences brings us back to governance. It has of course been with us all the time. However, looking at governance with an ARP focuses research on the ways in which governance modifies authority relations concerning research content, and on the ways in which this authority is exercised by governance. The experiments with the ARP in this book demonstrate that the analytical emphasis on authority relations is capable of producing interesting results.

References

Benkler, Yochai (2002) 'Coase's Penguin, or, Linux and *The Nature of the Firm'*, *Yale Law Journal*, 112: 369–446.

Benz, Arthur, Lütz, Susanne, Schimank, Uwe, and Simonis, Georg (2007) 'Einleitung', in A. Benz, S. Lütz, U. Schimank, and G. Simonis (eds.), *Handbuch Governance: Theoretische Grundlagen und empirische Anwendungsfelder* (Wiesbaden: VS Verlag für Sozialwissenschaften), 9–25.

Braun, Dietmar, and Merrien, Francois-Xavier (1999) 'Government of Universities and Modernisation of the State: Analytical Aspects', in D. Braun and F.-X. Merrien (eds.), *Higher Education Policy: Towards a New Model of Governance for Universities?* (London: Jessica Kingsley), 10–33.

Clark, Burton R. (1983) *The Higher Education System: Academic Organization in Cross-National Perspective* (Berkeley, Calif.: University of California Press).

—— (1998) *Creating Entrepreneurial Universities: Organizational Pathways of Transformation* (New York: Pergamon Press).

Curry, Dennis J., and Fischer, Norman M. (1986) 'Public Higher Education and the State: Models for Financing, Budgeting, and Accountability', paper presented at the Annual Meeting of the Association for the Study of Higher Education, San Antonio, Tex., Feb. 1986: http://www.eric.ed.gov/ERICDocs/data/ericdocs2sql/content_storage_01/0000019b/80/2f/38/be.pdf (accessed Nov. 2009).

French, John P., and Raven, Bertram (1959) 'The Bases of Social Power', in Dorwin Cartwright (ed.), *Studies in Social Power* (Ann Arbor: University of Michigan Press), 150–67.

Gläser, Jochen (2006) *Wissenschaftliche Produktionsgemeinschaften: Die soziale Ordnung der Forschung* (Frankfurt a. M.: Campus).

—— and Lange, Stefan (2007) 'Wissenschaft', in A. Benz, S. Lütz, U. Schimank, and G. Simonis (eds.), *Handbuch Governance: Theoretische Grundlagen und empirische Anwendungsfelder* (Wiesbaden: VS Verlag für Sozialwissenschaften), 437–51.

Hayek, Frederick A. (1991) 'Spontaneous ("Grown") Order and Organized ("Made") Order', in G. Thompson, J. Frances, R. Levacic, and J. Mitchell (eds.), *Markets, Hierarchies and Networks: The Coordination of Social Life* (London: Sage), 293–301.

Lange, Stefan, and Schimank, Uwe (2007) 'Zwischen Konvergenz und Pfadabhängigkeit: New Public Management in den Hochschulsystemen fünf ausgewählter OECD-Länder', in K. Holzinger, H. Joergens, and C. Knill (eds.), *Transfer, Diffusion und Konvergenz von Politiken: Sonderheft der Politischen Vierteljahresschrift* (Wiesbaden: VS Verlag für Sozialwissenschaften), 522–48.

Latour, Bruno (1988) *The Pasteurization of France* (Cambridge, Mass.: Harvard University Press).

Mayntz, Renate (2004) 'Governance im modernen Staat', in A. Benz (ed.), *Governance: Regieren in komplexen Regelsystemen* (Wiesbaden: VS-Verlag für Sozialwissenschaften), 65–75.

Rhodes, Roderick A. W. (1997) *Understanding Governance: Policy Networks, Reflexivity, and Accountability* (Buckingham: Open University Press).

Schimank, Uwe (2005) '"New Public Management" and the Academic Profession: Reflections on the German Situation', *Minerva*, 43: 361–76.

Van Keersbergen, Kees, and Waarden, Frans van (2004) '"Governance" as a Bridge between Disciplines: Cross-Disciplinary Inspiration Regarding Shifts in Governance and Problems of Governability, Accountability and Legitimacy', *European Journal of Political Research*, 43: 143–71.

Van Vught, Frans A. (1994) 'Autonomy and Accountability in Government/University Relationships', in J. Salmi and A. M. Verspoor (eds.), *Revitalizing Higher Education* (Oxford: IAU Press), 323–62.

Von Blumenthal, Julia (2005) 'Governance: Eine kritische Zwischenbilanz', *Zeitschrift für Politikwissenschaft*, 15: 1149–80.

Index